Addressing Environmental Challenges Through Spatial Planning

Athar Hussain
Netaji Subhas University of Technology, India

Kailash Chandra Tiwari
Delhi Technological University, India

Alpana Gupta
Indian Institute of Technology, Delhi, India

A volume in the Advances in Geospatial Technologies (AGT) Book Series

IGI Global
PUBLISHER of TIMELY KNOWLEDGE

Published in the United States of America by
IGI Global
Engineering Science Reference (an imprint of IGI Global)
701 E. Chocolate Avenue
Hershey PA, USA 17033
Tel: 717-533-8845
Fax: 717-533-8661
E-mail: cust@igi-global.com
Web site: http://www.igi-global.com

Library of Congress Cataloging-in-Publication Data

Names: Hussain, Athar, 1976- editor. | Tiwari, Kailash Chandra, 1964- editor. | Gupta, Alpana, 1996- editor.
Title: Addressing environmental challenges through spatial planning / Athar Hussain, Kailash Chandra Tiwari, and Alpana Gupta, editors.
Description: Hershey, PA : Engineering Science Reference, [2022] | Includes bibliographical references and index. | Summary: "This edited book is devoted to addressing environmental concerns and technology innovations in domains such as pollution, water insecurity and resources management, loss of green cover, energy efficiency, reducing carbon footprint, climate change, and disaster management with an aim to bridge the gap between engineering considerations and spatial aspects of planning"-- Provided by publisher.
Identifiers: LCCN 2021026051 (print) | LCCN 2021026052 (ebook) | ISBN 9781799883319 (h/c) | ISBN 9781799883326 (s/c) | ISBN 9781799883333 (ebook)
Subjects: LCSH: Land use--Environmental aspects. | Environmental protection. | Environmental policy. | Urban policy. | Land use--Planning. | Sustainable architecture. | Geospatial data.
Classification: LCC HD108.3 .A33 2022 (print) | LCC HD108.3 (ebook) | DDC 333.73/13--dc23
LC record available at https://lccn.loc.gov/2021026051
LC ebook record available at https://lccn.loc.gov/2021026052

This book is published in the IGI Global book series Advances in Geospatial Technologies (AGT) (ISSN: 2327-5715; eISSN: 2327-5723)

British Cataloguing in Publication Data
A Cataloguing in Publication record for this book is available from the British Library.

Advances in Geospatial Technologies (AGT) Book Series

Nilanjan Dey
Techno India College of Technology, India

ISSN:2327-5715
EISSN:2327-5723

Mission

The geospatial technology field is a fast-paced, high growth industry that is involved in a variety of fields, including military planning, public health, land use, environmental protection, and Google Earth mapping. With such a diverse body of applications, the research in geospatial technologies is always evolving and new theories, methodologies, tools, and applications are being developed. **Advances in Geospatial Technologies (AGT) Book Series** is a reference source and outlet for research that discusses all aspects of geographic information, including areas such as geomatics, geodesy, GIS, cartography, remote sensing, and other areas. Because geospatial technologies are so pervasive in such a variety of areas, AGT also includes books that address interdisciplinary applications of the technologies.

Coverage

- Public Health Applications of Geospatial Technologies
- Disaster Informatics
- Surveying Technologies
- Geomatics
- Geovisualization
- Geographical information systems
- Military Security and Geospatial Technologies
- Digital Terrain Modeling
- Environmental Monitoring
- Global Positioning Systems

IGI Global is currently accepting manuscripts for publication within this series. To submit a proposal for a volume in this series, please contact our Acquisition Editors at Acquisitions@igi-global.com or visit: http://www.igi-global.com/publish/.

The Advances in Geospatial Technologies (AGT) Book Series (ISSN 2327-5715) is published by IGI Global, 701 E. Chocolate Avenue, Hershey, PA 17033-1240, USA, www.igi-global.com. This series is composed of titles available for purchase individually; each title is edited to be contextually exclusive from any other title within the series. For pricing and ordering information please visit http://www.igi-global.com/book-series/advances-geospatial-technologies/73686. Postmaster: Send all address changes to above address.

Titles in this Series

For a list of additional titles in this series, please visit: www.igi-global.com/book-series

Methods and Applications of Geospatial Technology in Sustainable Urbanism
José António Tenedório (Universidade NOVA de Lisboa, Portugal) Rossana Estanqueiro (Universidade NOVA de Lisboa, Portugal) and Cristina Delgado Henriques (Universidade de Lisboa, Porugal)
Engineering Science Reference • © 2021 • 684pp • H/C (ISBN: 9781799822493) • US $195.00

Interdisciplinary Approaches to Spatial Optimization Issues
Sami Faiz (University of Tunis El Manar, Tunis, Tunisia) and Soumaya Elhosni (University of Tunis El Manar, Tunis, Tunisia)
Engineering Science Reference • © 2021 • 271pp • H/C (ISBN: 9781799819547) • US $195.00

Geospatial Technologies for Effective Land Governance
Moha El-Ayachi (Institut Agronomique et Vétérinaire Hassan II, Morocco) and Loubna El Mansouri (Institut Agronomique et Vétérinaire Hassan II, Morocco)
Engineering Science Reference • © 2019 • 302pp • H/C (ISBN: 9781522559399) • US $205.00

Emerging Trends in Open Source Geographic Information Systems
Naveenchandra N. Srivastava (Bhaskaracharya Institute for Space Applications and Geo-Informatics, India)
Engineering Science Reference • © 2018 • 302pp • H/C (ISBN: 9781522550396) • US $195.00

Geospatial Technologies in Urban System Development Emerging Research and Opportunities
Alok Bhushan Mukherjee (North-Eastern Hill University Shillong, India) Akhouri Pramod Krishna (Birla Institute of Technology Mesra, India) and Nilanchal Patel (Birla Institute of Technology, India)
Engineering Science Reference • © 2018 • 179pp • H/C (ISBN: 9781522536833) • US $145.00

Spatial Analysis Techniques Using MyGeoffice®
João Garrott Marques Negreiros (University of Saint Joseph, China & Macau Polytechnic Institute, China)
Engineering Science Reference • © 2018 • 336pp • H/C (ISBN: 9781522532705) • US $225.00

Handbook of Research on Geospatial Science and Technologies
Joyce Gosata Maphanyane (University of Botswana, Botswana) Read Brown Mthanganyika Mapeo (University of Botswana, Botswana) and Modupe O. Akinola (University of Lagos, Nigeria)
Engineering Science Reference • © 2018 • 457pp • H/C (ISBN: 9781522534402) • US $345.00

IGI Global
PUBLISHER of TIMELY KNOWLEDGE

701 East Chocolate Avenue, Hershey, PA 17033, USA
Tel: 717-533-8845 x100 • Fax: 717-533-8661
E-Mail: cust@igi-global.com • www.igi-global.com

Table of Contents

Detailed Table of Contents

Climate change is a multidimensional observable fact and is regarded as one of the greatest challenges human society is facing in the 21st century. Urban researchers advocate that well formulated urban spatial planning policy has the ability to mitigate climate change and adapt to the adverse impacts of climate change. However, there has been limited research on analysing the extent to which spatial planning policies address climate change mitigation and adaptation. This chapter presents a qualitative evaluation of urban spatial planning polices in India by assessing planning policies against an evaluation framework. The analysis highlights that there are limited climate change mitigation and adaptation indicators in planning documents, and these indicators have been very limitedly integrated in the planning documents. This research supports the use of spatial planning policy as an effective tool in addressing climate change mitigation and has an implication for mainstreaming climate change mitigation and adaptation in urban planning.

The goal of this chapter is to demonstrate the classification of natural and man-made objects from multisensory remote sensing data. The spectral and spatial features play an important role in extracting the information of natural and man-made objects. The classification accuracy may be enhanced by fusion technique applied on feature knowledge database. A significantly different approach has been devised using spatial as well as spectral features from multisensory data, and the classified results are enhanced by majority voting fusion technique. The author concludes by presenting extensive discussion at each level and has envisaged the potential use of multisensory data for object-based land cover classification.

The River Yamuna is one of the most prominent and sacred rivers of India. Due to the unaccounted discharge from many anthropogenic activities and industrialization in the basin, the deterioration of the river with respect to time is at its maximum. Periodical analysis is essential for remediation of the causes and the source. The study has been undertaken using GIS technology to understand the variations being taken place in the past decades. The transformation in the physical and geographical characteristics, during the years since 1980-81, of the Yamuna River Basin has been analyzed and converted into digital GIS domain in terms of maps and charts. The reference condition of the basin has been set up and utilized in order to identify the change in land use and land cover of the river premises around Delhi and its neighbouring states. The vegetation analysis with respect to climatic conditions and water quality deterioration has also been assessed in the present study.

Bangladesh has a long history of physical planning. It started with blueprint-based master planning approach in the 1950s. Since the late 1970s, it has slowly shifted to the rational planning approach following the path laid down by the United Kingdom. Lately, infusion of the North American planning thoughts is observed. Thus, Bangladeshi physical planning never had its own identity. Structure plans are loosely tied to the other tiers of planning packages that are used for development management. Though most of the structure plans address environmental issues, lower tiers of plans often lack their reflection. Often policies are laid down based on popular belief rather than scientific evidence. To establish these arguments, the evolution of the planning practices of Bangladesh is illustrated briefly in the context of the evolution of planning theory. Additionally, environmental sensitivity of a physical plan covering one of the most ecologically sensitive and economically potential areas of Bangladesh, namely Mongla master plan (MNP) is evaluated, as an example, considering the spatial dispersion of sulfur dioxide to be discharged by the coal-fired thermal power plant at Rampal. Unfortunately, MMP failed to stand out from all the rest to prove its soundness.

Basic infrastructure services like clean water, proper sewage and sanitation, etc. are important attributes in urbanization of a city to make it healthy, livable, and sustainable. Goal 6 of the Sustainable Development Goals (SDG) aims to ensure clean water and sanitation for all. Access to clean drinking water and basic sanitation facilities are crucial requirements for good health. Issues like intermittent supply, poor water quality, and low-pressure cause unreliable situations due to which community members resort to various coping strategies. Coping strategies are influenced by the household's level of education, income, and the level of unreliability of the service. The coping mechanisms used by the households are costly, and the poor people end up paying more because of the hidden cost. The poor water supply services result in induced costs by the residents in terms of the time costs for water collection and illness due to disease and water treatment costs. Because of these induced costs, the urban poor end up spending more for a clean and reliable water supply.

With the rapid pace of urbanization, land-use change is essential for economic and social progress; however, it does not come without costs. With such rapid urbanization, there comes pressure on the land and its resources, like that of food and timber production with a significant impact on the livelihood of millions of people. With the loss of agricultural land due to developmental activities, future agriculture would be very intensive. Therefore, it is likely with the existing pattern of allocating land uses for future development that we may lose the ecosystem services and highly productive agricultural lands. The value of these ecosystem services to agriculture is enormous and often underappreciated. The study focuses on identifying underlying causes of the land-use change, ecosystem services affected due to land-use change in peri-urban areas of Chandigarh using spatial mapping of affected ecosystem services and suggesting proposals for promoting agricultural ecosystem values using economically-informed policy instruments.

The Aravalli range is one of the oldest mountain ranges in the world. Many lakes are sustained by the streams generated from the Aravalli range. In the last few years, hill systems are facing changes affecting the catchment of water bodies. Many waterbodies that were earlier able to provide water in peak summers are now dried up before the summer. The study aims to analyse land use land cover change in the Aravalli range on Sambhar Lake. Sambhar Lake is the largest inland saltwater lake in India and was selected as a "Ramsar Site." A catchment area of the lake is delineated, and micro-watersheds are selected for detailed analysis. A primary survey is done to know the impacts and causes of changes in the lake. The study indicates significant green cover changes in three seasons over the years. Because of the loss in green cover, the soil is losing its moisture, affecting the cropping pattern and agriculture production in the area. Factors causing the drying of the lake are affecting its ecosystem and the number of birds coming to it.

Currently, cities are home to more than half of the world's population. The increasing urbanization rates create an unprecedented urban sprawl that worsens the urban climate situation. Urban areas modify their local climate and face the consequent urban climate impacts, which are particularly exacerbated by global climate change. This chapter shares scientific knowledge on how cities affect their climate and how urban spatial planning can mitigate the negative impacts of urban climate. Focus is given on the urban heat

Ecological areas are the network of protected areas that contribute to the ecosystem's productivity and services. With increased human demands, towns and cities are blooming with changes in landuse patterns around their peripheral areas or in the immediate vicinity. Land intensification and disproportionate urbanization have inflicted various challenges such as qualitative and quantitative depletion of natural resources, ecosystem services, and degradation of environmental quality in and around ecological areas. The chapter will focus on the circumstantial elucidation of ecological areas, their recorded challenges caused by urbanization, and the need for their conservation in previous research studies. The authors explore reported challenges encompassing ecological areas by urbanization. This will help understand the various aspects of urban transformation, like physical, social, cultural, and economic change in and around the ecological areas, and develop measures and strategies for ecosystem conservation and environmental restoration in ecological areas.

Integrated watershed management is a holistic approach aimed at linking upstream and downstream areas. It encompasses preventing the watershed from being polluted and maintaining the quality of water and prevention of soil erosion. Remote sensing and GIS techniques play a crucial role in assessing watershed conditions through modeling. These techniques are also very helpful in acquiring the spatial information of the land use, land cover, vegetation, water resources, etc. With the available data, characteristics of aquifers can easily be deciphered. Remote sensing aids in identifying and assessing the topographical features and alienates the unnecessary ones. Integration of GIS with remote sensing along with hydrological, hydrogeological, geomorphologic studies will make the practice of integrated watershed management more effective. The primary aim of this chapter is to encompass the remote sensing and GIS technologies currently in use for watershed monitoring and management.

Anthropogenic greenhouse gas emissions have increased since the beginning of the Industrial Revolution. This triggered an unnatural warming and the Earth's temperature which has increased dramatically in a short period of time. This brought many problems such as natural disasters, tropical cyclones, heavy precipitation events, extreme heat events, and drought. It is an urgent need to make all processes "sustainable" in order to protect the environment and human life. Since urban areas are estimated to be the source of approximately 80% of global greenhouse gas (GHG) emissions and a significant proportion of these greenhouse gas emissions in urban areas originate from buildings, buildings should be considered as one of the first stops on the road to sustainability. The purpose of this chapter is to highlight the importance of sustainable architecture in addressing environmental challenges and ensuring sustainable urban development.

Preface

In the past decade, two developments have brought environment management issues to the fore. First has been the increased dependence of human beings on natural resources, not only for key operational purposes but also for gaining strategic advantage. Second, abetted by urbanization and globalization, the whole ecosystem model for has been transformed. Whereas in the past population rely on confining themselves to a particular geographical area to conduct their lifestyle. Today anthropocenes are increasingly becoming location independent and are finding themselves to be strategically disadvantaged if they are confined to a particular place. The consequence of advances in technologies and the changing boundaries helps planners realize their objectives and helps policy makers to take adequate decisions. Today because the environmental challenges and scope of information processing has evolved, managing environment is not just restricted to maintaining integrity and availability. Rapid deterioration and consumption of our planets' natural resources, immense waste production, environmental decline, and climate change are just some of the environmental issues threatening human health and the well-being of planet Earth. Environmental considerations are assuming an increasingly important role in planning and development of human settlements. Alarming consequences of ozone layer depletion, global warming, desertification, and air pollution, pollution of the oceans, species' extinction, deforestation and flooding are some of the manifestations of nature's reactions to human activities on the planet Earth. Implications of such environmental degradations have been the subject of growing concern to people all over the world, and has resulted in various initiatives and actions taken at global, regional, national and international levels.

It starts by defining a field of action for development assistance, and ends with a review of urban environmental strategies. In between, it examines a range of policy relevant issues, from how environmental improvements can reduce poverty to how local governments can work with private enterprises and civil society groups to address environmental problems. Environmental problems are not purely scientific; some of the major problems deal with poor management and the inability to involve people in environmental decision making process. The approach taken in this book is to review environmental problems as they are affected by poor planning and management. Understanding of management issues involved will help to get top management to buy into environmental management. The tendency is for top management to view environmental management efforts as expensive and wasteful to an organization. However, when top management is exposed to the high cost of doing nothing and the lack of competitiveness as a result of poor environmental quality, it is more likely to buy into the idea of environmental quality and work towards achieving sustainable goals.

Spatial Planning provides an integrated, quantitative, and interdisciplinary approach to establish the interaction between environmental issues and human response and to find solutions to resolve the problems associated with it for sustainability. To explore the issues, innovations and integrated approaches towards environmental sustainability, an initiative has been taken under an umbrella covering following range of topics: Water insecurity and resources Management, Loss of Green Cover, Energy Efficiency, Reducing Carbon Footprint, Climate Change, and Disaster Management. Due to scattered pattern of settlement and their more adaptability to urban culture, environment is degrading. Climate Change which is a huge buzz today is only because of changing anthropogenic interventions, which is directly or indirectly correlated with human behavior. There should be a consolidated balance between human, nature, and spatial aspects. As urbanization is giving rise to environmental concerns like, urban flooding which generally occurs due to the construction of house in the low lying areas, loss of green cover leading to disturbance in ecological cycle, water scarcity due to growing needs and deforestation leading to habitat fragmentation, wildlife corridors disturbance, forest fires, and climate change. In the fields of Environment Engineering and Planning, Technical as well as spatial solutions needs to be integrated on the common platform. Therefore, there exists a need for an edited collection of original research in this area.

OBJECTIVE

The proposed book is a collection of various authors from academic and industrial backgrounds. The compilation presents a structured view of the important concepts, considerations, and procedures involved in environmental planning. While numerous books focus on Geographic Information Systems (GIS) capabilities of computer mapping and spatial database management, few provide an understanding of its analytical potential and practical realities in a non-technical manner. The unique character of this book draws from the author's ability to convey seemingly simple as well as complex concepts of spatial data and GIS operations in words that resonant with others less technically versed. The result is a book that engages the reader to "think spatially" and formulate new spatial, policy driven and innovative solutions to complex spatial problems. Key to this process is a paradigm shift that extends the traditional paper map perspective of "where is what" to the modern perspective of "why and so what." In just three decades, GIS technology has radically changed our perspective on both what constitutes a map and the information it contains. The edited book is devoted to addressing environmental concerns and Technology innovations in domains such as Pollution, Water insecurity and resources Management, Loss of Green Cover, Energy Efficiency, Reducing Carbon Footprint, Climate Change, and Disaster Management. The main aim is to bridge the gap between engineering considerations and spatial aspects of planning. It will cover the different aspects of environmental engineering, science and planning related to all four critical elements of life cycle i.e. water, air, land and food. The main elements are to conserve our ecosystem and to maintain balance interaction between environment and human beings. The main objective of editorial is to enhance and communicate knowledge in different aspects of environmental issues and providing their precise measures in both technical and spatial manner to use it for individuals, groups, organizations, and communities and also focus on the practical, technical, social and political problems in the planning and management of the environment.

A case study approach is used with each topic area describing the application of a set of related analysis techniques. The discussion is followed by a series of hands-on exercises providing practical experience in applying the techniques. Planning theorists are often criticized for being insufficiently concerned with the needs of practitioners. The author of this book takes a view of planning which centres on the decision-making process and offers a theoretical approach which takes practice as its starting point. Going beyond previous 'generic' approaches, the work culminates in a consideration of theory and practice in the planning of all forms of environmental intervention. Environmental Informatics is a fast growing field which deals with all methods from computer science, environmental planning, ecology and related subjects. As well as being an interdisciplinary area, Environmental Informatics provides an interface between all involved professional groups. Monitoring the state of the environment, analysing existing data, presenting the data to scientists and the public, as well as providing decision support are only some of the topics involved. Environmental Informatics is therefore a good foundation for the computer-assisted protection of the environment.

TARGET AUDIENCE

If there is one area of environmental science in which GIS or Geographic Information Systems/Science is vital, it is Environmental Planning and Design. This is true whether we are looking at urban or rural landscapes, both of which provide issues of their own with different sets of data and considerations. The role of an Environmental Planner or Environmental Designer is to make decisions based on geographic data; GIS makes available to full range of information in an easy to use database and map that can be manipulated by each issue and for various interested parties. An Environmental Planner will use it to plan and organize a transport network and examine its impact on the environment, manage natural habitats and study the impact that other elements of the landscape may have. Environmental Planners and Designers work with government policy on the management of the built and the natural landscape; GIS presents and allows us to work more freely and quickly with the sort of geographic data you will be handling all the time. It is more suitable than any other form than digital cartography because it is concerned with the data extracted from the map rather than the features of an environment. GIS specialists are highly skilled with expert knowledge of the concepts of geographic data.

- Environmental Engineering, Planning Researchers, and Faculty Members
- Environmental and Civil Administrators, Architects, Planners, and Consultants
- Environmental Engineering Scientists, Environmental Science Researchers, Landscape Architects, and Civil Engineers
- Environmental Activists
- Town and Country Planning Organization, Environmental Statutory Bodies
- Professionals in all industries who are aspiring to have environment friendly atmosphere and to provide sustainable way of dealing with environment in their respective domains for process efficiency, cost optimization, and so on.

ORGANIZATION OF THE BOOK

The book is organized into 14 chapters. A brief description of each of the chapters follows:

Chapter 1 identifies climate change as a multidimensional observable fact and is one of the greatest challenges to the human society in the 21st century. The chapter contributes significantly by developing empirical evidence that highlights the limited integration of climate change mitigation and adaptation indicators in planning documents.

Chapter 2 establishes the need of spectral and spatial features and their role in extracting the information of natural and man-made objects. A significantly different approach has been devised using spatial as well as spectral features from multisensory data and the classified results are enhanced by majority voting fusion technique.

Chapter 3 takes the River Yamuna into consideration. Due to the unaccounted discharge from many anthropogenic activities and industrialization in the basin, the deterioration of the river with respect to time is at its maximum. The transformation in the physical and geographical characteristics, past during the year since 1980-81, of the Yamuna River Basin has been analyzed and converted into digital GIS domain in terms of maps and charts. The reference condition of the basin has been set up and utilized in order to identify the change in land use, land cover of the river premises around Delhi and its neighbouring states.

Chapter 4 reviews the evolution of the planning practice of Bangladesh and is illustrated briefly in the context of the evolution of planning theory. For in-depth analysis, Mongla Master Plan (MMP) is evaluated considering the spatial dispersion of Sulfur Dioxide discharged by the coal-fired power plant at Rampal.

Chapter 5 revolves around the basic infrastructure services like clean water, proper sewage and sanitation, etc. which are important attributes in urbanization of a city to make it healthy, livable and sustainable. Goal six of Sustainable Development Goals (SDG) aims to ensure clean water and sanitation for all. Coping strategies are influenced by the household's level of education, income and the level of unreliability of the service. The coping mechanism used by the households are costly and the poor people end up paying more because of the hidden cost.

Chapter 6 presents an analysis of rapid urbanization and pressure on the land and its resources, like that of food, timber production with a significant impact on the livelihood of millions of people. With the loss of agricultural land due to developmental activities, future agriculture would be very intensive. Therefore, it is more likely with the existing pattern of allocating land uses for future development we may lose the ecosystem services and highly productive agricultural lands. The value of these ecosystem services to agriculture is enormous and often underappreciated. The chapter focuses on identifying underlying causes of the land-use change, ecosystem services affected due to land-use change in Peri-urban areas of Chandigarh using, spatial mapping of affected ecosystem services along suggesting proposals for promoting agricultural ecosystem values using economically informed policy instruments.

Chapter 7 presents an analysis of land use land cover change in the Aravalli range on Sambhar Lake. Many lakes are sustained by the streams generated from the Aravalli's. In the last few years, hill systems are facing changes affecting the catchment of water bodies. Many waterbodies that were earlier able to provide water in peak summers are now dried up before the summers. The chapter indicates significant green cover changes in three seasons over the years. Because of the loss in green cover, the soil is losing its moisture, affecting the cropping pattern and agriculture production in the area. Factors causing the drying of the lake are affecting its ecosystem and the number of birds coming to it.

Chapter 8 shares scientific knowledge on how cities affect their climate and how urban spatial planning can mitigate the negative impacts of urban climate. Focus is given on the urban heat island, the most documented aspect of urban climate, directly linked to cities spatial characteristics and functions. This phenomenon is explained and tools and methods to assess it and mitigate its intensity are introduced in an attempt to help urban planners and designers to use climatic knowledge in urban planning, to build more sustainable and climate-resilient cities.

Chapter 9 emphasizes to prioritize flood risk management in the urban planning process in India. Kochi, the commercial capital of Kerala, India has been frequently affected by flooding events. The chapter focuses on comprehending significant factors attributed to the vulnerability of settlements in the study region and promoting a way forward based on lessons learned and good practices across the world. This is achieved by analyzing significant databases, and computations using GIS. The research outcome would help define strategies for sustainable land-use-based development, promoting effective flood management in the Kochi urban area.

Chapter 10 describes water as an essential component for the survival of mankind and for balancing the ecosystem and livelihood. Analysis of satellite images for estimating the quality and quantity of natural water has become an accepted tool for better spatial planning.This chapter reviews the work of several researchers worldwide who have used and demonstrated the capability of satellite images with other geo-big data such as elevation, landcover etc., for water resource management on the GEE platform. It can be concluded from the review work, that GEE can help in estimating the water quality parameters with reasonable accuracy, comparable to the in-situ measurement, albeit quickly.

Chapter 11 presents the notion of ecological areas as the network of protected areas that contribute to the ecosystem's productivity and services. Land intensification and disproportionate urbanization have inflicted various challenges such as qualitative and quantitative depletion of natural resources, ecosystem services, and degradation of environmental quality in and around ecological areas. The chapter will focus on the circumstantial elucidation of ecological areas, their recorded challenges caused by urbanization, and the need for their conservation in previously done research studies. The authors are aimed to intricate the reported challenges encompassing ecological areas by urbanization. This will help understand the various aspects of urban transformation, like physical, social, cultural, and economic change in and around the ecological areas, and develop measures and strategies for ecosystem conservation and environmental restoration in ecological areas.

Chapter 12 aims to encompass the remote sensing and GIS technologies currently in use for watershed monitoring and management. Integrated watershed management is a holistic approach aimed at linking upstream and downstream areas. Remote sensing aids in identifying and assessing the topographical features and alienates the unnecessary ones. Integration of GIS with remote sensing along with hydrological, hydrogeological, geomorphologic studies will make the practice of integrated watershed management more effective.

Chapter 13 highlighted the importance of sustainable architecture in addressing environmental challenges and ensuring sustainable urban development. Anthropogenic greenhouse gas emissions have increased since the beginning of the Industrial Revolution. It is an urgent need to make all processes "sustainable" in order to protect the environment and human life.

Chapter 14 concludes the integration of construction industry share in Sustainable Development Goals and their targets to test the built environment sustainability and the need to develop the most adequate frameworks for commencing the topic. Sustainable Development is a requisite for future generation, as increasing urbanization; destruction of natural resources by anthropic activities, degrading ecosystems for the sake of present economic development at cost of environmental exploitation is increasing by each passing day on earth. The goal of this chapter is to provide meaningful insights for policy-makers and decision-makers towards sustainable development in the construction industry. Two possible frameworks are also suggested for future research recommendation needed for Indian context to enhance sustainability within construction industry based on LCSA and MCDM analysis.

Alpana Gupta
Indian Institute of Technology, Delhi, India

Acknowledgment

Individually we are one drop. Together we are ocean -Ryunosuke Satoro

Addressing Environment Challenges Through Spatial Planning has been depicting the journey of various authors across the world. Like flowing of river, it is natural that we should acknowledge their hard work. First, I would like to thank each one of the authors for their contributions. I wish to express my gratitude to my co-editor Dr. (Col) Kailash Chandra Tiwari, Professor & Head, Multidisciplinary Centre for Geoinformatics, Delhi Technological University for initiating the preparation of the book. The editor would also like to thank the IGI Global Publishers staff for their assistance in the publication of the book.

I am extremely thankful and indebt of Ms. Alpana Gupta, Innovator at Indian Institute of Technology, Delhi who did everything that was foisted on her and left no stone unturned in pursuit of excellence- your pivotal contributions are gratefully acknowledged.

Chapter 1
Ability of Urban Planning Policy in Addressing Climate Change Adaptation and Mitigation

Kiranmayi Raparthi
School of Architecture and Planning, Anna University, Chennai, India

ABSTRACT

Climate change is a multidimensional observable fact and is regarded as one of the greatest challenges human society is facing in the 21st century. Urban researchers advocate that well formulated urban spatial planning policy has the ability to mitigate climate change and adapt to the adverse impacts of climate change. However, there has been limited research on analysing the extent to which spatial planning policies address climate change mitigation and adaptation. This chapter presents a qualitative evaluation of urban spatial planning polices in India by assessing planning policies against an evaluation framework. The analysis highlights that there are limited climate change mitigation and adaptation indicators in planning documents, and these indicators have been very limitedly integrated in the planning documents. This research supports the use of spatial planning policy as an effective tool in addressing climate change mitigation and has an implication for mainstreaming climate change mitigation and adaptation in urban planning.

INTRODUCTION

In most countries, urban planning refers to the spatial distribution of residential, commercial, industrial and recreational activities over a widespread area and generally involves integration of various components of planning namely land use planning, transportation planning and environmental planning (Levy, 2010). Through urban planning, regional and local governments have an ability to respond to the changing social, economic and environmental conditions by controlling the distribution of various land uses and planning decisions related to local development (Agyeman, Evans, & Kates, 1998). Hence, urban planning policies by incorporating coordination among land use, transportation and environmental planning might have an ability to reduce carbon-di-oxide (CO_2) emissions thereby mitigate climate change.

DOI: 10.4018/978-1-7998-8331-9.ch001

Although, the Intergovernmental Panel on Climate Change (IPCC) and various United Nations (UN) agencies have repetitively acknowledged that urban planning practices can significantly reduce CO_2 emissions. However, lack of empirical evidences that emphasize the impact of urban planning policies on CO_2 emissions, to some extent has hampered the mainstreaming of CO_2 reduction policies in urban planning. Nevertheless, International negotiations such as U.S. Mayors Climate Protection Agreement and Local Governments for Sustainability (IPCC, 2007a) have put forth noticeable efforts to reduce CO_2 levels in the atmosphere by developing local climate change action plans which consist of activities that are relatively consistent in reducing CO_2 emissions. However, these climate change action plans lack the regulatory framework and institutional capacity to achieve their goals and objectives (Raparthi, K. 2018). So, it is essential to formulate policy discussions related to CO_2 reduction in accordance with their applicability at the regional as well as at the local level.

Wilbanks (2003) advocate that any effort taken to mitigate climate change must take into account its multi dimensions and can be effectively mitigated at the urban level by incorporating a comprehensive approach. However, Wilbanks (2003) highlights that a comprehensive approach requires urban areas to mainstream CO_2 reduction strategies into respective land use, transportation, environmental and developmental policies to mitigate climate change. Furthermore, Betsill and Bulkeley (2007) identified that, individual commitment is one of the main reason to effectively address CO_2 reduction at the regional as well as at the local level. Acknowledging the fact that regional authorities with their regulatory power have an ability to promote behavioral changes. Urban planning can be used as a proactive tool to influence individual and organizational behavior and to incorporate comprehensive approach. Thereby effectively reduce CO_2 emissions.

This chapter focuses on a developing country like India. The goal of this chapter is to evaluate the ability of urban planning policies in reducing CO_2 emissions. Accordingly, this chapter presents a qualitative and quantitative evaluation of urban spatial planning polices in India by assessing planning policies against an evaluation framework consists of a set of indicators at the local level and the regional level for addressing climate change. Later, the impact of those policies in addressing climate change mitigation is analyzed.

The analysis highlights that there are limited climate change mitigation and adaptation indicators in planning documents and these indicators have been very limitedly integrated in the planning documents. However, there is a scope to address climate change by formulating and aligning the spatial planning policies towards certain identified indicators and by undertaking frequent evaluation of the planning policies to ensure implementation of the policies at the local level.

BACKGROUND

Global climate change is possibly one of the greatest threats human society is facing in the 21st century. According to the National Aeronautics and Space Administration's (NASA), Surface Temperature Analysis Report (2007), since 1850s the warmest years that have been recorded all occurred from the year 1998, with the year 2005 being noted as the warmest year. Based on improved scientific data and general observations such as increase in atmospheric temperatures, increased rainfalls and rising sea levels, The Intergovernmental Panel on Climate Change (IPCC) affirms that climate change is unequivocal (IPCC 2007a).

Global warming which is the main cause of climate change is a result of increased concentration of atmospheric greenhouse gases (GHG) (IPCC, 2007e). GHG are mainly composed of carbon dioxide, methane, nitrous oxide and fluorinated gases (IPCC, 2007b). Among all these GHG, CO_2 emitted through anthropogenic activities, is considered to be mainly responsible for climate change, since it accounts for about 84% of total GHG emissions in the atmosphere (Hunt, 2004). According to the National Oceanic and Atmospheric Administration's (NOAA's) State of the Air quality Report 2010, CO_2 level in the atmosphere has risen drastically from 280 parts per million (ppm) during the 1700s to 397 parts per million (ppm) by the year 2005.

According to the Environmental Planning Agency (EPA), the major sources of CO_2 emissions are electricity accounting for 38%, transportation emitting 31%, industrial activities accounting for 14% of CO_2 emissions followed by residential and commercial activities emitting 10% and non-fossil fuel combustion accounting to 6% (IPCC, 2007c). This increase in CO_2 level is mainly due to burning of fossil fuels, trees, solid waste, and emissions from industrial processes that have occurred since the industrial revolution (IPCC, 2007a; Hughes et al., 2003; Parmesan & Yohe, 2003).

Almost 80% of the world population lives in urban areas (Kates & Torrie, 1998). Urban activities consume huge amounts of energy to improve the quality of life and as a result emit about 78% of CO_2 gas into the atmosphere (Grimm et al., 2008; Stern & Taylor, 2007). Hence, urban areas significantly increase CO_2 level in the atmosphere thereby, contribute to climate change (Dodman, 2009). However, urban areas as huge emitters of CO_2 not only contribute to climate change, but also experience the resulting impacts of climate change (Bell & Batterson, 1978).

Impacts of climate change such as loss of ecosystem, heat related illness, increased mortality rates and high economic losses are expected to have an effect on almost all aspects of human life (Post & Altman, 1994). Recently, scientific evidence reveals that all over the world, almost 160 million people residing in 20 coastal cities are expected to be affected by the rising sea levels (Nicholls, 1995). Hence, the impacts of climate change are very alarming and it is necessary to address climate change.

URBANIZATION AND CLIMATE CHANGE

Urbanization is happening rapidly. It is analyzed that by 2007, almost half of the world's population live in urban areas. In India, urban areas as hubs of economic growth 24 contribute to about 64% of the country's GDP (Ministry of Urban development, 2009). This productivity is mainly dependent upon the availability of infrastructure services. Infrastructure such as roads, mass transportation, roads, water supply and solid waste management not only increase the productivity of urban areas but also improves the quality of life in these areas.

Economic growth generally implies the need for more land to accommodate urban development (residential uses, industrial and commercial uses) and basically refers to the conversion of agricultural land into urban based industries and services. In developing countries, urbanization and per capita income increase at the same rate as productivity increases. So, Rapid productivity and high GDP implies improvements in industries and services, transportation production and information technologies thereby having a profound effect on urbanization (Quigley 1998, Romer 1986 and Lucas 2007; Engel, K. 2005). Rapid urbanization and improvements in technologies have spurred economic growth (Quigley 2009). Hence, Urbanization in simple terms means the development of industries which means increased CO_2

emissions and thereby contributes to climate change. So, it is important for both the developing and developed countries to address climate change.

Urban areas as concentrations of large economic activities encompass residential, commercial, industrial, and infrastructure services. Thereby has a significant impact on climate change. It is evident that urban areas through sprawl increase the usage of cars thereby contributing to climate change. This means that more the sprawl more is the CO_2 emissions. Similarly urban areas means increased residential and commercial buildings. More energy is required to heat and light the buildings (McCarney 2009). Increased consumption of energy leads to more generation of energy and eventually increased emissions thereby adding more CO_2 emissions to the atmosphere. 25 Increased area of a building requires more energy to heat and light. According to the Clinton Foundation, urban areas are responsible for about 75 percent of GHG emissions that are released into the atmosphere.

In this context, it is necessary to understand how urban areas in a developing country like India contribute to climate change. Grazi and Waisman (2009) analyzed that economic growth through urban agglomeration has external cost benefits on land use and transportation thereby contribute to climate change.

LOCAL CLIMATE CHANGE MITIGATION

The various global, regional and local impacts of climate change suggest that urban areas act as an interface between the components of climate change and urban activities. Various urban activities results in increased GHG emissions thereby, contributing to an increase in atmospheric temperatures and subsequently impacting the urban areas. Thus, urban areas not only result in causing climate change but are also adversely affected by climate change (Brown, 2001, 2008 & Nicholls, 1995). In order to overcome the increasing consequences of climate change there is a need for regions and cities to efficiently reduce their CO_2 emissions (mitigation action) and develop climate change resilient regions and cities (adaptation action) (Mcevoy, et al., 2006).

Opportunities of Climate Change Mitigation

In the literature of climate change, mitigation activities are broadly defined as strategies that tend reduce greenhouse gas emissions especially CO_2 gas emissions or promote greenhouse sinks (IPCC, 2007b). Climate change mitigation strategies are mainly developed to reduce CO_2 emissions into the atmosphere thereby not only safeguard the environment but also the humans who are most likely to be affected by the impacts of climate change CO_2 reduction activities mostly include but are not limited to promoting public transportation, improving energy efficiency, infrastructural investments and the implementation of renewable energy (Laukkonen et al., 2009). According to McCarney, climate change mitigation strategies mainly tend to target five sectors that are considered to be responsible for emitting large amounts of CO_2 emissions. They are: transportation, energy supply, industry, buildings, and waste management.

Generally, most researchers consider mitigation of climate change as a global, issue and responsibility. According to them, a major amount of GHG emissions are mainly generated from power generation industries. So, it is necessary for national level policies to target these industries and thereby significantly reduce their GHG emissions (Lim, Burton, Malone, & Huq, 2005). However, recently, there is an increasing research that focuses on analyzing the impacts of urban form, urban patterns and urban characteristics on carbon emissions. These studies conclude that by 28 altering the spatial patterns of urban development the CO_2 emissions of a community can be significantly reduced (Condon, 2010, Ewing, 2008, Raparthi, K. 2015, Hough, 2004).

Incorporating changes to the urban form by increasing the density within a community, by altering the design form, promoting regional accessibility and transit availability, and promoting the usage of low carbon energy statistically prove to be efficient in reducing CO_2 emissions to a considerable extent (Raparthi, K. 2018). For instance, transportation emissions which are primarily a result of car travel and one of the major sources of CO_2 gas can be reduced to some extent by reducing car travel through promoting better development patterns or considerable changes to existing patterns (Glaeser, 2008). In addition, individual behavioral choices to rely on public transport, car pooling and use bicycle mode or walking for short trips will all add up to reducing the transportation emissions which is one of the major contributor to GHG emissions. Hence, mitigation policies are no more limited to global and international level negotiations (Kyoto protocol), but can also be implemented at regional as well as local level. While mitigation is necessary and also proves to be beneficial in reducing CO_2 emissions, there are evidences that suggest that certain climatic changes are unavoidable and unexpected. So, there is a need to adapt to the changes by incorporating adaptation actions (Reilly et al., 2001).

Opportunities of Climate Change Adaptation

Climate change adaptation is referred to as actions that tend to prepare a region or a city to become accustomed to harmful effects of climate change. Unlike mitigation actions, which focus more on minimizing climate change, adaptation action focus more on acclimatizing to climate change (IPCC, 2007a). Unlike mitigation strategies, adaptation strategies are mostly short term, action–oriented solutions (Huq, 2006). According to IPCC, implementation of adaptation options lies on the ability of government 29 organizations, communities and individuals to transform strategies into actions (Tompkins 2005). Adaptation actions apart from being action oriented can also tend to facilitate social capital within a community, promote public works development and market processes (Smit, et al., 2000).

Adaptation actions similar to the mitigation actions can be addressed at the regional level and are mostly community based approaches. But unlike mitigation, they depend on the intensity of threat a local environment faces from a particular climate change impact and the inbuilt socio-economic capacity of the community to deal with the impacts (Mendelsohn, 2001). Hence, adaptation actions mostly depend on the ability of the region to respond to the impacts of climate change when they are experienced. The regional/community response to the impacts of climate change is almost similar to the way a region/ community responds to natural disasters. So, climate change adaptation actions require assessing both the risks and vulnerabilities that limit capacity building of impacts imposed by climatic changes (Adger, 1996). Hence, it is sensible to assess climate change vulnerability and plan for climate change by taking advantage of natural hazard vulnerability literature (Mendelsohn, 2000).

Synergies and Conflicts

In the climate change literature, it is evident that research on climate change adaptation is very limited when compared to the research on climate change mitigation. As there are increasing scientific evidences that acknowledge that climate change will continue to happen and it is unavoidable, it is necessary to protect the society from the impacts of climate change and hence requires more research (Stehr, 2005). However, researchers advocate that both mitigation and adaptation go hand in hand and there is a need to supplement mitigation actions with adaptation actions. According to Ager 2001, 30 by promoting climate change mitigation strategies in a community, the need for adaptive actions can be significantly reduced (Ager, 2001; Wilbanks, 2005).

Climate change mitigation policies mainly focus on reducing CO_2 emissions thereby eventually reducing adaption actions. Likewise, adaptation actions such as developing urban forests limits storm water run-off and eventually reduces CO_2 emissions in the atmosphere through carbon sequestration. Hence, synchronization between mitigation and adaptation actions can prove to be beneficial in addressing the impacts of climate change in a cost effective manner (Adger, 2001).

In contrary, mitigation and adaptation policies can also result to be counterproductive if they are implemented without addressing the co-relation between the climate change mitigation and adaptation strategies. Thereby, tend to undermine their ability in achieving their respective objectives (Wheaton, 1999). For example, emphasizing more on development of urban forests can lead to sprawling communities, resulting in more vehicular trips thereby increasing CO_2 emissions. Hence adaptation actions can prove counterproductive to mitigation action (Hamin, 2009). So, it is necessary to achieve a balance of mix between both mitigation and adaptation actions.

With the existing knowledge on climate change, researchers highlight that there is a need to develop a justifiable mix of mitigation and adaptation strategies to effectively address climate change (Klein, et al., 2005). The comprehensive approach requires urban areas to mainstream climate change mitigation and adaptation strategies into respective land use, transportation, environmental and developmental policies (Hamin & Gurran, 2009). This comprehensive approach mainly focuses and targets the CO_2 emissions sectors that emit huge amounts of CO_2 emissions such as transportation, buildings, energy supply industries, waste and industrial pollution activities and also promotes adaptation strategies such as pursuing innovative cooling, managing storm water and preserving trees and plants to adapt to climate change at the local level. Hence, mitigation and adaptation benefits are best achieved through this the community based approach, since it is not only manageable but also feasible (Kates et al., 2001).

However, most of the research and analysis related to climate change is qualitative rather than empirical and most of the studies are confined to only a single issue of climate change such as increase in sea levels (Sato, 2012, Markandya & Chiabai, 2009; Hunt and Watkiss 2007; Kreimer et al. 2003). The main challenge that is yet to be resolved in the literature of climate change is to develop statistical studies that analyze the relationship between climate change mitigation policies and CO_2 emissions. Later, integrate climate change mitigation policies in urban planning (IPCC, 2007e).

Reducing Climate Change Impact on Cities

Schlosberg, (2008) highlights that most of the climate change mitigation actions currently taking place to reduce CO_2 emissions at the local level are informed by urban planners. The various mitigation strategies can be broadly categorized into "compact city" approach and the "green city" approach. The 'compact city' approach emphasizes on high density development within a city (densification) and the 'green city' approach emphasizes the promotion of new developments as a result of the polycentric growth of a city (spatial expansion).

The "compact city" approach emphasizes that high density development can be ascertained by promoting future development of a city within the existing city limits. This requires an effective land use transportation planning approach and also incentives that encourage not only reducing distances between various activities and car travel demand but also reducing single family homes demand as they have a high demand for energy. Alternatively the "green city" approach emphasizes the urban design centered aspect of the developing urban sub-centers by promoting mixed land use developments. This involves the promotion of mass transit routes, and improvements in architectural designs.

PLANNING IN INDIA

Planning is the fundamental process required promoting development. In India, the "National Action Plan for Climate Change" under its eight missions' only highlights the aims and goals related to CO_2 emission reduction, which the country has to achieve (Biesbroek et al., 2009). For instance within the NAPCC, the aim of the "National Solar Mission is to promote the development and use of solar energy for power generation and other uses with the ultimate objective of making solar competitive with 39 fossil-based energy options." Likewise the goal of this mission is "to increase the use of solar thermal technologies in urban areas, industry, and commercial establishments". Therefore, the key role of the regional/local planning officials is to coordinate climate change (CO2 reduction) activities/policies that are dominant in reducing CO_2 emissions, and accordingly include such activities in the planning process to effectively reduce CO_2 emissions (Biesbroek et al., 2009).

In India, the idea of community participation and multi stakeholders in urban and regional planning is an accepted phenomenon in current day practice. The bottom up approach in Indian urban planning may be traced back to the late 1980's when Lindblom (1963) ideas of incremental planning, Davidoff's advocacy model and John Friedman's action-oriented model planning were widely welcomed among Indian urban planners, due to evident failures of the long-term master plans, concerning over-optimistic growth estimations and socio-economic problems of large scale development.

As such, India constitutional mandate (73rd & 74th Constitutional Amendment Act 1992) laid the foundation of citizen and stake holder's participation in urban / rural planning and also entrusted local Non-Governmental Organizations' (NGOs) and Community Design Centre (CDC) with the responsibility to intervene in the implementation process of plans with the participatory action of the public in great way. As such NGOs and CDCs advocates for participation in planning and they became the staging ground for professionals for such work.

Since, 1999, India has been actively engaging in multilateral negotiations in the UNFCCC. One of the main concerns of these negotiations is to reduce CO_2 emissions in the country. India as a developing country receives funding from the World Bank to focus on their emission reduction and the national government commissioned municipal governments to work towards emission reduction.

In regard to climate change, Betsill and Bulkeley (2007) identified that, lack of individual commitment to reduce CO_2 emissions is one of the main reason for not being able to effectively address CO_2 reduction. Hence, it is essential for communities to be effectively involved in the climate change planning process. As such planning for CO_2 emission reduction/climate change mitigation in Indian cities is necessarily a response to the environmental pressures and also to the pressures from the planning officials.

Challenges of Urban Planning

Most of the countries in the world were originally based on the European urban planning tradition. Accordingly most of the countries influenced by the European planning practice emphasized on land use planning and architectural urban design as a solution for most of their spatial urban problems, without much regard to the specificities of context (Friedman, 2005). Cohen (2006) describes the characteristics of urban planning by highlighting the enormous challenges urban planners' experience. One of the most common challenges is the unregulated growth and expansion of the cities due to urbanization. In many cities, unregulated growth has led to an increased pressure for land thereby, intensifying sprawl and congestion. Existence of inappropriate and obsolete planning regulations and codes and the dominance of the comprehensive long range planning as some of the challenges related to urban planning.

However, in the late 1980s, with the growing realization that planning can significantly influence the livability of cities if its environmental, socio-cultural and economic dimensions are fully explored many developing countries broadened their scope of urban planning by incorporating sustainable planning in their planning agendas. Accordingly, this has led to a shift from traditional planning concerns about land use, to concerns about the promotion of the sustainable development in many developing countries

India's Response to Climate Change

Impacts of climate change are very extensive and it is urgent to address these challenges. India has 17 per cent of the world's population and contributes nearly 18 per cent of the total global greenhouse gas emissions. While US contributes approximately about 30% and the European countries contribute nearly 23% of GHG emissions. In terms of per capita CO_2 emissions, India also has high per capita emissions of nearly 10 MT CO_2 (about 33% of the global average) compared with 24 MT CO_2 in US and 17 MT CO_2 in EU per capita emission. According to WRI (2009), India emits an estimated 1.7 T CO_2 emissions per capita per year (Tiwari, G. 2001; WRI, 2009).

India looks forward and participates in global negotiations on climate change. The country believes that developed countries are mainly responsible for the climate change problem, due to increased urban and economic growth. As growth will continue to happen, there is a need for developing countries to manage the growth and act accordingly by reducing their GHG emissions. In order to meet the demands of living and raise their living standards, total GHG emissions are bound to increase in India as well as in other developing countries.

India being committed to a pathway of sustainable development, recently adopted a National Action Plan on Climate Change (NAPCC) that consists of various urban CO_2 emission reduction missions. The NAPCC outlines national level policies that address climate change mitigation as well as adaptation. The action plan identifies eight core areas and refers them as national missions to effectively mitigate and adapt climate change. They are "National solar mission", "National Mission for Enhanced Energy Efficiency", "National Mission on Sustainable Habitat", "National Water Mission", "National Mission for Sustaining the Himalayan Ecosystem", "National Mission for a "Green India" "National Mission for Sustainable Agriculture" and "National Mission on Strategic Knowledge for Climate Change" (NAPCC, 2008).

The NAPCC through these eight missions emphasize the need to maintain high standards of air quality, living conditions and economic growth. Accordingly, identifies strategies that promote sustainable development objectives and at the same time addressing climate change. Being a member of the Asia Pacific Partnership on Clean Development and Climate, the country actively participates at the UN Framework Convention on Climate Change (UNFCCC) and various international conferences to collaboratively develop and promote the usage of clean and environment friendly technologies.

PUBLIC POLICY IN URBAN PLANNING

Comprehensive, development and master plans are regarded as a means of managing growth and directing development in order to achieve the desired goals of a community. These plans generally include elements of land use, urban design, transportation, housing, energy, environment, public facilities and economic development (Raparthi, K. 2021). This type of cross-thematic integration between development policies and various plan elements of a community along with the underlying regulatory framework are regarded as a viable tool in managing the growth and development of a community thereby help in accomplishing the goals and objectives of a community (Raparthi, K. 2018). Hence, these plans are regarded as the necessary policy document that promotes urban planning process in the city (Kaiser, 1995b).

Across the world, public participation in urban planning process is widely accepted to be efficient (Tauxe, 1995). The principles of public participatory planning in urban planning mainly emphasizes the stakeholders rights to be informed, to promote innovative issues and also assess alternative solutions during the planning process (Kaiser & Godschalk, 1995b). According to Barrett and Fudge (1981), implementation that develops as a response to actions is referred as an action-centered approach as such involves an action in planning. Research on urban planning and policy formation methods analyzes the effectiveness of decision making through public participation to be an effective practice for the implementation of planning goals (Barrett & Fudge, 1981).

Traditionally urban planning structure has been a regarded as public policy that is shaped by power (Kaiser & Godschalk, 1995b). Research emphasizes the role of urban planning policies in promoting sustainability (Raparthi, K. 2020), in environmental management (Brody, 2003), intergovernmental collaboration (Kaiser & Godschalk, 1995b) and promoting natural resiliency performance and mitigating natural hazards (Raparthi, K. 2016; Nelson and Steven, 2002; Berke and Beatley 1992; Brody 2005). This aspect of public policy within urban planning process that lies beneath the urban planning documents symbolizes the regulatory framework, the integrative, collaborative nature and the scope for public participation to develop innovative solutions in planning process. Accordingly, urban planning and public policy have the ability to mitigate climate change by reducing CO_2 emissions and also the potential to increase governance to mainstream climate change mitigation policies in urban planning respectively. Thereby providing the regions an opportunity to be more responsive towards climate change mitigation.

CLIMATE CHANGE PLANNING IN INDIA

Urban planning is a diverse field and can be viewed from various perspectives. This may involve perceptions which focus on urban planning policy issues while others may consider it in terms of urban planning practice. Urban planning involves a range of activities designed to promote the growth and development of the built environment (Kaiser, 1995b). Since the 1900s, the field of urban planning has undergone a gradual shift from an initial emphasis on the promotion of architectural beauty to urban reform movements with concern about improving the living conditions of people, focusing on issues of equity and social justice. Urban planning is a 'rational' process and must be approached from a scientific and technical perspective. Accordingly, rational planning mainly emphasized on identifying and focusing on the best methods for practicing planning and future courses of action with regard to the city.

Gradually in the mid-1970s with the introduction of political discourse into urban planning and later, in the 1980s with the launch of John Friedman's context-value laden nature of participatory, action oriented, communicative and radical approaches, planning was viewed as a mutual learning process involving consensus building rather than a rational decision making process. As such in the 1990s the field of urban planning witnessed a significant shift in its use of 'rationality' from one that views plan making 35 (providing an outline of the various development strategies, policies, and activities taken at the different planning scales) and plan implementation (putting plans into effect) in isolation to one that views plan making process and plan implementation as a single activity.

In the mid-1970s, implementation was described as an inseparable activity from plan making process, as such involving the inclusion of a wide-range of actions in the process of shaping and making of plans. Action model of planning highlights the political process involved in the implementation of the plans. He advocates that a planner must be like a facilitator or communicator trying to promote participatory planning among all interest groups, discussing issues with the public and developing a hierarchy of goals and workable strategies for plans with the public interest and also put them into action with the help of the public. Accordingly highlights that, planning is a form of undertaking action, and implementation means thoroughly carrying out the planning.

Planning is a mutual learning process and a participatory action process where in both the planner and the public participate together in the plan making process and put into action as such work towards the reformation of the project undertaken. Radical planning process is the best approach to practice planning since they are created with the public interests within the social context and is value laden (Taylor, 1998). Thereby highlights that for effective implementation of a plan, the planner must act as a facilitator, researcher and must be skilled at contacting, communication and negotiating with others (Raparthi. K. 2020; Forester, 1989; Healey, 1992; Taylor, 1998).

Later in the early 1990s, the phrase communicative planning was coined highlighting that negotiation was a pre-requisite skill for a planner. Since then, implementation was widely used to imply that decision-making processes for the city has been widened out, with a restructuring of the relationships between city government officials, planning professionals, non-profit organizations (NGOs), stakeholders and the public. As such, radical planning demands for more participation in decision-making and underscores the need for planners to involve clients in debate and dialogue to effectively implement urban plans.

From a public policy perspective, this approach to implementation is referred as the "bottom-up; action centered" approach of implementation. From this point of view implementation is seen as "action perspective" which is in contrast to the "policy centered" approach. The latter considers implementation to be "putting policy into effect" whereas action-perspective claims that instead of formulating policy first then going on with the implementation; policy could arise as the plan was implemented.

Mainstreaming Climate Change Concerns in Local Urban Planning

Discussion of integrating climate change concerns into urban planning and considering climate change as a new aspect of sustainability has started evolving. Themes of sustainability, smart growth, and new urbanism have been a part of planning concerns and have been integrated into master plans. Likewise, master plans may also have the ability to integrate climate change mitigation policies that reduces or moderates the impacts of climate change in the master plans (Raparthi, K. 2014).

Researchers who advocate mainstreaming climate change mitigation into urban planning have identified a set of planning actions that benefit in achieving CO_2 emission reduction. Climate change mitigation effective planning policies focus on promoting low building energy use, reducing vehicle miles, creating dense urban environments and green urban spaces. All these policies eventually reduce CO_2 emissions (Raparthi, K. 2014; Kahn, 1996). Cervero and Kockelman (1997) identified the 3 D's namely density, diversity and design as a model for sustainable urbanism. Later, Ewing (2008) in addition to the 3 D's adds two more components thereby puts forth five key components for climate change mitigation and addresses them as the 5 D's. These are: accessibility to destination, short distance between work and transit, increased density; increased diversity in land uses; and small block designs. The main benefit of all these aspects is the significant reduction in vehicle miles travelled and eventually the reduction of CO_2 emissions.

In addition to these aspects, Ewing (2008) also suggest urban growth boundaries, workforce housing within close proximity to jobs, pedestrian friendly surroundings and low energy efficient building design standards. Recently, compact urban form is also closely related to climate change mitigation (Ewing, 2008). The justification behind this is that low density development is associated with more sprawling areas that lead to more vehicles miles travelled and increased CO_2 emissions. Recently low density is also related to large single family dethatched units with large interior spaces that lead to more energy consumption and eventually more energy demand. This eventually increases GHG emissions (Raparthi, K. 2018; Ewing, 2008).

The Climate Protection Agreement recommends a set of both short and long term activities for mitigating climate change (ICLEI, 2005). A set of short-term activities that is relevant to planning are: planting shady street trees; maintaining urban forests, encourage car-pooling and mass-transit; promoting usage of green energy; strict residential and commercial building codes; promoting reuse and recycling programs and public education. Long-term actions include: promote high-density, relocating facilities to reduce travel time; ordinances to limit sprawl, infill development and Brownfield redevelopment; preserve open space; and infill development.

Climate change researchers agree that for policies to effectively mitigate climate change it must constitute a mix of climate change mitigation and adaptation actions. Policies and actions that emphasize in limiting or reducing CO_2 emissions contribute to the climate change mitigation activities.

QUALITATIVE AND QUANTITATIVE EVALUATION OF URBAN SPATIAL PLANNING POLICES

To understand the impact of urban planning policies on CO_2 emissions it was necessary to understand the extent to which the regional master plans included climate change mitigation policies and also the extent to which they are successfully being implemented. Hence, climate change mitigation policy evaluation protocol identified the extent to which master plans developed climate change mitigation policies and (2) climate change mitigation implementation evaluation protocol highlighted the extent to which master plans implement the climate change mitigation policies.

The two evaluation protocols consist of a set of indicators that tend to have a significant impact on climate change mitigation and ability of the policies to be successfully implemented. In specific planning policies related to land use, urban design, physical planning, building specifications, transportation, environment, incentive tools, educational tools, attainment tools and physical infrastructure were analyzed. Table 1 describes the protocol against which the master plans were evaluated for the presence of climate change mitigation policies.

Table 1. Policy evaluation protocol

Plan Component	Recommended Climate Change Mitigation indicators
Land use	Promotes: • Mixed use development • Brownfield (or Greyfield) redevelopment • Infill development • Limiting use or limits use of hazardous areas/ marginal areas (Overlay zones/ reduced densities)
Urban Design	Promotes: • High density development • Urban landscape development • Has proposals/actions to decrease urban heat island effect (urban forests, tree lined streets) • Has regulations on building height/ orientation guidelines, street width to building height ratios
Physical planning	Requires: • Site plan review for land suitability assessment • Setbacks/Buffers • Subdivision regulations
Building specifications	Requires/Promotes use of: • Solar PV panels and wind turbines on roofs • Storage, collection and recycling of wastes • Water-efficient construction • Recycling of grey-water • Rainwater harvesting • On-site water treatment • Building standards/code for enhanced protection.
Transportation	Requires/Promotes use of: • Creating /implementing /enhancing public transportation systems • Transit oriented developments • Car sharing and car pooling • Increased public transportation stops/nodes • Creation/ upgrading of bicycle paths • Creation/ upgrading of pedestrian facilities • Management of no traffic zones
Environment	Promotes: • Environmentally sensitive area protection (national/state parks) • Conservation of forests, vegetation, and riparian areas • Creating wildlife corridors • Preventing habitat fragmentation • Sediment and erosion control regulation • Wetlands restoration
Incentive tools	• Subsidized mass transit / incentives for car pooling • Impact fees for development in ecologically sensitive areas • Density bonuses
Educational tools	• Education and outreach program during plan implementation • Training/Technical assistance to developers or property owners
Attainment tools	• Land and property acquisition • Transfer/Purchase of development rights
Physical Infrastructure	• Maintenance of public Infrastructure • Capital improvements for developments

Source: (created by author, 2021)

In addition, it was necessary to evaluate the plans for their implementation capacity. Accordingly, table 2 describes the protocol against which the master plans were evaluated for their implementation capacity.

Table 2. Plan implementation protocol

Implementation Indicators
• Initiation of NGOs, CDCs, Research Institution
• Designation of responsibility for implementation
• Funding for implementation
• Timetable for implementation
• Sanctions for failure to implement
• Regular update of the projects
• Provisions for technical assistance
• Monitoring of environmental and human impacts
• Public participation process in monitoring and review
• Provision of plan response to new information/ data

Source: (created by author, 2021)

The climate change mitigation policy index was developed by evaluating the planning policies within the master plans against the policy evaluation protocol (table 1). Likewise, the climate change mitigation implementation index was developed by evaluating the planning policies against the plan implementation protocol (table 2). As such, these protocols evaluated the plans for their ability to address mitigation projects and their ability to successfully implement the addressed projects.

Climate Change Mitigation Index Calculation

The Climate change mitigation policy index in this chapter was calculated by using the "Developing Index method", which was widely used by many researchers in evaluating the quality of a plan. The climate change mitigation implementation index was calculated by using the "Plan Implementation Evaluation (PIE) method" which was used by planning researchers to evaluate the implementation success of plans.

The development of the index includes five steps: The first step was to assign the scores for each policy on a scale of 0-1-2. Scores will be assigned based on the presence of the indicators that are mentioned in the evaluation protocol. If the indicator was absent within a policy then, the score of the policy was "0". A score of '1' was assigned to policies which address the indicator but tend to be a suggestive policy. Words such as 'may', 'should', 'prefer', 'encourage', 'suggest' indicated the suggestive character of the policy. If the indicator was present in the policy and the policy was a mandatory policy then that policy receives a score of 2. Mandatory policies usually addressed keywords such as 'shall', mandated', 'must', and 'will'. The range of these scores was from 0 to 2 (Edwards, 2007; APA 2013). The second step was to sum all the indicator scores within each plan component. The third step involves calculating the fractional scores of each plan component. This was achieved by dividing the total of assigned scores for each plan component by the maximum possible scores of the plan component. (Maximum possible scores imply that all the indicators are mandatory. For example, if there were 5 indicators in a plan component, then the maximum possible score of that plan component was 10). In the fourth step, the fractional scores were standardized by multiplying the fractional score by 10. So that scores for each plan component can be scaled on a range between 0 and 10.

Climate Change Mitigation Implementation Index

In addition to the master plan's policy evaluation for climate change mitigation, the implementation potential of the master plan was also assessed using the plan implementation capacity evaluation protocol that consists of ten implementation indicators. Absence of the implementation indicator was coded as '0'. A score of '1' was given if the implementation indicator was "mentioned but not in detail" (suggestive). Indicators that were "mentioned in detail" (mandatory) were assigned a score of '2'. The implementation index" of a city was calculated as the ratio of the indicator capacity score received by each city's plan to the maximum possible score any city plan can achieve and multiplied it by 100 (In this research, the maximum possible points any city can achieve is 20 points (10x2) since there were 10 indicators against which the implementation capacity of a city was being evaluated.The climate change mitigation implementation index ranges from 0 to 100. Assuming that for a city's plan all the 10 implementation indicators were "mentioned in detail" then the Implementation Score is (10 x 2)/20 multiplied by 100 which is 100.

Accordingly, by calculating both the climate change mitigation policy index and implementation index for each city master plan, it is able to deduce whether planning polices within the master plan have an ability to influence climate change mitigation within the city.

CLIMATE CHANGE MITIGATION POLICIES IN MASTER PLANS

Generally, most of the master plans in India are organized in two sections. The first section provides the baseline data and an analysis of the population and employment growths in the city thereby provides the rationale of urban planning within the city and lays down the basis for future developments. The second section of the master plans describes the vision and planning goals of the city. The master plans comprised of ten components/sections (land use, urban design, physical planning, building specifications, transportation, environment, incentive tools, education tools, attainment tools and physical infrastructure) that guide the development of a city. The objectives and policies pertaining to each of the plan components/ section are addressed individually under their respective sections. As anticipated, none of the sampled master plans had considered climate change mitigation as a planning goal or had any planning section exclusively focusing on climate change mitigation. Therefore, the climate change mitigation evaluation protocol that was developed to analyze the presence of climate change mitigation policies in the master plans consists of a set of common indicators that tend to have a significant impact on CO_2 emissions.

To facilitate analysis, climate change mitigation indicators were categorized into 10 sections of the master plans namely Land use, urban design, physical planning, building specifications, transportation, environment, incentive tools, education tools, attainment tools and physical infrastructure.

Descriptive statistics of the overall climate change mitigation policy scores and the total policy index are presented in table 3. Each plan component of the climate change mitigation evaluation protocol consisted of indicators ranging from 2 to 7. The transportation and building specifications components have maximum of 7 indicators each, while the educational tools component, attainment tools component and the physical infrastructure component have a minimum of 2 indicators. The environment component has 6 indicators. Land use zoning component and urban design components have 4 indicators each. The physical planning component and incentive tools component includes 3 indicators each. The maximum possible score an indicator can receive is 2 and this is possible only if the indicator is addressed to be mandatory in the master plan.

Table 3. Climate change mitigation policy index scores

Plan Component	Number of policies	Maximum Possible score	Mean Minimum Score	Mean Maximum Score	Mean Standardized Score
Land use	4	8	2	7	6.26
Urban design	4	8	4	6	6.12
Physical planning	3	6	2	4	6.29
Building specifications	7	14	2	12	5.30
Transportation	7	14	2	12	8.36
Environment	6	12	2	10	7.10
Incentive tools	3	6	1	5	5.39
Education tools	2	4	1	3	5.68
Attainment tools	2	4	1	2	4.24
Physical Infrastructure	2	4	2	3	3.80
Total Policy Index	40	80	15	64	58.54

Source: (created by author, 2021)

For instance, maximum possible score of the transportation component will be14 if all seven indicators are addressed as mandatory policies (7 indicators multiplied by 2). Likewise a minimum possible score of 0 indicates that the plan did not address any of the indicators that have been included in the plan component within the evaluation protocol. The indicator score for each plan component is a measure of the extent to which that indicator is included in the planning component. The climate change mitigation policy index of a city's master plan is the measure of the extent to which climate change 75 mitigation policies are promoted by that city. Assuming that all the 10 plan components of a city's master plan achieved a standardized score of 10, then the mitigation index of such a city will be 100.

Results show that each of the 64 master plans had at least one planning policy related to climate change mitigation (had a significant impact on CO_2 emissions) within their planning components. 76 As mentioned earlier, the climate change mitigation policy index for a city's master plan ranges from 0-100 points. The mean climate change mitigation index for all the sampled plans (64 master plans) is 58.54 points, as such representing 60% of the maximum climate change mitigation policy index. The mean maximum and minimum index ranges from 64 to 15 thereby suggesting that the climate change mitigation policies and its associated details vary within the sampled communities. Among the various components, transportation planning component has the highest mean climate change mitigation standardized score 8.36 (83.6%) of the possible maximum component score 10 (100).

This means that the sampled plans have made efforts to reduce their regions CO_2 emissions by addressing transportation related issues, thereby emphasizing more on transportation planning policies in their master plans. Comparatively, the mean mitigation standardized score for physical infrastructure component is only 3.80 (38%) of the possible maximum component score, indicating a lack of attention to these policies in the sampled master plans. On the whole, the mean standardized scores for most of the plan components ranges from 4.0 to 6.0 (40 - 60%) of their respective possible maximum component scores (10; 100%).

This suggests that a moderately fair amount of climate change mitigation related urban planning policies are present in their master plans. Thereby, indicating that the sampled master plans include policies that target climate change mitigation.

CONCLUSION

Ability of master plans in sustainable development, ecological and environmental conservation has been widely analyzed. Research by (Kaiser & Godschalk, 1995b) supports the use of master plans as an effective tool in reducing disaster losses thereby promoting disaster resilient communities through urban planning. Other research studies emphasize the importance of master plans in managing and promoting sustainability, ecosystem management and natural hazards (Laurian, et al., 2004; Brody, 2003).

However, studies on urban planning policies and their ability in mitigating climate change are limited. Hence, outcomes of this research may tend to complement the earlier studies that depict the ability of urban plans. This chapter builds on the literature of the diverse role of master plans in urban planning, and contributes to the existing literature by developing empirical evidence that supports the role of master plans in mitigating climate change. This chapter finding emphasize the ability of master plans to mitigate climate change and supports the use of master plans as an effective tool in mitigating climate change thereby promoting low-carbon regions through urban planning.

Overall, this research supports the use of master plans as a practical tool in mitigating climate change. Findings of this research emphasized the relation between CO_2 emission and urban planning policies contained in the master plans. As a result, this research has an implication for mainstreaming climate change mitigation in urban planning. The phase 1 of this chapter (Development of climate change policy and implementation indexes for the master plans in the sample) addressed the first research question which was to identify whether master plans promoted and included policies that target climate change mitigation. The research hypothesis was that higher climate change mitigation policy Index and higher climate change mitigation implementation index for a master plan indicated that, the master plan included and promoted planning policies that mitigate climate change.

Thereby, they have an ability to support climate change mitigation. Content analysis of the policy framework identified that 60% of the sampled master plans had policies that addressed climate change mitigation and 58% of the sampled master plans had implemented their policies towards climate change mitigation. Besides, higher climate change mitigation indexes will significantly have lower CO_2 emissions thereby have an ability to mitigate climate change.

This research has an implication for mainstreaming climate change mitigation in urban planning and for promoting low-carbon communities through urban planning. Thereby encourages planners to integrate climate change mitigation policies in local master plans. This chapter significantly contributes to the field of urban planning and public policy by developing empirical evidence that analyzes the relationship between urban planning policies and CO_2 emissions. Thereby emphasizes the application of urban planning as a policy tool for mitigating climate change.

FUTURE RESEARCH DIRECTIONS

This chapter makes a significant first step approach towards understanding and analyzing the ability of urban planning policies and other socio economic characteristics in mitigating climate change by reducing CO_2 emissions. Future research is necessary to enable a more detailed understanding of the relationship between urban planning policies and climate change mitigation.

This chapter only focused on evaluating the impact of climate change mitigation policy indexes and implementation indexes on CO_2 emissions. Future studies can be conducted by undertaking a disaggregated discriminant analysis of each planning component and identify which of the planning components within master plans contributes more to climate change mitigation. Discriminant analysis research will generate a greater evaluation of the strength and weaknesses of each plan component and policies with respect to climate change mitigation. Undertaking detailed local level analysis by conducting, case studies and/or in-depth content analysis of metropolitan regions master plans that had sections on details of their urban plans related to important city's /neighborhoods within the metropolitan regions/urban agglomeration will help in identifying the intricacies and best practices for effectively reducing CO_2 emissions.

Qualitative content analysis also identified the need to develop alternative governance structures that focus on collaboration and sharing power with rather than structures that are built to enforce or control from a climate change mitigation perspective. Accordingly undertaking detailed case study analysis will help in identifying barriers and also opportunities that hinder mainstreaming of climate change mitigation in urban planning and in gaining valuable insights that can help in developing an integrated framework for climate change mitigation.

REFERENCES

Agyeman, J., Evans, B., & Kates, R. (1998). Greenhouse gases special: Thinking locally in science, practice and policy. *Local Environment*, *3*(3), 245–246. doi:10.1080/13549839808725563

Barrett, S., & Fudge, C. (1981). *Policy and Action. Essays on the Implementation of Public Policy*. Metheun.

Berke, P. R. (1995). Natural-hazard reduction and sustainable development: A global assessment. *Journal of Planning Literature*, *9*(4), 369–382. doi:10.1177/088541229500900404

Betsill, M., & Bulkeley, H. (2007). Looking back and thinking ahead: A decade of cities and climate change research. *Local Environment*, *12*(5), 447–456. doi:10.1080/13549830701659683

Bulkeley, H. (2006). 'A Changing Climate for Spatial Planning' in 'Is the Issue of Climate Change too big for Spatial Planning?'. *Planning Theory & Practice*, *7*(2), 201–230. doi:10.1080/14649350600681875

Cervero, R., & Kockelman, K. (1997). Travel demand and the 3ds: Density, diversity, and design. *Transportation Research Part D, Transport and Environment*, *2*(3), 199–219. doi:10.1016/S1361-9209(97)00009-6

Engel, K. (2005). Mitigating global climate change in the United States: A regional approach. *New York University Environmental Law Journal, 14*, 54–85.

Ewing, R., Bartholomew, K., Winkelman, S., Walters, J., & Che, D. (2008). *Growing cooler: Evidence on urban development and climate change*. ULI.

ICLEI. (2005). *International council for local environmental initiatives*. Retrieved Oct 2005, from http://www.iclei.org/

IPCC. (2007a). *Climate change 2007: Impacts, adaptation and vulnerability*. Cambridge University Press.

IPCC. (2007b). *Climate change 2007: Mitigation of climate change*. Cambridge University Press.

IPCC. (2007c). *Climate change 2007: Summary for policymakers of the synthesis report of the IPCC fourth assessment report*. Cambridge University Press.

IPCC. (2007e). *Climate change 2007: The physical science basis*. Cambridge University Press.

Kahn, M. (1996). New evidence on trends in vehicle emissions. *The RAND Journal of Economics, 27*(1), 183–196. doi:10.2307/2555798

Kaiser, E. J., & Godschalk, D. R. (1995b). Twentieth century land use planning: A stalwart family tree. *Journal of the American Planning Association, 61*(3), 365–385. doi:10.1080/01944369508975648

Laukkonen, J., Blanco, P. K., Lenhart, J., Keiner, M., Cavric, B., & Kinuthia-Njenga, C. (2009). Combining climate change adaptation and mitigation measures at the local level. *Habitat International, 33*(3), 287–292. doi:10.1016/j.habitatint.2008.10.003

Nelson, A. C., & Steven, P. (2002). Plan quality and mitigating damage from natural disasters: A case study of the north ridge earthquake with planning policy considerations. *Journal of the American Planning Association, 68*(2), 194–207. doi:10.1080/01944360208976265

Raparthi, K. (2014). Assessing smart growth strategies in Indian cities: 1441 A grounded theory approach to planning practice. *J. Urban Plann., 1442*. ,05014031 doi:10.1061/(ASCE)UP.1943-5444.0000267

Raparthi, K. (2015). Assessing Smart Growth Strategies in Indian Cities: A Grounded Theory Approach to Planning Practice. *Journal of Urban Planning and Development, 141*(4), 1–10. doi:10.1061/(ASCE)UP.1943-5444.0000267

Raparthi, K. (2015). Analyzing the Relationship between Environmental Planning Policies and Climate Change: Multinomial Logit Regression Model Evaluation of Tarrant County, Texas. *Current Urban Studies, 3*(1), 1–10. doi:10.4236/cus.2015.31001

Raparthi, K. (2016). Assessing the Initiatives in Climate Responsive and Energy Efficient Architecture: Bridging the Gap between Architectural Research and Practice". *International Journal for Science and Advance Research in Technology, 2*(6), 94–101.

Raparthi, K. (2016). Assessing Climate Change Planning in Indian Cities: Bridging the Gap between Climate Change Research and Practice. *Social Science and Humanities Journal, 1*(3), 160-174.

Raparthi, K. (2018). Assessing the Role of Urban Planning Policies in Meeting Climate Change Mitigation Goals in Indian Cities. *Journal of Urban Planning and Development, 144*(4), 1–16. doi:10.1061/(ASCE)UP.1943-5444.0000440

Raparthi, K. (2020a). *Investigating Open Space Reservation as the New Urban Commons: A Case of Chennai City* (Vol. 52). NAGARLOK Quarterly Journal of Urban Affairs.

Raparthi, K. (2020b). Ability of Nature Based Solutions Through Urban Green Space Management in Building Urban Resilience. *International Journal of Urban Design., 3*(2), 46–56.

Raparthi, K. (2021). 'Assessing the Relationship between Urban Planning Policies, Gender and Climate Change Mitigation: Regression Model Evaluation of Indian Cities. *Journal of Urban Planning and Development, 147*(2), 05021007. Advance online publication. doi:10.1061/(ASCE)UP.1943-5444.0000677

Tiwari, G. (2001). Traffic flow and safety: Need for new models in heterogenous traffic. In D. Mohan & G. Tiwari (Eds.), *Injury prevention and control* (pp. 71–88). Taylor & Francis.

Wheeler, D., & Hammer, D. (2010). *The Economics of Population Policy for Carbon Emissions Reduction in Developing Countries. CGD Working Paper 229.* Center for Global Development. https://www.cgdev.org/content/publications/detail/1424557

Wilbanks, T. J. (2003). Integrating climate change and sustainable development in a place-based context. *Climate Policy, 3*(Supplement 1), S147–S154. doi:10.1016/j.clipol.2003.10.013

Wilbanks, T. J. (2005). Issues in developing a capacity for integrated analysis of mitigation and adaptation. *Environmental Science & Policy, 8*(6), 541–547. doi:10.1016/j.envsci.2005.06.014

Chapter 2
Object-Based Land Cover Classification Using Multisensor Remote Sensing Data

Rubeena Vohra
Bharatividyapeeth's College of Engineering, India

Kailash Chandra Tiwari
Delhi Technological University, India

ABSTRACT

The goal of this chapter is to demonstrate the classification of natural and man-made objects from multisensory remote sensing data. The spectral and spatial features play an important role in extracting the information of natural and man-made objects. The classification accuracy may be enhanced by fusion technique applied on feature knowledge database. A significantly different approach has been devised using spatial as well as spectral features from multisensory data, and the classified results are enhanced by majority voting fusion technique. The author concludes by presenting extensive discussion at each level and has envisaged the potential use of multisensory data for object-based land cover classification.

INTRODUCTION

Ever since the advent of satellite technology, remote sensing has grown in leaps and bounds simultaneously keeping pace with the development in sensor technology. Today, these integrated technologies equip us with the capability to remotely acquire information about the ground surface and study various Earth features and phenomena occurring on it such as water, soil, minerals, agriculture, urban growth and various other geographical features and phenomena.

Remote sensing sensors can now obtain high-resolution imageries in different specific bands which seem useful in identifying and categorizing various land-cover classes. The categorization of land-cover classes provides significant information which is used in applications related to land use planning such as, suburban sprawl, urban growth trends, management policy regulations and inducements. These ap-

DOI: 10.4018/978-1-7998-8331-9.ch002

plications require authentic information on Land Use / Land Cover (LULC) classification patterns which in turn help in facilitating the balanced interaction between the humans and the environment through developers of LULC information (Varlyguin, 2001). Not limited to this, the information appreciably develops the association of geospatial technologies along with the development of technological product with the available resources in least possible time.

One of the major issues affecting mankind and its continued progression as well survival on planet Earth today is the growth in urbanization that has led to various environmental and climatic changes. Land-cover planning and management using remotely sensed images has therefore become a very helpful tool in identification and classification of various features of the urban land surfaces. In many studies, high spectral resolution data has been explored for classification of land features in urban environment. The high spectral resolution data is largely characterized by the radiance parameters captured from different land-covers in a large number of bands of the electromagnetic spectrum. Many a times, it may not have an optimal spatial resolution; therefore, it may be inadequate for a specific identification and classification task despite of its good spectral resolution. On the other hand, in a high spatial resolution remote sensing data, while geometric features can easily get recognized but shadow of various land features and the spatial adjacencies between the features may lead to ambiguities and lower the classification accuracy. The classification of physical features of land surfaces using only multispectral remote sensing data may lead to uncertainty as the use of single source remote sensing data may arise the ambiguities due to spatial adjacencies between the various land cover features. The spectral similarities shown by many land-covers features may also contribute in uncertainty if only a single source multispectral data is considered.

Similarly, while all types of remote sensing data such as Panchromatic (PAN), Multispectral (MS) and Hyperspectral (HS) data are suitable for urban land feature classification; all of them have their advantages and limitations. Most physical urban land- cover features either show similar spectral radiance or are combination of mixed pixels, therefore the classification of land features becomes difficult to be handled by low resolution or single source remote sensing data, it may require either data with fine spatial resolution or data with high spectral resolution. The desired data may not always be available and where available, there may be a requirement of analytical tools to facilitate urban land cover feature classification. For example, HS data is known to be rich in information content due to its high spectral resolution but traditional classification algorithms using HS data are still under different stages of investigation.

Classification unit in the algorithms are either pixels or a group of pixels and in practice, most land-cover features are known to exhibit either spectral or spatial similarity. Therefore, there is a requirement of exploring efficient ways of handling the HS data for urban land feature classification. Similarly, while most land features of land surface may now get captured in different electromagnetic spectrums simultaneously and the information is available in wider range, yet classification algorithms deal with the spectral range information for land cover features separately and this may lead to lower classification accuracy. The issue again merits investigation. Remote sensing data may be useful in classification of various physical features on urban land surfaces. For example, the idea of combining the multisource remote sensing data seems to strengthen the urban land-cover classification. Similarly, long wave infrared thermal HS data may possibly aid in detection of natural and man-made land features. There, however, appears limited indigenous effort in exploitation of spectral and spatial features from multisource airborne and satellite based remote sensing data. The motivation of this chapter originates from an urge to explore some of these possibilities discussed here.

In view of the foregoing discussion on the identification and extraction of objects in an urban area, the present chapter focuses on the potential use of Long Wave Infrared HS (LWIR) and Visible (VIS) RGB data for selection of spectral and spatial features and delineation of urban land-cover objects using advanced classifier *i.e.,* SVM. The use of Thermal Infrared (TIR) imaging is scarcely used in fusion techniques though the data is highly relevant for urban land-cover classification. TIR imagery can describe surface-emitted energy differences caused by human activity (Galiano, 2012), therefore, helpful in acquiring the information about the location and shape of objects even in darkness. For this reason, TIR imagery is getting popular and can be used for identification of natural or man- made objects in an urban area.

The chapter highlights the extraction of natural and man-made objects by computing features from coarse resolution HS-1-m LWIR and fine resolution -20-cm VIS RGB data. The spectral and spatial characteristics are preserved from HS and high-resolution spatial data for classifying five classes under natural and man-made objects. Under the category of natural objects; the classes that have been extracted are, vegetation, trees and soil, and under the category of man-made objects; roads and buildings have been extracted. Feature oriented object-based classification is performed to generate the classification map that highlights the natural (soil, trees, and vegetated area) and man-made objects (roads and buildings) in the given dataset.

The chapter discusses various tasks that have been formulated and investigated. Following are the investigations/tasks presented,

- Explore the possibility of multi-sensor data for preserving the spectral and spatial characteristics of various land-cover objects for classification.
- Explore the effectiveness of fusion in multi-sensor data i.e., thermal infrared hyper spectral and VIS RGB data.
- Extract the spatial and spectral features for extraction of natural and man-made objects.
- Segment the features on bounding box of each class under investigation.
- Classify the data based on the feature analysis using SVM classification scheme.
- Generate the classification maps of the land-cover classes under natural and man-made objects.

REMOTE SENSING DATA ACQUISTION

Remote sensing is the science and art which involves the acquisition of remote sensing data and understanding it for extracting the relevant information. The acquisition of data in an appropriate wavelength range is highly important for extracting the information of various objects on the ground. Most objects on ground reflect a part of the electromagnetic energy radiated towards it from the sun or emit some energy due to their own temperatures (Jong, 2004). The remote sensing sensors that are classified as passive and active sensors record the information about an object by measuring the electromagnetic energy in the form of reflectance or emittance. The data acquisition by the remote sensing sensors is illustrated in Figure 1.

Figure 1. Remote sensing systems for data acquisition
Source: Jong, 2004

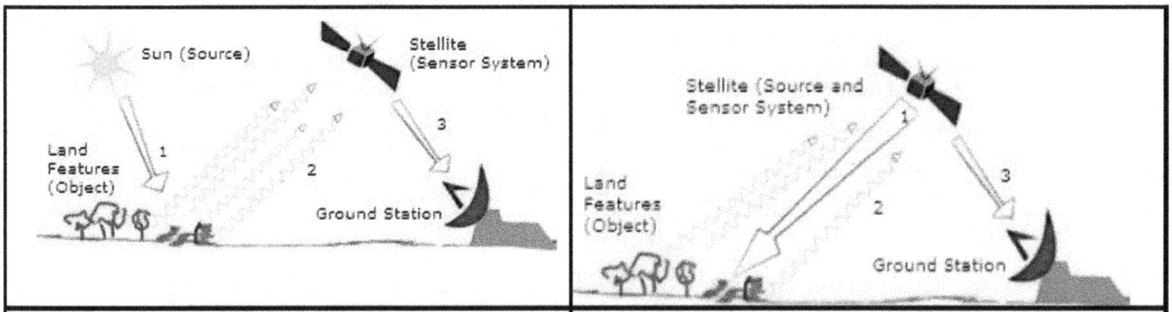

(a) Data acquisition – Passive sensor | **(b) Data acquisition – Active sensor**

The extraction of information using remote sensing data depends upon sensor resolution which is described broadly in four domains *i.e.,* spatial, spectral, radiometric and temporal (Liang, 2012). The characteristics of these four types of sensor resolutions are explained in Table 1 (Malgorzata, 2010). These resolutions are characterised by the sensors deployed on the platform for capturing the wavelength bands of the objects of interest on ground. Certain salient details of sensors based on spatial and spectral resolutions are given in Table 2 and Table 3 (Malgorzata, 2010, Ilcev, 2019).

Table 1. Characteristics of sensors

Types of Resolution	Explanation
Spectral resolution	Represents the wavelength interval acquired by panchromatic/multi-spectral or hyper-spectral sensor. Narrow wavelength range of a channel or band symbolizes better spectral resolution.
Radiometric resolution	Illustrates the potential of the sensor to discriminate the slightest differences in the energy. The more the sensitivity of detecting the reflected or emitted energy differences, the finer is the radiometric resolution.
Spatial resolution	The area/region of the Earth is represented by the number of pixels deployed to detect the size of the smallest possible feature.
Temporal resolution	Represents the repeated observations over the surface of Earth with respect to the time interval between consequent observations.

Source: (Malgorzata,2010)

Table 2. Classification of imaging systems in terms of spatial resolution

Types of Imaging Systems	Range of Spatial Resolution
Low resolution sensor	Nearly 1-km or more
Medium resolution sensor	Nearly 100-m to 1-km
High resolution sensor	Nearly 5-m to 100-m
Very-high resolution sensor	Nearly 5-m or less

Source: (Malgorzata,2010, Ilcev,2019)

Table 3. Classification of imaging systems in terms of spectral resolution

Types of Imaging Systems	Number of Spectral Bands
Optical imaging sensors (including electromagnetic ranges in visible, near infrared, and shortwave infrared regions)	§ Panchromatic 1-band – IKONOS PAN, SPOT PAN *etc.* § Multispectral - 3-10 bands– Landsat TM, IKONOS MS *etc.* § Hyperspectral - 100s or 1000s– MODIS, MERIS
Thermal imaging sensors	
Synthetic Aperture Radar (SAR) imaging system	Radarsat -C-band, Terra SAR - X-band

Source: (Malgorzata,2010, Ilcev,2019)

MULTI-SENSOR DATA

Multisensory data has been used in this chapter to carry the relevant tasks discussed above. The multisensory data consists of two airborne data sets of an urban area at Thetford Mines, Canada acquired at different spectral and spatial resolutions. The data *i.e.,* LWIR HS and VIS RGB is shown below in Figure 2 (IEEE, 2014). The data set includes the residential and commercial buildings, roads and green vegetated area of an urban area.

Figure 2. Hyper spectral and Multi spectral Data
Source: (IEEE,2014)

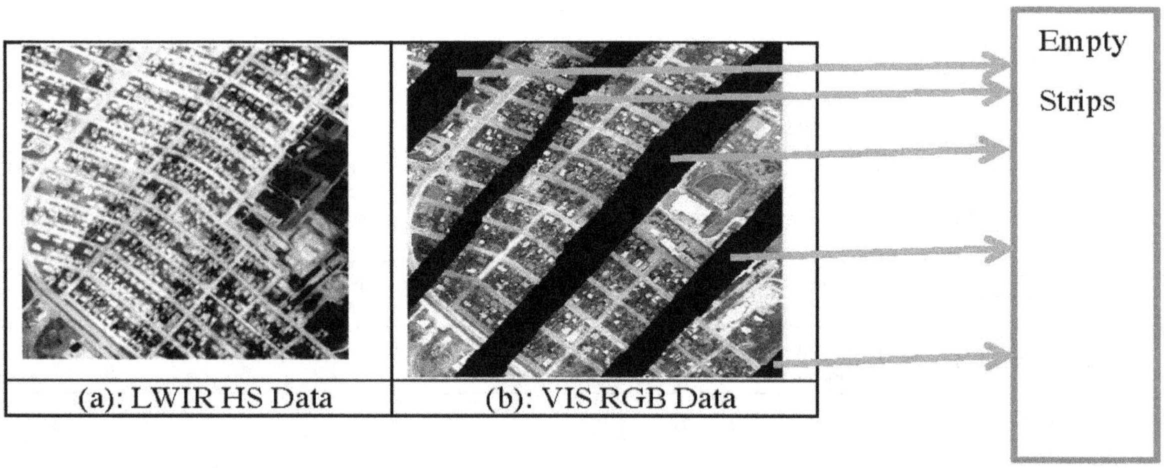

The brief summary of salient features of the two data sets is given in Table 4.

Characteristics of Multisensory Data Used

For object-based classification, it is necessary to explore the two datasets *i.e.,* LWIR thermal HS and VIS RGB data to understand the challenges posed by them so that the proper procedural processing can be done in land-cover classification. From the two datasets following important information is drawn,

1. LWIR thermal HS data is comparatively a low spatial resolution data with more spectral channels *i.e.,* 84 spectral bands. Due to large number of spectral bands, the data is likely to have higher spectral redundancy, that is, the linear correlation between the spectral bands may exist which may lead to erroneous results in depicting the objects.
2. As the thermal data shows the radiant energy of man-made objects, the spectral response curves are required to be studied. It is found that the maximum radiant energy is quiet low, which is less than 0.12 that is shown in Figure 3. Lower radiant energy of man-made objects may lead to lesser precision in extraction of these objects.
3. The challenge posed by VIS RGB dataset is that there are large gaps between the given strips of the entire scene which are highlighted in Figure 2(b). The spatial descriptors performance can be corrupted due to these gaps in between the strips.
4. Furthermore, some of the spectral bands of LWIR data show little information about the scene; therefore, it is necessary to choose the useful spectral bands for studying the spectral signatures of the objects for classification. Only the useful spectral bands can enhance the spectral characteristics in defining the objects in the classification process.

Table 4. Summary of multisensory data

Data Set	Data set Type	Features	Parameters
2(a)	LWIR Hyperspectral Data	Samples and Lines	751 x 874
		Bands	84
		File Type	ENVI
		Spectral Range	7800 -11500 nm
		Spatial Resolution	1m
2 (b)	Visible RGB Data	Samples and Lines	3769 x 4386
		Bands	3
		File Type	3
		Spectral Range	ENVI
		Spatial Resolution	0.2m

Source: (IEEE,2014)

CHOICE OF FEATURE EXTRACTION

From the characteristic study of multisensory data, it becomes easy to identify the spectral and spatial features which may aid in forming the database for identification of various classes in the data.

Spectral Feature Set

The acquisition of information regarding vegetation is captured in the form of electromagnetic wave reflectance spectra (Baofeng, 2017). The spectral reflectance or emission characteristics of vegetation are determined by biophysical and morphological characteristics of the surface of plants or leaves (Baofeng, 2017). The important reflectance bands that capture the various categories of vegetation are: (i) the visible spectra, which are composed of the blue:450-495-nm, green:495-570-nm, and red:620-750-nm wavelength regions; and (ii) the near and mid infrared band: 850-1700-nm (Mahlein, 2012, Oerke, -2014). Many vegetation studies are dependent on the reflectivity peaks or overtones of reflectance spectra within the visible and near/mid infrared regions (Karnieli, 2010, Sripada, 2005). Plant reflectivity in the thermal infrared spectral range: 8-14-µm follows the blackbody radiation law (Zhang, 2012), which interprets the plant emission in relation to plant temperature. Hence, indices obtained from this spectra range can be used as a proxy to assess the dynamics of various categories of the vegetation in an urban area such as ground grass, trees and shrubs.

Figure 3. Spectral response curve
Source:(Created by Author, 2015)

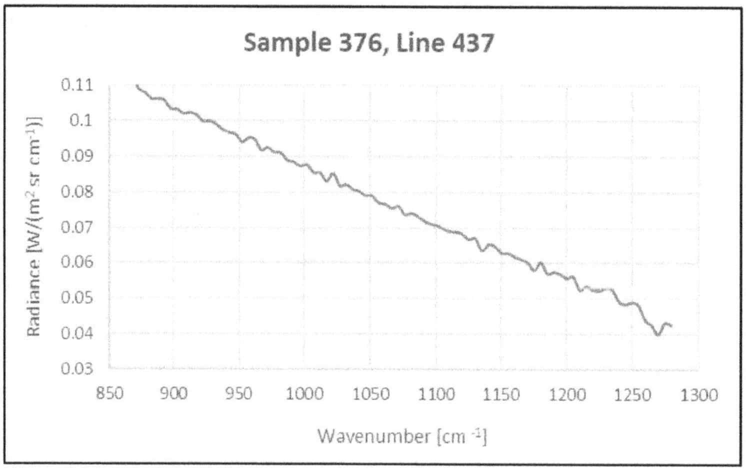

Since, LWIR HS imagery comprises of spectral channel band ranging from 7800- nm to 11500-nm and VIS RGB band ranging from 400-nm to 700-nm, therefore, NDVI spectral feature is computed using the bands from the multi-sensor data. From the particular multisensory data discussed in this chapter, 4 classes out of 5 are either the types of vegetation, soil or building and the morphological and biophysical characteristics of buildings, soil and green plants is high in infrared spectral bands, therefore, it becomes relevant to compute NDVI. The average of LWIR channels are extracted from LWIR HS data and red channels from VIS RGB data to compute NDVI feature (as given in equation Eq.1).

$$NDVI = \frac{R_{Average\ LWIR\ Channels(LWIR\ HS)} - R_{Red\ Channels(VIS\ RGB)}}{R_{Average\ LWIR\ Channels(LWIR\ HS)} + R_{Red\ Channels(VIS\ RGB)}} \quad (1)$$

Spectral – Structural Feature

The categorization of man-made objects in an urban area is getting popular with the availability of high-resolution imageries. This is so because it has been reported in various literature surveys that the spectral information of man-made objects exhibits similar characteristics, thereby, making it inefficient in distinguishing the classes. The knowledge of spectral-structural characteristics of such objects from high resolution imageries is highly useful in extracting the important attributes that may lead to correct identification. MBI is recently developed algorithm which is used for the extraction of buildings from high resolution imageries (Huang, 2011). Since, the present study involves the classification of buildings, roads and soil; therefore, it becomes relevant to extract MBI to have the information in spectral-structural domain of the mentioned classes as they exhibit similar spectral characteristics.

The visible channels in the data provide significant contribution of spectral information of man-made objects, particularly, for buildings. Therefore, the brightness of pixel is computed from all channels of VIS RGB data. The brightness of pixel from each band is recorded as shown in Eq.2,

$$(x) = max1 \leq k \leq K\ (band_k\ (x)) \tag{2}$$

where, $band_k(x)$ is the spectral value of pixel x at the kth spectral band and K is 3 expressing RGB bands. Next, local contrast information for man-made objects is captured through Differential Morphological Profiles (DMP) of the White Top-Hat by reconstruction (W- TH) using a series of linear Structuring Elements (SE), describing the size and directionality of objects, such as buildings, roads and soil. Mathematically, it is defined as shown in Eq.3,

$$W-\ (d, s) = b - \gamma^{re}{}_{b}\ (d, s) \tag{3}$$

where, s and d indicate length and direction of a linear SE, respectively. Morphological Profile (MP) of the white top-hat is defined mathematically as shown in Eq.4 and Eq.5,

$$MP_{W-TH}\ (d, s) = W - TH\ (d, s) \tag{4}$$

$$MP_{W}-(d, 0) = b \tag{5}$$

The DMP of the white top-hat is given in Eq.6,

$$DMP_{W}-(d, s) = |MP_{W-TH}(d, s + \Delta s)) - MP_{W-TH}(d, s)| \tag{6}$$

where, Δs is the interval of the profiles and $s_{min} \leq s \leq s_{max}$.

Finally, MBI is constructed as the average of DMP of the white top-hat as shown in Eq.7,

$$MBI = \frac{\sum DMPW - TH\left(d.s\right)}{\left(D * S\right)} \tag{7}$$

where, D and S denote the numbers of directionality and scale of the profiles, respectively. Four directions are considered in this study i.e., D is kept 4, since increase of D did not result in higher accuracy for building detection. The sizes of SE (s, s_{min}, *and* Δs should be determined according to the spatial characteristics of buildings and the spatial resolution, in this study, $s_{min} = 2$, $s_{max} = 52$, and $\Delta s = 5$. The number of scales is calculated by $((s_{max} - s_{min})/\Delta s)) + 1$. The larger values obtained in MBI vector refer to man-made structures since these structures show larger values in most of the directions of DMP.

Spatial Feature

From VIS RGB data co-occurrence statistics is calculated as the spatial resolution of the data is 0.2-m. The statistics are calculated using GLCMs (Zhang, 2017). The contrast and homogeneity features are extracted from each band of VIS RGB data. The feature vector using GLCM characterizes the texture of the objects in the scene by calculating the number of pixel pairs in the neighbourhood with the specified spatial intensity and then extracting the statistical measures from the matrix consisting of specific spatial relationship.

SEGMENTATION

The regions of interest polygons for all classes are obtained from VIS RGB data which is high spatial resolution imagery. Since, VIS RGB data has five strips which do not contain any information; therefore, no polygons have been obtained from the strips. The connectivity of the polygons is retrieved through 8-pixel neighbourhood rule followed by assigning a label to the connected *object* of all regions of interest classes. Bounding box created around the connected objects leads to the object identification for image analysis in classification of objects in an urban area.

CLASSIFIER NETWORK

SVM classifier network is a popular method of supervised classification. SVM with one-vs-one approach is built because of its effective processing and the ability to handle unbalanced training datasets. SVM classifiers (Melgani, 2004) of the form $f(x) = w.\Phi(x)+b$ are made to learn from the data $\{ x_i , y_i \}$, where x_i is an n- dimensional input feature vector, f(x) denotes the hyper-plane that separates the class labels on the basis of the support vectors on the hyper-plane $y_i = \pm 1$ on each side, w and b are the parameters i.e., weights and bias of the hyper-plane. The hyperplane calculation can be optimized as follows in Eq.8,

$$\min_{w,b,\varepsilon} \frac{1}{2} w^2 + C\sum_{i=0}^{N} \varepsilon_i \qquad (8)$$

For a given a set of training data, support vectors for the concerned objects/classes are generated near or far from the hyper plane. The larger the distance of the support vectors from the hyper plane the better is the separation of the objects/classes from the other set of objects/classes (Huang, 2013, Luo, 2015). Few distinct features of using SVMs in the present study for multi-resolution and multi-sensor data classification is that the SVM classifier networks are capable of using the HS data, LWIR-84 spectral channels, without considering dimensional reduction algorithms and are also insensitive to the size of training samples and quality of the input data.

IMAGE ANALYSIS FOR CLASSIFCATION OF AN URBAN AREA

A series of image processing steps have been framed for implementing various tasks in the present study for object-based classification of natural and man-made objects in an urban area using data shown in Figure 2. The flow chart of the implementation steps is given in Figure 4. The implementation includes careful investigation of two types of datasets i.e., LWIR HS and VIS RGB data for the extraction of significant attributes in spectral as well as spatial domain. In this regard, for the classification of natural and man- made objects from the data of an urban area, three sets of features are calculated i.e., spatial, spectral and morphological features to generate the knowledge data base vector. In spatial domain, the textural second order statistical feature is computed for defining the contrast and homogeneity in the three bands of VIS RGB data. In addition to the textural features, spectral feature is computed using Normalised Difference Vegetation Index (NDVI) which is formulated by exploring the infrared bands from LWIR and red band from VIS RGB data. The mean of near infrared bands from LWIR data and red bands from VIS RGB data is used to compute the spectral index. Besides, textural feature and spectral feature index, spectral-structural feature is computed using Morphological Building Index (MBI) to highlight the characteristics of built up objects in high resolution VIS RGB data. Once, the feature vectors are built up, which contains the information of pixels either in the form of spatial intensities or the spectral values, they are extracted to form a knowledge data base for all classes in a unique way. For this purpose, regions of interest for all natural and man- made objects/classes are generated in the form of polygons. This is created to from the ground truth for all objects/classes from very high spatial resolution VIS RGB data. Connected component analysis is performed on each polygon of every object so that the polygon only consists of the information of those pixels that hold certain level of connectivity with the neighbourhood pixels. This connectivity of pixels is labelled, and this way forms the connected object for each class of polygons. The features that are calculated consist of information in the form of contrast, homogeneity, spectral reflectance and spectral-structural values. These feature vector values are then extracted in the bounding box around each connected polygon for all objects/classes. The feature vector values in a bounding box is a knowledge data base of each object/class. Bounding boxes of all classes of connected objects are of irregular size as the polygons (region of interest) captured have varied sizes, therefore, for computational purposes and for minimizing the mixed pixel issue, all bounding boxes are fixed to single size window i.e., 25x25. Bounding box not only helps in identifying the objects of interest in an image but also provides the knowledge of position of the object in an image. The various spatial-spectral features extracted onto the bounding boxes have appreciable information on the edges/position of different objects/classes and each image object feature of the bounding box has feature knowledge too of respective object/class. The feature segmented bounding box for each object/class is divided into training and testing samples. In this study, the ratio of training data is kept 60% and the ratio for testing data is 40%. It is chosen as it seems to be an optimal ratio in concern to the classification accuracy of distinguished classes. The training samples are uniquely fed to Support Vector Machine (SVM) classifier for classification purpose. To classify natural and man-made objects/ classes like building, vegetation, road, soil and trees multi-class SVM is applied. The SVM used here is non- linear SVM which separates the data in two classes by means of a classification hyper-plane drawn in higher dimensional feature space. Classifier training uses the labelled data of bounding box feature samples for learning. For classification of n-class problem, ${}^{n}C_{2}$ pairs of classifiers are created. An object feature thus gets ${}^{n}C_{2}$ confidence measures for belonging to various classes. In post-processing stage, fusion of features is done with classifiers decision, such that each classifier votes for a class label, and the majority class label is extracted.

Figure 4. Flowchart of object-based classification using LWIR HS and VIS RGB data
Source: (Created by Author,2014)

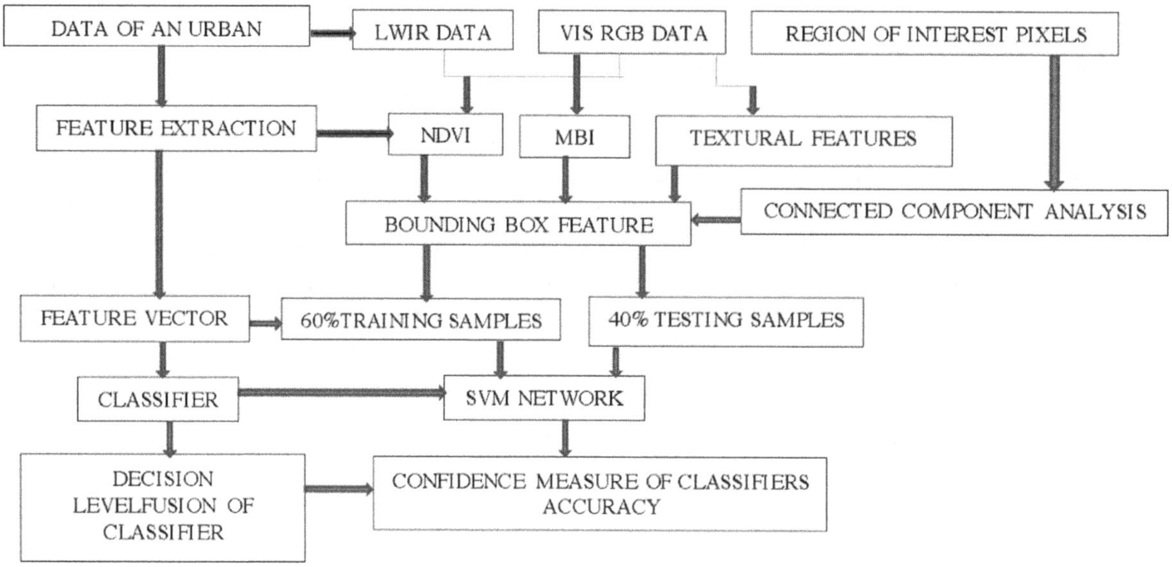

Spectral-Spatial Feature Extraction

The images of the calculated features are shown below in Figure 5 (which represents NDVI spectral image) and Figure 6 (which represents MBI image). The calculated features form the learning knowledge data base for SVM classifier. Following is a brief discussion on the results obtained from three sets of features.

1. NDVI calculations are done using LWIR HS and VIS RGB data. HS data has bands in infrared range that extends from nominal red edge of the visible spectrum at 7800- nm to 11500-nm. The mean of all the bands is computed in order to gather the average spectral information of near infrared range for the computation of healthy vegetation. As LWIR data supports the infrared range information, therefore, the red band information for computing the vegetation index is extracted from VIS RGB data. For depiction of healthy vegetation, the maximum absorption is captured in red band and maximum reflectance captured in infrared band by the sensors, therefore, the spectral information of both bands is required for understanding the reflectance pattern for vegetation, hence, the fusion of average infrared bands from LWIR and red band from VIS RGB data has complemented the extraction of vegetation index by normalised difference of the spectral band information. The value of NDVI feature extracted lies in the range from -0.549 to +1, which is in the range -1 to +1 defined for NDVI in the literature. The values ranging from -0.1 to +0.1 are indicated as soil in NDVI class identification range. Similarly, for other classes, the same information is observed and is highlighted in Figure-5 which is the false color composite image of the NDVI spectral index.

Figure 5. False color composite image of NDVI spectral index
Source: (Created by Author,2016)

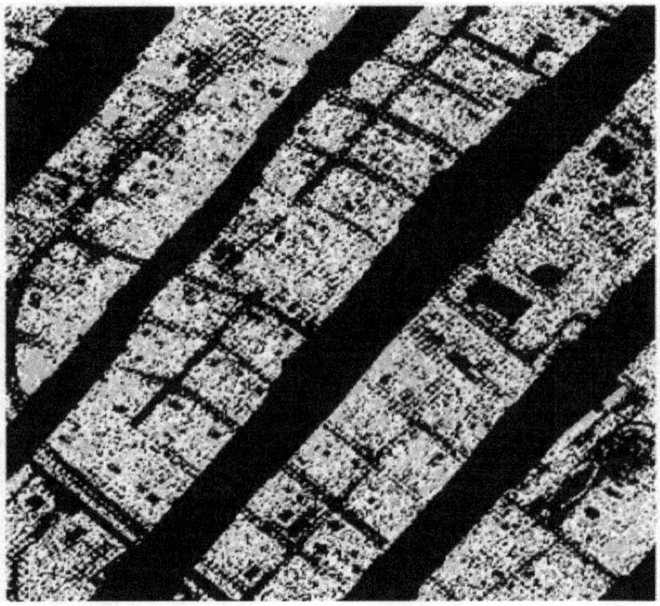

Figure 6. MBI image
Source: (Created by Author,2016)

2. Another feature that is extracted contains the information of man-made structures such as, buildings in an urban area. The extraction of such structures is a challenging task because of its complexity in terms of shape, size and brightness; and its spectral similarities with the surroundings with respect to roads, parking lots and open areas. MBI is therefore used which aims at describing the spectral-spatial properties of buildings (e.g., contrast, size and directionality) by a set of morphological operators. The low and high contrast intensity numeric values in MBI matrix helps in identifying the buildings with respect to natural objects/classes. Figure 6 shows the high low contrast image obtained from MBI calculations.

The bright structures seen in an image are probable man-made structures. A subset of values obtained from MBI matrix is shown in Table 5 which highlights the sudden change in the values showcasing the low and high contrast values. The larger values indicate the pixel intensities for man-made structures as man-made structures tend to have larger values in most of the directions of white top- hat DMP, since they show high local contrast in these directions.

Table 5. Subset of MBI matrix

Pixel Intensity Values						
4.500	17.568	17.068	14.431	10.477	12.727	10.795
4.840	18.272	18.090	16.909	12.931	12.045	0.886
6.590	18.113	18.250	19.545	18.977	0	0
9.227	18.204	19.568	21.113	2.799	0	0
10.295	19.454	21.113	21.113	0	0	0
12.295	18.068	18.590	0.954	0	0	0
12.409	16.863	0	0	0	0	0
13.068	13.704	0	0	0	0	0
15.818	0	0	0	0	0	0
18.045	0	0	0	0	0	0

Source: (Created by Author,2016)

3. The third set of features extracted is the spatial feature which is second order textural features. Two spatial statistical parameters are calculated- contrast and homogeneity for each visible band. A total of 486 statistical values are obtained, in which from each VIS band a textural feature vector space of 81 GLCMs are of contrast and another 81 GLCMs are homogeneity values. In the experiment, minimum value of contrast is 0 where the constant intensity values of the objects are seen. Maximum value of contrast is 0.667 which defines the difference in brightness levels of the man-made structures such as, buildings and roads within the same field of view. For homogeneity, the minimum value observed is 0.667 and the maximum value observed is 1. Homogeneity value is 1 for diagonal GLCMs define the closeness of the distribution of the pixel elements for buildings and roads.

Regions of Interest

The regions of interest pixels for natural and man-made objects from very high- resolution VIS RGB data are generated. Since, VIS RGB data contains five strips which do not contain any information; therefore, no regions of interest have been extracted from that portion of the data. The regions of interest pixels extracted for each class under natural and man-made objects are irregular shaped polygons which are shown in Figure 7. VIS RGB data has 3769x4386=16530834 pixels and the polygons representing regions of interest for natural and man-made objects are captured from the entire site of an image. Table 6 shows the total number of pixels captured for each class under the category of natural and man-made objects.

Figure 7. Regions of interest polygons of natural and man-made objects
Source: (Created by Author,2017)

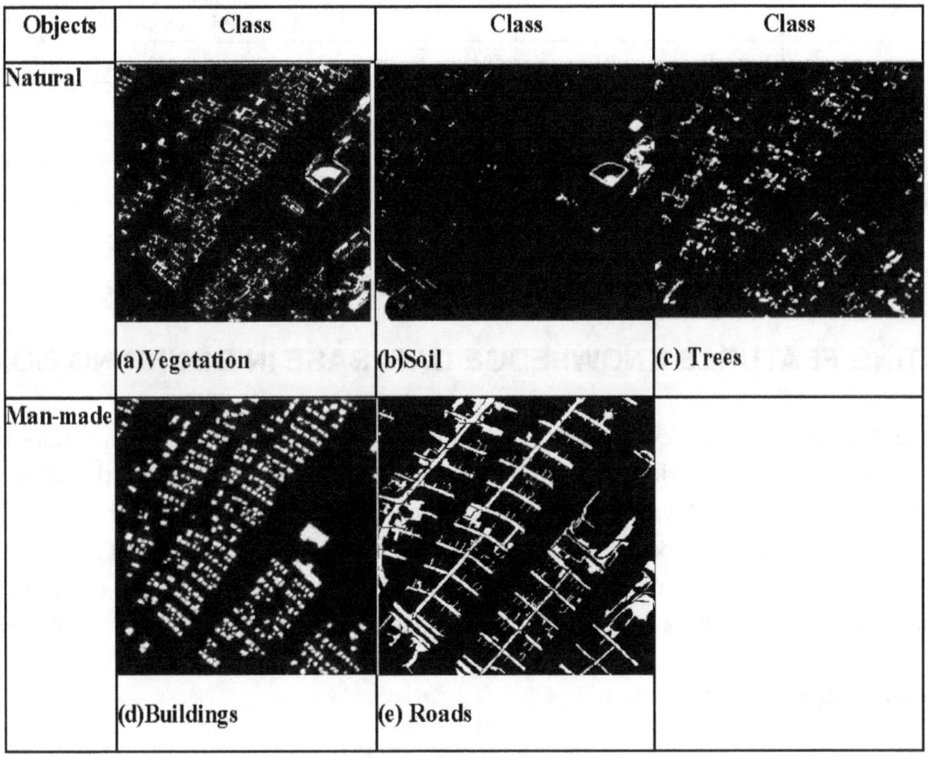

Table 6. Total number of pixels in each class of natural and man-made objects

Objects	Class	Total number of pixels
Natural	Vegetation	1021428
	Soil	316798
	Trees	727620
Man-made	Buildings	1590235
	Roads	2514842

Source: (Created by Author,2017)

Connected Component Analysis and Bounding Box

Connected component analysis is performed to know the count of the connected components in each polygon of natural and man-made objects, which is shown in binary image Figure 7, by 8-point neighbourhood system. The region property of the connected component which is used in this study is the bounding box. It is a small rectangle which not only identifies the objects of interest in an image but also their coordinate positions. Table 7 shows the total number of connected components obtained in the polygons containing the region of interest pixels for each class of natural and man-made objects.

Table 7. Total number of connected components in each class of natural and man- made objects

Objects	Class	Total number of connected components
Natural	Vegetation	594
	Soil	140
	Trees	552
Man-made	Buildings	456
	Roads	363

Source: (Created by Author,2017)

EXTRACTING FEATURES KNOWLEDGE DATABASE IN BOUNDING BOX

The spatial and spectral features computed from already existing mathematical formulations are then incorporated in the connected regions of bounding box. Hence, in this manner all the connected components in the bounding box will have all feature index values for each class. This shows that the spatial intensity relationship of the neighbourhood pixels and spectral reflectance values of each feature is now extracted in the bounding box of connected components for all classes under the category of natural and man-made objects forming a complete knowledge data base in spectral as well as spatial domain.

Resizing Bounding Box

The bounding box of connected objects of each class has different rectangular size due to irregular pattern of polygons. Therefore, for computational reasons the size of the bounding box is kept same and, in our experiment, the optimal size of 25x25 is taken. Further, the total number of connected objects/ components in the bounding box containing the spatial and spectral feature knowledge is then divided into training and testing samples. From the total number of bounding boxes for each class, 60% of them are training samples and the remaining 40% are testing samples. Table 8 shows the number of training and testing bounding boxes for each class. As the feature window size of each connected component is 25x25 i.e. 625, therefore, the training and testing sample size for all classes is 625xnumber of connected components in training and testing sample. Table 9 shows the total size of feature sets for training and testing samples.

Table 8. Total number of training and testing bounding boxes for each class

Objects	Class	Training Bounding Box	Testing Bounding Box
Natural	Vegetation	356x4	238x4
	Soil	84x4	56x4
	Trees	331x4	221x4
Man-made	Buildings	274x4	182x4
	Roads	218x4	145x4

Source:(Created by Author, 2017)

Table 9. Feature size for training and testing samples for each class

Objects	Feature	Class	Training Feature Sample Size	Testing Feature Sample Size
Natural	NDVI	Vegetation	625x356	625x238
		Soil	625x84	625x56
		Trees	625x331	625x221
Man-made		Buildings	625x274	625x182
		Roads	625x218	625x145
Natural	MBI	Vegetation	625x356	625x238
		Soil	625x84	625x56
		Trees	625x331	625x221
Man-made		Buildings	625x274	625x182
		Roads	625x218	625x145
Natural	TEXTURAL	Vegetation	486x356	486x238
		Soil	486x84	486x56
		Trees	486x331	486x221
Man-made		Buildings	486x274	486x182
		Roads	486x218	486x145

Source: (Created by Author, 2017)

SVM Classifier

Each training feature bounding box is given the class label and the system is trained using non-linear SVM classifier. For each pair of class, one SVM network is constructed. For a 5-class problem, 5C2 SVMs are trained to distinguish the samples of one class from the samples of another class. Classification of an unknown test bounding box samples is done according to maximum voting, where each SVM classifier network votes for one class. Following Tables 10 to 12 shows the confidence measure of each class against the other class from SVM generated network which is based upon spectral and spatial feature set analysis.

Decision Level Fusion on Multi-Class SVM Classifier

To calculate the classification accuracy of SVM classifier network appropriate post- processing is required. In this study, decision level fusion of the classifiers is done. The decision of all testing class labels is done by the majority vote of each class of the testing sample from each feature set analysis. The resultant of this decision accounts for the best match or the maximum vote the class label obtained from learning the training set of the concerned feature set. Hence, the decision is done for one-against-one class by the majority vote rule. Each classifier votes for a class label, and the majority class label is chosen. Table 13 highlights the confidence measure of each class achieved from multi-class SVM classifier by decision level fusion of classifiers for each feature set.

Table 10. Confidence measure of one class against another class using NDVI feature

Accuracy % Class	Class			
Building *Row1*	Road *Column 1*	Soil *Column 2*	Tree *Column 3*	Vegetation *Column 4*
	69.2	63.1	100	59.3
Road *Row2*	Building	Soil	Tree	Vegetation
	62.0	36.5	67.5	66.9
Soil *Row3*	Building	Road	Tree	Vegetation
	82.1	64.2	73.2	75.0
Tree *Row4*	Building	Road	Soil	Vegetation
	73.3	75.5	57.9	57.9
Vegetation *Row5*	Building	Road	Soil	Tree
	87.3	90.3	74.7	65.1

Source: (Created by Author, 2017)

Table 11. Confidence measure of one class against another class using MBI feature

Accuracy % Class	Class			
Building *Row 1*	Road *Column 1*	Soil *Column 2*	Tree *Column 3*	Vegetation *Column 4*
	100.0	100.0	93.4	100.0
Road *Row 2*	Building	Soil	Tree	Vegetation
	53.7	36.5	46.9	53.7
Soil *Row 3*	Building	Road	Tree	Vegetation
	76.7	66.0	73.2	75.0
Tree *Row 4*	Building	Road	Soil	Vegetation
	69.6	68.7	47.0	60.1
Vegetation *Row 5*	Building	Road	Soil	Tree
	77.7	75.6	70.1	68.9

Source: (Created by Author, 2017)

A few important observations emerging from Tables 10 to13 have been discussed and analysed below,

In Table 10, row-2, column-2, it is seen that the classification of road against soil presents the least confidence measure i.e., 36.5%. This may be because tr*opica*l soils and built environment generally have red band contribute to its spectral signature. Since, classes such as road and soil show high reflectance in red spectral band, therefore, there spectral separability may be difficult to be addressed by NDVI feature alone and this may lead to low classification accuracy.

From Table 10, row-1, column-3, it can be seen that the confidence measure of class building against tree is 100%, the confidence measure of class soil against building- row-3, column-1, is 82.1%, the confidence measure of road against building and road, row-4, column-1 and column-2, is 73.3% and 75.5% and the confidence measure of class vegetation against building and road, refer row- 5, column-1 and column-2 is 87.3% and 90.3%. The inference from the results show that SVM classification using NDVI feature analysis is recommended for classifying man-made objects and vegetation of an urban area using multisensory imageries.

From Table 10, row-4, column-1 and column-2, the confidence measure of natural vegetation against man-made objects i.e., tree against building and *road* is 73.3% and 75.5%. The results show that the use of thermal data in NDVI feature calculation has given the advantage in classification of natural vegetation and metropolitan area. Since, thermal infrared uses emitted radiation instead of reflected radiation and man-made objects are significantly warmer than the natural vegetated areas, hence, this can be well interpreted by using thermal bands for NDVI feature calculations.

Table 11 shows MBI feature analysis for classifying the data. From the results, it is seen that confidence measure of class building against all other classes is high, row-1, column-1, column-2, column-3 and column-4. This shows that high resolution VIS RGB imagery is useful in depicting the class building from MBI feature.

Table 12. Confidence measure of one class against another class using textural feature

Accuracy % Class	Class			
Building *Row 1*	Road *Column 1*	Soil *Column 2*	Tree *Column 3*	Vegetation *Column 4*
	63.7	73.6	100.0	54.4
Road *Row 2*	Building	Soil	Tree	Vegetation
	0.0	62.7	56.5	75.1
Soil *Row 3*	Building	Road	Tree	Vegetation
	51.7	19.6	39.2	100.0
Tree *Row 4*	Building	Road	Soil	Vegetation
	75.5	73.7	61.5	58.3
Vegetation *Row 5*	Building	Road	Soil	Tree
	72.6	64.2	60.0	0.02

Source: (Created by Author,2017)

Table 13. Confidence measure of natural and man-made objects by decision level fusion of SVM classifiers using NDVI, MBI and textural features

Class	Confidence Measure using NDVI feature (%)	Confidence Measure using MBI feature (%)	Confidence Measure using Textural feature (%)
Column 1	*Column 2*	*Column 3*	*Column 4*
Building	100.0	100.0	100.0
Road	79.3	66.9	66.8
Soil	91.0	85.7	57.1
Tree	82.8	75.1	83.7
Vegetation	91.6	84.8	75.6

Source: (Created by Author, 2017)

From the low confidence measure of class road against building and soil, Table 11, row-2, column-1 and column-2, it is seen that MBI is subject to commission errors due to the similar characteristics between buildings, soil and roads. Hence, lower classification accuracy of roads is observed against building and soil.

From Table 12, row-5, column-1 and column-2, the confidence measure of vegetation against buildings -72.6% and for roads -64.2% and also the confidence measure, row-4, column-1 and column-2, of trees against buildings - 75.5% and roads- 73.7% highlights that TEXTURAL features are beneficial for classification of vegetated and man-made objects. A high contrast intensity value in textural statistics indicates the bright vegetated areas around buildings and roads which helps in delineating the vegetated portion from the man-made objects.

From Table 13, it is seen that majority voting decision fusion generated from NDVI feature analysis, column-2, has resulted in best classification accuracy for natural and man-made objects as compared to results obtained from majority voting fusion of MBI, column-3 and textural, column-4 features.

Classification Map

Based upon the above calculations, the classification map of natural and man-made objects is best obtained from the results of NDVI feature set analysis and is shown in Figure 8. The classification of natural and man-made objects is done as per the color legend chosen shown in Figure 8 (f)) for all classes. Certain inferences from Figure 8 and Tables 9 and 13 can be made which are discussed below.

The high confidence measure of vegetation which is 91.6%, refer Table 13 column-2 shows the maximum regions of interest pixels have been mapped correctly. The red marked area in the image, Figure 8 (a) shows that the vegetated area is not identified in the classified results. The classification algorithm shows high confidence measure of soil which is 91% refer Table 13 column-2 which shows almost all the regions of interest pixels as shown in Figure 7 (b)) have been correctly identified. The confidence measure attained for trees is 82.8% (refer Table 13 column-2. The classification map shows certain mixed pixel classes mapped under this category which is highlighted in the image with orange circles, shown in Figure 8 (c). The misclassifications are supported from the calculations obtained in Table 10 (Confidence measure of one class against another class using NDVI feature). The confidence measure of tree obtained against soil and vegetation is quite low which is 57.9% for both the classes, row-4, column-3 and column-4. Hence, certain regions of pixels containing soil and vegetation have been identified as trees (marked as orange circles in Figure 8(c)) lowering the confidence measure. The confidence measure of building is 100%, refer Table 1.13 column-2) which shows all regions of interest pixels are correctly identified and the objects identified as buildings in classification map can be seen and matched from Figure 7(d). The confidence measure achieved for road is not highly appreciable i.e., 79.3%, refer Table 13 column-2. The classification map of road, Figure 8(e) shows mixed pixel mapping under this category.

Figure 8. Classification maps for natural and man-made objects
Source: (Created by Author,2017)

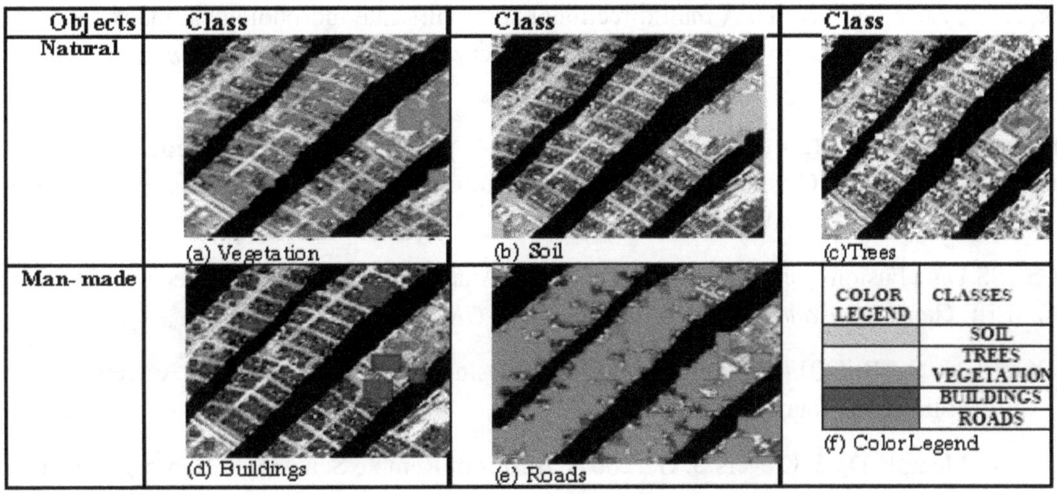

CONCLUSION

In the present Chapter, the potential use of multi-sensor data sets for classification of natural and man-made objects is presented. It has been seen that spectral and spatial features extracted from multi-sensor data have significant contribution in identification of objects in an urban area. The use of mean infra-red channels in thermal range for computing NDVI features has shown better effects not only in identification of natural objects but also contributed in obtaining good identification of man-made objects. In the experiment, MBI features have given an appreciable percentage in identification of soil i.e., 85.7% along with effective depiction of buildings against other classes in the scene. The results show that the construction of MBI exhibits the spectral-structural characteristics of buildings and soil and these objects have shown larger values in the directions of the differential morphological profiles for their identification using a high-resolution VIS RGB data. Multiclass SVM classifier network built using textural features has shown good discrimination of natural and man-made objects because of the local contrast and homogeneity values in respect to the neighbourhood pixels, thereby, highlighting the edges and enhancing the discrimination.

Many classification algorithms are followed by segmentation processes but here in this study, a significantly different approach has been evolved. For classification of natural and man-made objects using multi-sensor and multi-resolution data, both spatial and spectral features are explored using connected component analysis and thereby, incorporating the knowledge feature set onto the bounding box of connected components of each class of interest. The knowledge feature set is used to train SVM classifier network and the validations for the testing samples are done using one-against-one class classification approach. Further, the decision of multi-class SVM classifier networks which are established using spectral and spatial features are fused together to explore the best suitable feature for classification. This approach has envisaged the potential use of high- resolution data for building up the region of interest pixels for identification of objects and exploring spectral and spatial features for object-based classification approach for an urban land-cover.

REFERENCES

Huang, X., & Zhang, L. (2011). A multidirectional and multiscale morphological index for automatic building extraction from multispectral GeoEye-1 imagery. Photogrammetric E*ngineering and Remote Sensing, 77(7), 721–732.*

*H*uang, X., & Zhang, L. (2013). An SVM Ensemble Approach Combining Spectral, Structural, and Semantic Features for the Classification of High-Resolution Remotely Sensed Imagery. IEEE Transactio*ns on Geoscience and Remote Sensing, 51(1), 257–272.*

IEEE GRSS Data Fusion Contest, Presented to Image Analysis and Data Fusion Technical Committee. (2014). IEEE Geoscie*nce and Remote Sensing Society. GRSS.*

*J*inru Xue, J., & Su, B. (2017). Significant Remote Sensing Vegetation Indices: A Review of Developments and Applications. Journal *of Sensors.*

Jong, S. M., Meer, F. D., & Clevers, J. G. (2004). Basics of Remote Sensing. In Remote Sensing Image Analysis: Including the Spatial Domain. Springer.

Karnieli, A., Agam, N., & Pinker, R. T. (2010). Use of NDVI and land surface temperature for drought assessment: Merits and limitations. Jou*rnal of Climate, 23(3)*, 618–633.

Liang, S., Li, X., & Wang, J. (2012). *Advanced remote sensing: terrestrial information extraction and applications*. Academic Press.

Luo, S., Wang, C., Xi, X., Zeng, H., Li, D., Xia, S., & Wang, P. (2015). Fusion of Airborne Discrete-Return LiDAR and Hyperspectral Data for Land Cover Classification. *Remote Sensing*, *8*(1), 3–12.

Mahlein, A. K., Oerke, E. C., Steiner, U., & Dehne, H. W. (2012). Recent advances in sensing plant diseases for precision crop protection. *European Journal of Plant Pathology*, *133*(1), 197–209.

Malgorzata, V. W., & Árpád, B. (2010). *Data acquisition and integration* (Vol. 6). Remote Sensing, Nyugat-magyarországi Egyetem.

Melgani, F. (2004). Classification of Hyperspectral Remote Sensing Images with Support Vector Machines. *IEEE Transactions on Geoscience and Remote Sensing*, *42*(8), 1778–1790.

Oerke, Mahlein, & Steiner. (2014). *Proximal sensing of plant diseases, Detection and Diagnostics of Plant Pathogens*. Academic Press.

Rodriguez-Galiano, V. F., Ghimire, B., Pardo-Igúzquiza, E., Chica-Olmo, M., & Congalton, R. G. (2012). Incorporating the downscaled Landsat TM thermal band in land-cover classification using random forest. *Photogrammetric Engineering and Remote Sensing*, *78*(2), 129–137. doi:10.14358/PERS.78.2.129

Sripada, R. P., Heiniger, R. W., White, J. G., & Weisz, R. (2005). Aerial color infrared photography for determining late-season nitrogen requirements in corn. *Agronomy Journal*, *97*(5), 1443–1451.

Stojče Dimov Ilčev. (2019). *Global Satellite Meteorological Observation (GSMO) Applications* (Vol. 2). Springer International Publishing.

Varlyguin, D. L., Wright, R., Goetz, S. J., & Prince, S. D. (2001). *Advances in Land Cover Classification for Applications Research: A Case Study from The Mid-Atlantic*. RESAC.

Zhang, X., Cui, J., Wang, W., & Lin, C. (2017). A Study for Texture Feature Extraction of High-Resolution Satellite Images Based on a Direction Measure and Gray Level Co-Occurrence Matrix Fusion Algorithm. *Sensors (Basel)*, *17*.

Chapter 3
An Integrated Study of the Yamuna River Basin to Set Up a Reference Condition Using an Archival Data Set of the 20th Century by Geospatial Technology

Uma Shanker
Delhi Pollution Control Committee, India

Peeyush Gupta
National Mission for Clean Ganga, Ministry of Jal Shakti, India

Athar Hussain
https://orcid.org/0000-0002-7187-5977
Netaji Subhas University of Technology (West Campus), India

ABSTRACT

The River Yamuna is one of the most prominent and sacred rivers of India. Due to the unaccounted discharge from many anthropogenic activities and industrialization in the basin, the deterioration of the river with respect to time is at its maximum. Periodical analysis is essential for remediation of the causes and the source. The study has been undertaken using GIS technology to understand the variations being taken place in the past decades. The transformation in the physical and geographical characteristics, during the years since 1980-81, of the Yamuna River Basin has been analyzed and converted into digital GIS domain in terms of maps and charts. The reference condition of the basin has been set up and utilized in order to identify the change in land use and land cover of the river premises around Delhi and its neighbouring states. The vegetation analysis with respect to climatic conditions and water quality deterioration has also been assessed in the present study.

DOI: 10.4018/978-1-7998-8331-9.ch003

INTRODUCTION

Water is essential for sustaining life and development. However, the variation in rainfall, rapid growth in population, increased urbanization, intensive use of fertilizers in agriculture, unplanned industrial growth, all have contributed in putting great stress on the availability of water both in terms of quantity and quality. Planning of water resources for the future development, thus, calls for careful monitoring of this precious resource both in qualitative and quantitative terms for various uses e.g. drinking, agricultural, recreational, industrial etc. Lesser availability of fresh waters and entry of larger quantum of pollutants into the river systems in the recent past have put considerable strain on their self-assimilation capacity. The status of pollution and the effect of pollutants change with the seasonal flows. Evaluation of fitness of water for various uses, therefore, requires monitoring of water quality at various locations of the rivers. For any nation to progress, it is imperative that its resources are optimally used with a goal on long-term utilization. Like other resources, the water resources also play a very vital role in the progress of a country. Proper planning should always precede the utilization and similarly, the monitoring of resources precedes the planning. India, in totality is blissfully endowed with large water resources of good potable surface water in the shape of large rivers and lakes. In spite of their uneven distribution in time & space, these resources have greatly contributed to the development in India. Considerable tapping of river water resources has been done after the independence.

The tributaries of Yamuna contribute 70.9% of catchment area and balance 29.1% accounted for direct drainage into the Yamuna River or to the smaller tributaries. On the basis of area, the catchment basin of Yamuna amounts to 40.2% of the Ganga Basin and 10.7% of the total land mass of the country. However, the deleterious effects on water quality, the potential of pollution contributes by runoff water originating from rural communities, cattle farming and agricultural drainage is the major source of polluting a river water body and should not be underestimated. Intensive use of fertilizers, pesticides and other chemicals in agricultural use is bound to create impact on stream water quality during floods through surface wash off and during non-monsoon period through leaching and seepage. For developing country like India an extremely important aspect is formulating of the water pollution control programs and policies. Considering the above mentioned facts in view a study was being planned with the objective in order to digitalize the archival data of duration from 1980 – 1981 of the entire Yamuna basin and information pertaining to water use and water pollution has also been considered. The data and information includes physiography, geology, meteorology, hydrology, demography, land use, agriculture, industry, water use and waste water discharge along the river basin has been transformed on GIS domain.

BACKGROUND

The river Yamuna, a major tributary of river Ganges, originates from the Yamunotri glacier near Banderpoonch peaks in the Mussourie range of the lower Himalayas at an elevation of about 6387 meters above mean sea level in district Uttarkashi (Uttarakhand). In its first 170 km stretch, the tributaries Rishi Ganga Kunta, Hanuman Ganga, Tons and Giri join the main river. Arising from the source, river Yamuna flows through a series of valleys for about 200 Kms, in lower Himalayas and emerges into Indo-Gangetic plains. In the upper reaches, the main valley is overlooked by numerous hanging valleys, carved by glaciers during the last ice ages. The gradient of the river is steep here and the entire geomorphology of the valley has been influenced by the passage of the river. In the upper stretch of 200 Km, it draws water from several major streams. The combined stream flows through the Shivalik range of hills of Himachal Pradesh and Uttarakhand states of India and enters into plains at Dak Pathar in Uttarakhand where the river water is regulated through weir and diverted into canal for power generation. From Dak Pathar it flows through the famous Sikh religious place of Poanta Sahib. On the right side of the Yamuna basin is the Mussourie spur-along which, lies sprawled, the hill station of Mussourie. Flowing through Poanta Sahib it reaches Hathnikund/Tajewala in Yamuna Nagar district of Haryana state, where the river water is again diverted into Western Yamuna canal and Eastern Yamuna canal for irrigation. During dry season, no water is allowed to flow in the river downstream to Tajewala barrage and the river remains dry in some stretches between Tajewala & Delhi. The river regains water because of ground water accrual and contributions of feeding canal through Somnadi (seasonal stream) upstream of Kalanaur. It enters Delhi near Palla village after traversing a route of about 224 Km. the River is again tapped at Wazirabad through a barrage for drinking water supply to Delhi. Generally, a very small quantity of water is allowed to flow beyond Wazirabad barrage in dry season, as the available water is not adequate to fulfill the demand of water supply of Delhi. Whatever water flows in the downstream of Wazirabad barrage is the untreated or partially treated domestic and industrial wastewater contributed through several drains along with the water transported by Haryana Irrigation Department from Western Yamuna Canal (WYC) to Agra Canal via Nazafgarh Drain and the Yamuna. After 22 Km downstream of Wazirabad barrage there is another barrage, Okhla barrage, through which Yamuna water is diverted into Agra Canal for irrigation. A very small quantity of water is allowed to flow through Okhla barrage during dry season. Whatever water flows in the river beyond Okhla barrage is mostly contributed through domestic and industrial wastewater generated from East Delhi, Noida and Sahibabad and joins the river through Shahdara drain. The Yamuna after receiving water through other important tributaries joins the river Ganga and the underground Saraswati at Prayag (Allahabad) after traversing about 950 Km. Thus, Yamuna river cannot be designated as continuous river particularly in dry seasons (almost 9 months), but can be segmented in five distinguished independent segments due to characteristic hydrological and ecological conditions. The catchment of Yamuna river system covers parts of Uttar Pradesh, Uttarakhand, Himachal Pradesh, Haryana, Rajasthan, Madhya Pradesh & Delhi states.

METHODOLOGY

The physiographical, water quality, climate condition and land use land cover study have been conducted over the basin of river Yamuna. In order to determine the present situation of the basin of river Yamuna, the required data for utilizing the formation of policy planning and implementation has been analyzed using geospatial technology. Physiographical data was extracted from Cartosat 1 stero satellite data using photogrammetry approach. Stream flow was calculated with the help of Carto DEM & and arc hydro tool. Furtherance to this, Land use land cover map was derived from Recourse Sat (LiSS IV) data using onscreen visual interpretation technique. Precipitation map was derived from Tropical Rainfall Measuring Mission data. Historical water quality data was downloaded from Bhuvan Ganga Geoportal. All the prepared layer of physiographical data, LULC data, Climate data & water quality data have been applied in order to calculate the factors responsible for the transformation of the basin.

Figure 1. Study area
Source: Google Map, 2018

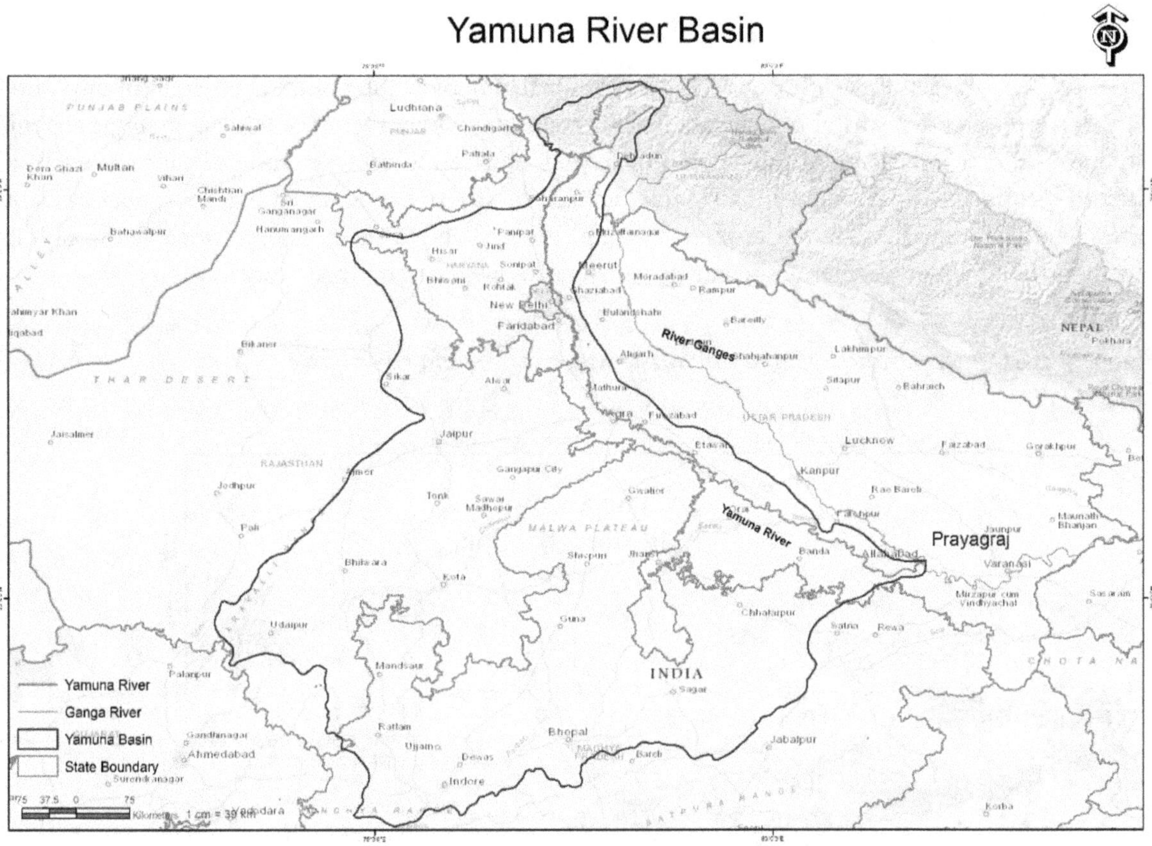

Figure 2. Methodology adopted
Source: Created by the author, 2018

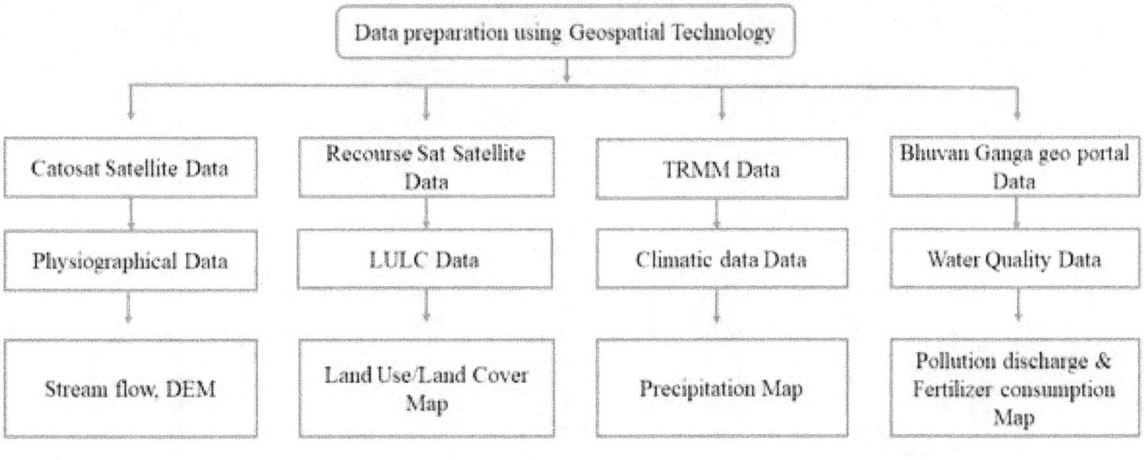

RESULTS AND DISCUSSION

Pollution Discharge

The Biological oxygen demand (BOD) is a parameter that depicts the pollution strength of any water body or wastewater in terms of organic load and is contributed by all types of human or anthropogenic activities. BOD is a measure of the amount of oxygen required to remove waste organic matter from water in the process of decomposition by aerobic bacteria (those bacteria that live only in an environment containing oxygen). The waste organic matter is stabilized or made unobjectionable through its decomposition by living bacterial organisms which need oxygen to do their work.

Figure 3. BOD concentration of domestic waste water in Yamuna river basin
Source: Created by the author, 2018

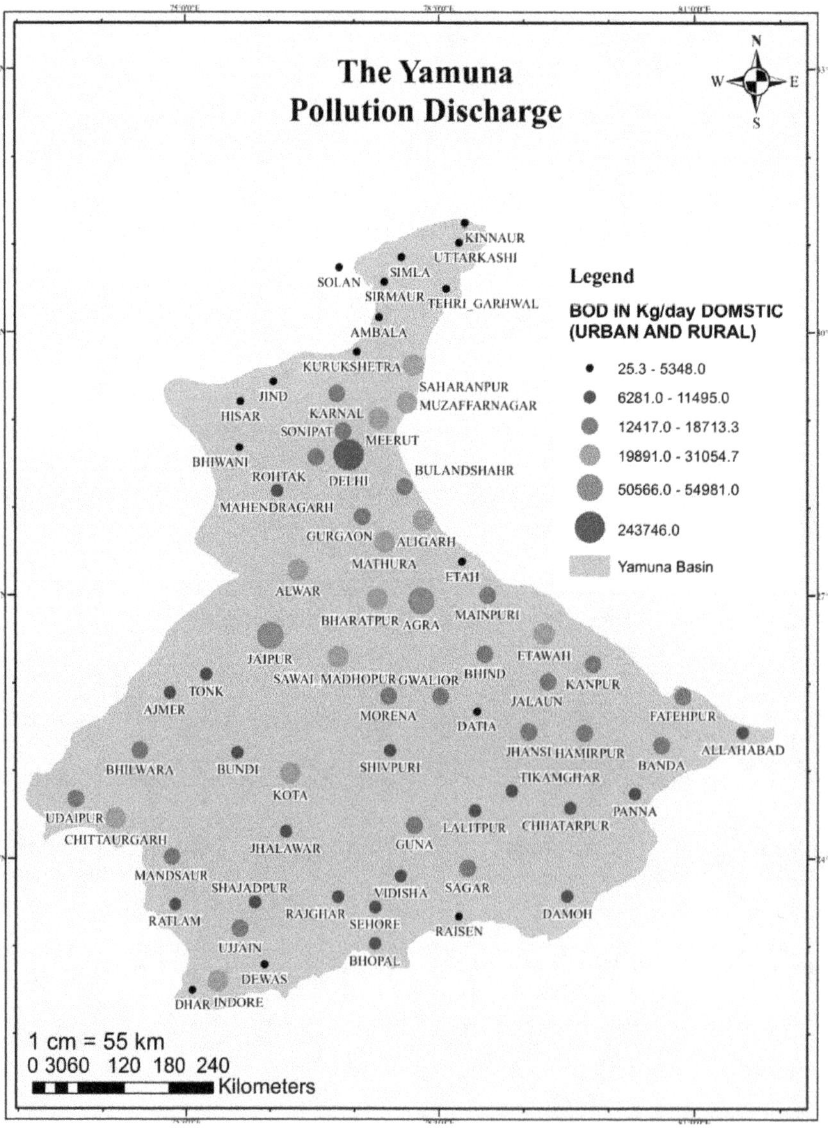

Figure 4. BOD concentration of industrial waste water in Yamuna river basin
Source: Created by the author, 2018

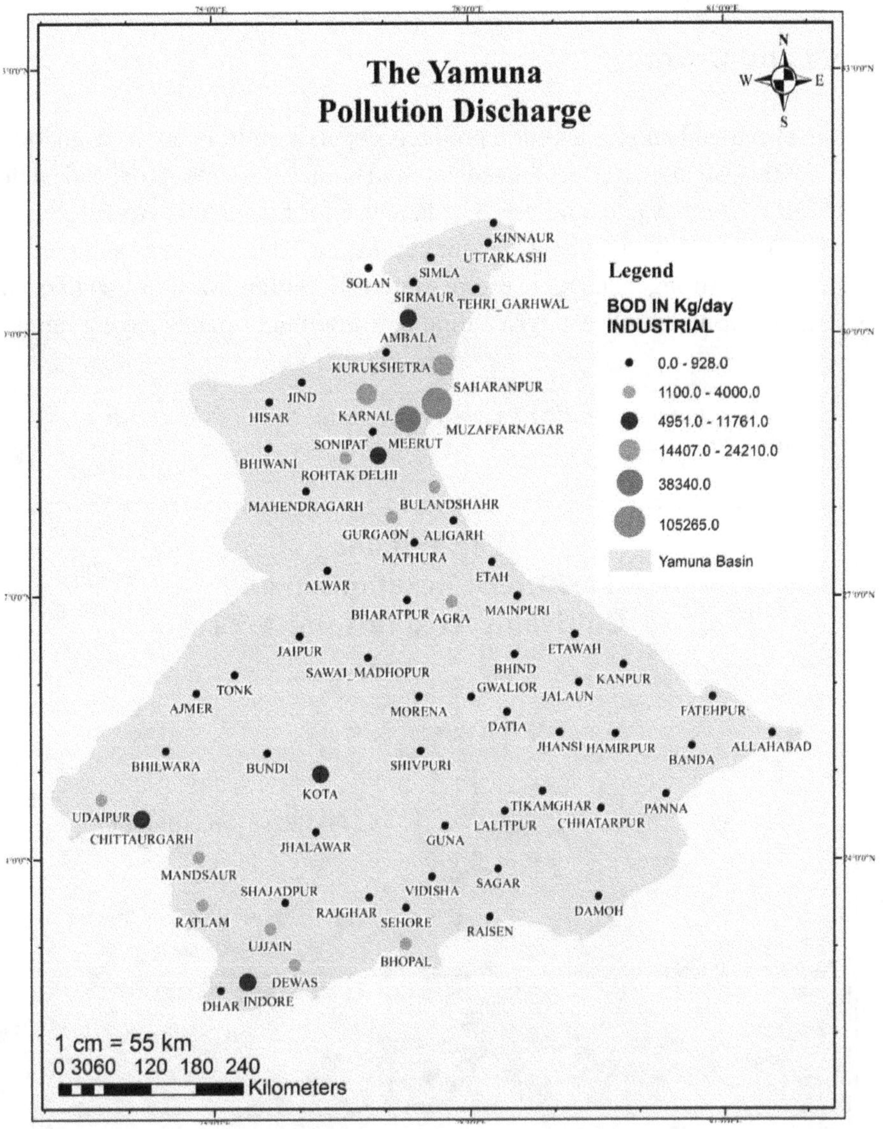

In this study, Haryana and Northern UP areas draining into the Yamuna (upstream of Etawah) and the only major tributary joining in this reach, viz. the Hindon in Delhi region, contribute the largest pollution loads on a per km stream length or per sq. km catchment area basis. The remaining three-fourths of the basin are so far low in pollution loads – as it in economic activities. The generation of the pollution loads in the case of the latter is sporadic near the urban industrial centres like Ratlam – Nagda, Indore – Ujjain, Bhopal, Gwalior, Kota, and Udaipur. The discharge of untreated domestic and industrial effluents has severely affected the quality of Yamuna River, which makes it fit only for recreation and industrial cooling, completely ruling out the possibility for underwater life and domestic supply. Biochemical oxygen demand (BOD) load has increased by 2.5 times between 1980 and 2005: from 117 tons per day (TPD)

in 1980 to 276 TPD in 2005 (Misra, A. K., 2010). The condition deteriorates further due to significant water abstraction which reduces the dilution capacity of the river (Sharma, D., & Kansal, A, 2006).

Land Use and Land Cover

Primary production of mineral and agricultural products depends entirely on the availability and use of suitable land. Besides this, all industrial establishments and human settlements depend on the availability of certain infrastructure facilities and land use. The land use for raising any type of plants should have a certain depth of arable top soil capable of retaining moisture. The soil types suitable for silviculture (forests) with plants having stronger and more extensive root system and slower growth rates are different from soil types suitable for more delicate agricultural or horticultural plants having rapid growth rates.

Figure 5. Map representing the area cultivable area to the total area of the basin
Source: Created by the author, 2018

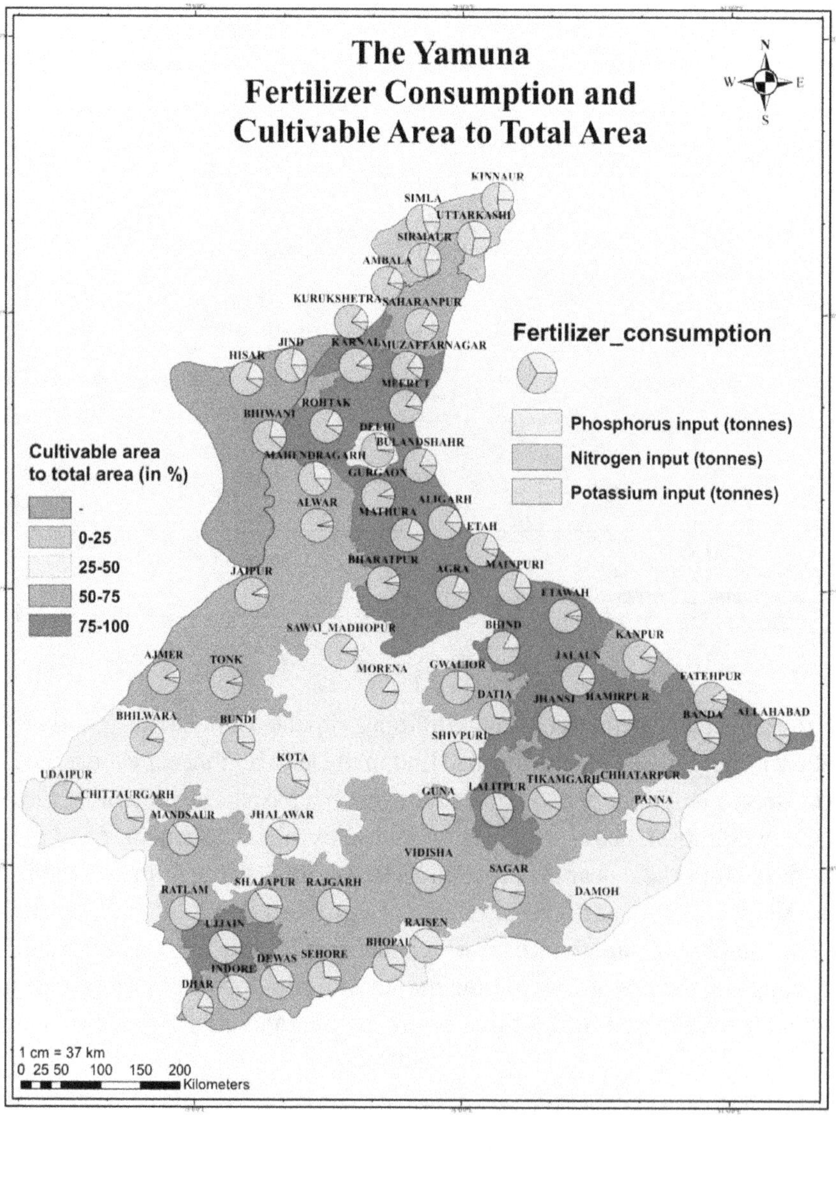

Soil type data was derived from Bhuvan Ganga geoportal. Those parts of the land in the basin which are too steep or rocky or are covered with sands as also those occupied by the flood plains of the rivers are unfit for agriculture, though sometimes they may be amenable to afforestation with special care techniques. Such non-arable lands can, of course, support mining activities in case economic minerals exist underneath. Thus almost all the mining activities in the Yamuna basin including the mining of limestone in the Dehradun district in Uttarakhand, building stones in Delhi, Jhansi, Lalitpur and several districts of Rajasthan and MP, of zinc and copper in Udaipur and Khetri and diamonds in Panna, are generally confined to use of non-arable lands.

Figure 6. Total fertilizers consumption along the Yamuna basin area
Source: Created by the author, 2018

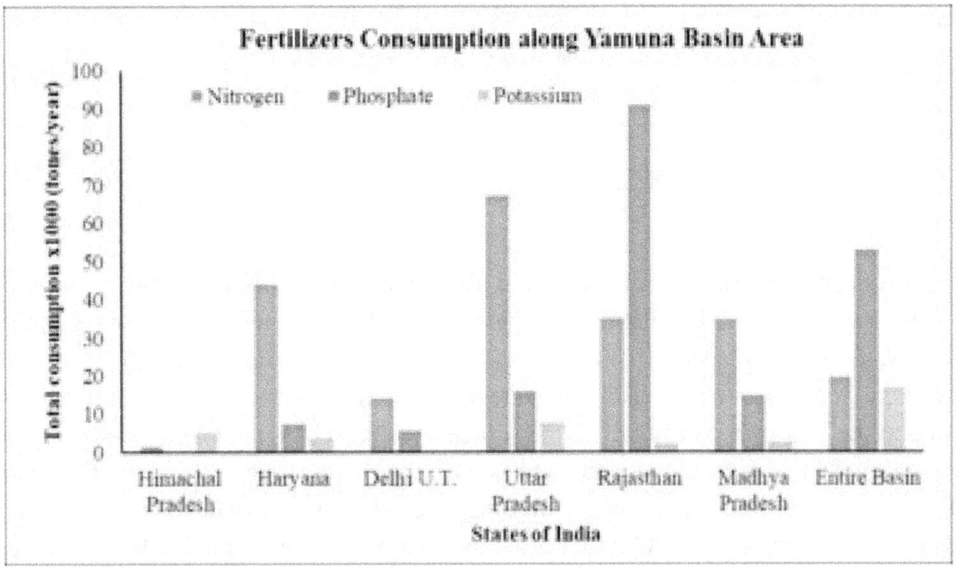

As much as 60 percent of the land in the basin classified as cultivable, viz., that which is arable and is not under forest. UP and Haryana have close to 80 percent of the total land classed as cultivable, MP has 56 percent while Rajasthan has only around 50 percent. In the Union Territory of Delhi, only 48 percent of the total land is cultivable. The district having the lowest fraction of land cultivable at 14 percent in Uttar Kashi is UP where most of the land (as much as 82 percent of the land) is covered with forests. The Muzaffarnagar district in UP is naturally endorsed with the largest percentage of cultivable land which is about 90.8 percent and is closely followed by Gurgaon and Rohtak district in Haryana (90 and 89.7 percent respectively). The above classification gives only an idea of the total available cultivable land in the Yamuna basin. But not all of it is under cultivation. In MP and Rajasthan, 90 percent of the cultivable land is reported to be actually under the plough. The district wise annual consumption of nitrogen, phosphatic and potassic fertilizers are shown in the map of figure no. 5. the rate of annual application of fertilizers per hectare of the gross cultivated area is given in graph no. 1 & 2.

Figure 7. Rate of fertilizers application along the Yamuna basin area
Source: Created by the author, 2018

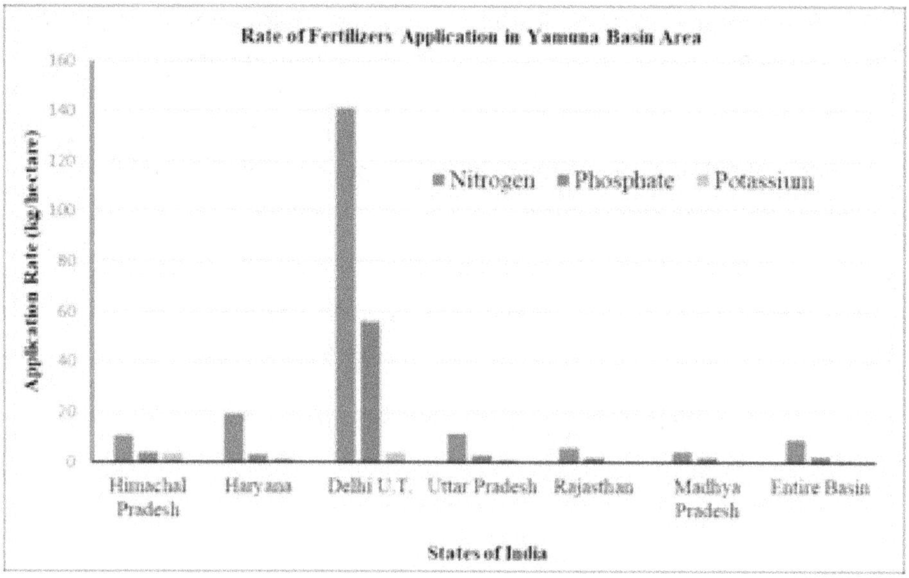

The incongruously high figures for the Union Tcrritory of Dclhi sccms to bc because significant quantities of fertilizers are easily available there and are, therefore, taken for use in the neighboring areas of UP and Haryana. The steep decline in the rate of application of chemical fertilizers from Haryana to UP, Rajasthan and MP indicates the declining state of agricultural development in these states which is primarily due to the lack of proximity from the source. Even in Haryana which has the highest rate of application among the major crop production states in India. However, the increase in the application of chemical fertilizers in the other states has not been lag behind with the expansion of irrigation facilities. Delhi land cover in the duration of 2008 to 2012 shows major changes in the landscape as there is high growth in the fallow and built-up area. Agriculture land decreased by 4.5% and forest area has reduced marginally by 1.5%, built up increased by 1.8% and water body is showing almost a constant condition over time (S.P. Pattanayak and S.K. Diwakar, 2013).

Physiography of the Yamuna Basin

The ground levels in the basin vary from about 6387 m above mean sea level (MSL) near Yamunotri Glacier to around 100 m above MSL near the confluence of the Yamuna with the Ganga basin at Prayagraj.

The basin can be classified into three groups as shown in the map in figure no. 6. Thus of out of the total basin area, about 3% area is classified as hilly, and the remaining is almost equally divided between plains and plateaus. It has been observed that the area under the plains of Rajasthan is 25% and Madhya Pradesh is 18%, while a significant fraction comprises the highly undulating Chambal Ravines. In general, it might be said that 40 per cent area of the basin with ground levels within 300 m above MSL has unconsolidated alluvial soils of recent origin and restricted to good groundwater availability. The plateau regions covering most of MP and Rajasthan on the other hand have black to red soils overlying older and much more consolidated rock formations and have very little groundwater potential. The hilly regions of the North are, of course, a separate category by themselves.

Figure 8. Map representing the mean sea level of the specific area of the basin classification of the canals
Source: Created by the author, 2018

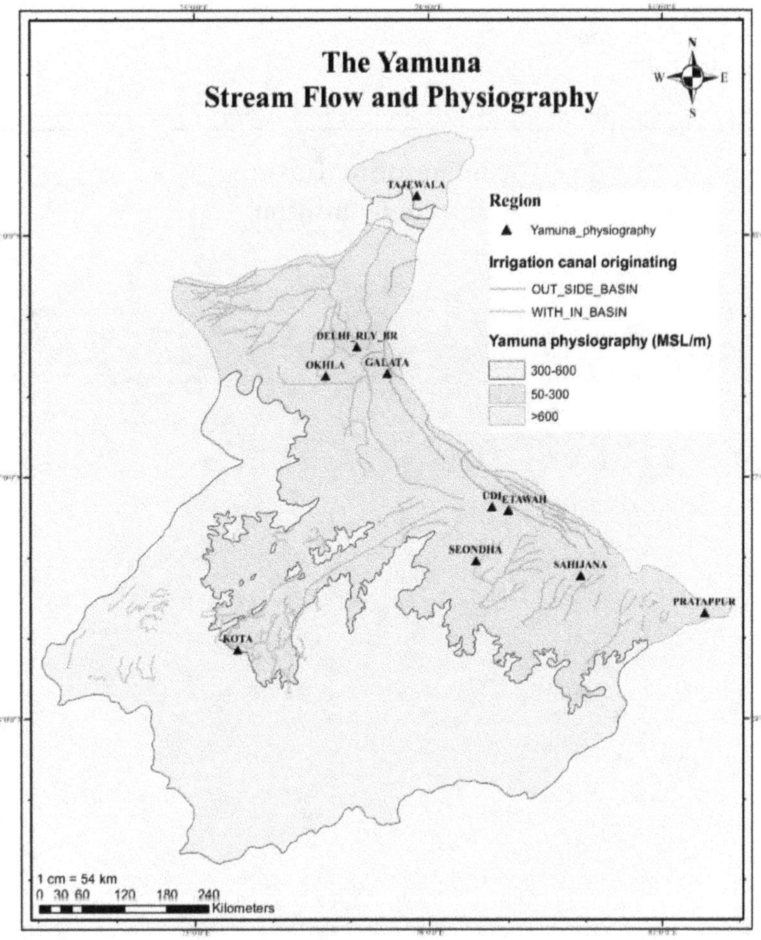

Table 1. The approximate coverage of the various categories

Classification	Height above mean sea level (m)		
	Hilly	Foot Hills & Plateau Region	Plains & Valleys
State	600	300-600	100-300
Uttar Pradesh	4900	4400	64908
Himachal Pradesh	5200	599	-
Haryana	-	800	20465
Rajasthan	1600	55610	45673
Madhya Pradesh	-	111508	28700
Delhi U.T.	-	-	1485
Total	11700	172917	161231

Source: (Created by the author, 2018)

Climate Condition of the Yamuna Basin

Figure 9. Map representing the annual rainfall data
Source: Created by the author, 2018

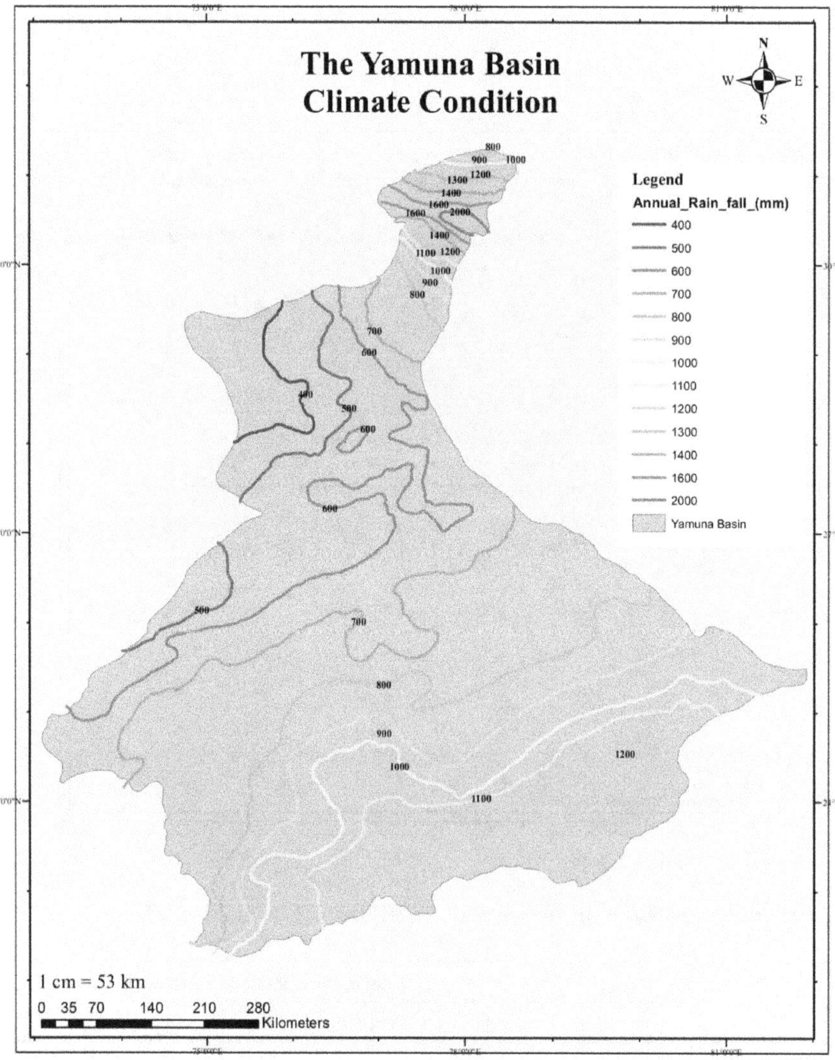

Identification of the precise nature and attributes of the time series of climatological data is very important and is usually the first step for water resources planning and management (Rai, R. K., 2013). The climate of the upper Yamuna river basin is subtropical monsoonal and is marked by strong seasonality. The rainfall is concentrated over the short period of the monsoons (June to September) when up to 90% of the total annual precipitation occurs (Gopal, B. and Chauhan, M. 2007). Isohyetal lines for mean annual rainfall in mm are shown in the map of figure no. 7 given above. The maximum precipitation in the basin, 1600 mm or more, the rainfall steeply decreases both towards the north as also towards the

South to reach 800 mm (half of the maximum) within a distance of about 100 km. The minimum annual rainfall occurs in the westernmost parts of the basin: 400 mm or even less in Mahendergarh and Bhiwani districts and 500 mm or even less in Ajmer, Jaipur, Gurgaon, and Rohtak districts. The rainfall intensity increases gradually as one move towards the East and more steeply as one move towards the South-East. The South-Eastern part of the basin running from Dewas to Prayagraj via Bhopal, Sagar, Damoh and Panna receives over 1000 mm of rainfall annually, while the Northern half of the basin (excluding the hills and foot-hills) receives only 500-800 mm annually.

CONCLUSION

From the present study using the geospatial technology for map composition and analysis/ processing of the data, it is concluded that the total annual run-off from the Yamuna basin is around 100 billion m^3. It is estimated to be around 80% of the total runoff that occurs during monsoon season. The fair weather flow condition is beneficial for the improvement of climatic conditions in terms of pollutant dilution, dispersion, and assimilation. It has also been observed that 27.5% is non-arable land while 53% is under actual cultivation and the other 3% is under habitational and urban usage. Therefore, as per the future land use policy, it can be recommended that (a) the agricultural area should not be expanded further, (b) the land under forests should not be utilized for other purposes, and (c) Major efforts should be made to reclaim the non-arable land for appropriate vegetation cover.

Analysis of the data in the present study indicates the adverse effect of agricultural drainage on stream in terms of rise of mineral constituents. Nutrients and biocides reaching water bodies from agricultural sources seem to be quite insignificant because of the low application rate of these chemicals in the Yamuna basin. The pollution discharge from domestic and industrial sources is estimated to be 200-500 kg BOD/ day per km. In the case of the Hindon and the reach of Yamuna from Okhla to Etawah, the estimated linear pollution load ranges between 700 and 900 kg BOD/day per km of the river length respectively. From the present study, it can be concluded that the magnitude of pollution load as estimated in the present study is well within the assimilative capacity of the river.

REFERENCES

Gopal, B., & Chauhan, M. (2007). River Yamuna from Source to Delhi: Human Impacts and Approaches to Conservation. In *Restoring River Yamuna: Concepts, Strategies and Socio-Economic Considerations* (pp. 45-69). National Institute of Ecology.

Joshi, R. R., Warthe, M., Dwivedi, S., Vijay, R., & Chakrabarti, T. (2011). Monitoring changes in land use land cover of Yamuna riverbed in Delhi: A multi-temporal analysis. *International Journal of Remote Sensing*, *32*(24), 9547–9558. Advance online publication. doi:10.1080/01431161.2011.565377

Misra, A. K. (2010). A River about to Die: Yamuna. *Journal of Water Resource and Protection*, *02*(05), 489–500. Advance online publication. doi:10.4236/jwarp.2010.25056

Pattanayak & Diwakar. (2013). District-Wise Change Analysis of Land Use-Land Cover In Delhi Territory Using Remote Sensing & GIS. *Journal of Urban and Environmental Engineering, 10*(2), 201-213. doi:10.4090/juee2016.v10n2.201213

Rai, R. K. (2013). *Temporal Variability of Climatic Parameters of Yamuna River Basin: Spatial Analysis of Persistence, Trend and Periodicity. The Open Hydrology Journal*. doi:10.2174/1874378101004010184

Sharma, D., & Kansal, A. (2006). *Current condition of the Yamuna River - An overview of flow, pollution load and human use.* https://fore.yale.edu/files/Current_Condition_of_Yamuna_River.pdf

Chapter 4
Environmental Sensitivity of Planning Practices in Bangladesh vis-à-vis Evolution of Planning Theories:
A Case Study on a Thermal Power Plant at Rampal, Bangladesh

Kazi Saiful Islam
https://orcid.org/0000-0002-4137-5904
Urban and Rural Planning Discipline, Khulna University, Bangladesh

ABSTRACT

Bangladesh has a long history of physical planning. It started with blueprint-based master planning approach in the 1950s. Since the late 1970s, it has slowly shifted to the rational planning approach following the path laid down by the United Kingdom. Lately, infusion of the North American planning thoughts is observed. Thus, Bangladeshi physical planning never had its own identity. Structure plans are loosely tied to the other tiers of planning packages that are used for development management. Though most of the structure plans address environmental issues, lower tiers of plans often lack their reflection. Often policies are laid down based on popular belief rather than scientific evidence. To establish these arguments, the evolution of the planning practices of Bangladesh is illustrated briefly in the context of the evolution of planning theory. Additionally, environmental sensitivity of a physical plan covering one of the most ecologically sensitive and economically potential areas of Bangladesh, namely Mongla master plan (MNP) is evaluated, as an example, considering the spatial dispersion of sulfur dioxide to be discharged by the coal-fired thermal power plant at Rampal. Unfortunately, MMP failed to stand out from all the rest to prove its soundness.

DOI: 10.4018/978-1-7998-8331-9.ch004

INTRODUCTION

As a country, Bangladesh has a long history of physical planning. Since the beginning of the 19th century, Bangladesh has traveled through several stages of physical planning. Bangladeshi plans started with the Master Planning Approach in the 1950s. Gradually it shifted towards Rational Comprehensive Planning (strategic planning) Approach. After the liberation of Bangladesh at 1971, urbanization triggered up in metropolitan cities and environmental issues started getting their much deserving attention among the urbanists, environmental activists, civil society, policymakers, and urban planners. As a response, in the 1990s, Bangladeshi plans started addressing environmental issues. These plans have been prepared under different names "Master Plan", "Detailed Area Plan", "Strategic Development Plan", "Urban Development Plan" etc. Most of these plans are mainly criticized through mass media on three grounds, *firstly*, lack of public participation in the plan-making process; *secondly*, errors in the plan proposals which is also linked to the first criticism, and *lastly*, lack of ecological and environmental considerations. Lack of implementation of these plans is another ground that is, in fact, linked to policy formulation based on popular belief rather than scientific evidence, failure to translate policies spatially, lack of ownership of the plan among the stakeholders, etc.

Instead of learning from mistakes and act upon them, Bangladesh has been continuously trying to adopt westernized ideas. In doing so, Bangladeshi Planning is losing its own identity. In fact, Bangladeshi planning never had its own identity. Traditional patron-client networks, political instability, Westminster style of democracy, etc. are often held responsible for such stagnancy and non-adaptive nature of urban planning in Bangladesh (Khan & Swapan, 2013). Unfortunately, Bangladeshi urban plans failed to transform themselves from Technocratic to Sociocratic planning. Since 1950s, Bangladesh followed the planning culture of the United Kingdom (UK). More recently, infusion of the North American planning thoughts are observed.

According to Albrechts (2004), strategic spatial planning should consider "*a limited number of strategically important issues, reactive to uncertainties, bridge the gaps among the competing values, critically analyze strengths and weaknesses in the context of opportunities and threats (both in terms of space and issues, present and anticipated future), acknowledge the resource constraints, be politically inclusive and participatory, be vertically integrated with higher-level plans, and incorporate provision for monitoring, feedback, and revisions*". All these issues are localized in nature. Thus, to justify the efficacy and soundness of a plan, one should not compare the same with other plans. Instead, one should take the theoretical underpinnings as primary point of departure, then considering local situation at the time of plan preparation and previous experience, one should evaluate a plan.

To validate these arguments, in the first part of the chapter, the evolution of urban planning practice of Bangladesh is illustrated briefly in context of the evolution of planning theory. Detailed and critical review of neither of these topics is within the scope of this chapter. Local-level planning (e.g., Detailed Area Plan, Action Area Plan etc.) is also not included, as it deserves special attention. For the sake of in-depth analysis, Mongla Master Plan (MMP) is evaluated to establish the arguments. There are several reasons for choosing this specific plan. *Firstly*, the plan area is located by one of the environmentally and ecologically most sensitive areas of Bangladesh, the Sundarbans Mangrove Forest. *Secondly*, Coal Based Thermal Power Plant at Rampal is now under implementation, which is in the jurisdiction of this plan. Thus, it would be easier to assess the environmental sensitivity of the plan. This will make the arguments objectively focused.

MASTER PLANNING IN BANGLADESH (1950-1990)

The concept of the master plan took shape after the Second World War, when city plan became synonymous to physical plan. This is also the period when planning started taking distinctive form. This approach emerged out of the physicalist conception of town planning. Meaning, the planners tended to view towns and their problems only in physical (and aesthetic) terms. It was also assumed that social and economic issues can be addressed through physical plans. Arguably, physically distinct residential neighborhoods with all necessary amenities would help developing social neighborhoods (increased social bonding and sense of kinship among the residents). This is called "Physical determinism" (Swinburne, 1969). This assumption is somewhat correct but highly questionable. However, let's focus on the Bangladeshi perspective of the master planning approach first.

The term "Master Plan" is often used quite ambivalently in most of the statutes and planning documents in Bangladesh. Master Plan was basically a large-scale urban design based 'blueprint' plan portraying spatial configuration of landuses and infrastructures. The 'blueprint' was essentially the end-state of a city or the future image of the same. Often the 'blueprint' used to accompany 'programming' plans showing the importance and phasing of different parts of the city and physical development projects for implementation.

The first plan for Dhaka was prepared by Patrick Geddes in 1917. I do not want to shift our focus on this plan, as it was long before the emergence of the planning theory as an individual field of knowledge. Iin 1958, Dhaka got its second master plan. Aims and objectives of this plan were rather vague, where landuse got the highest priority. The plan was built upon some assumptions, such as the dominance of the Buriganga River as the main transport route, constant population growth rate (1.75% per annum), expansion of the city in the northern direction (flood free high land), commercial agglomeration at the central part (old Dhaka) of Dhaka, etc. Based on these assumptions and projected population, the plan laid down proposals (landuse) for different sectors like transport, government buildings, housing, education, recreation, etc. The plan also laid down a few proposals for planning administration and implementation.

Long after the second master plan of Dhaka, in 1982, the Urban Development Directorate (UDD) initiated a project to prepare Landuse/Master Plans for 460 Upazila towns and 60 Zilla (district) towns. The planning process for the Upazila towns started after the completion of survey work. Based on the projected population, using standard (e.g., for 1.5-acre land should be allocated for commercial purpose for every 1000 population; 5.00-acre land should be allocated for the purpose of health complex for every 20000 populations, etc.) land requirement per provision (landuse) is identified. Justification of these standards cannot be found anywhere in these plans. Poignantly, these standards were used for all Upazila towns without considering any local conditions. Total amount of required land is then allocated over the project area considering certain planning principles. In doing so, the planners considered the functionality of the city. The plan usually concluded through identifying programming where a few projects or infrastructures were identified for implementation in different phases. Please see figure 01 for the simplified process of the master planning approach. Some suggestions were also laid down for development control and monitoring of the plan implementation.

Like Upazila towns, the master plans of Zilla towns also started with primary and secondary data collection to conduct several sectoral studies. Based on all these studies, the planners tried to develop future image of Zilla towns which determined the proposed landuse composition and infrastructure provision of the project area.

Figure 1. Simplified process of master planning approach of Upazilla Sahar

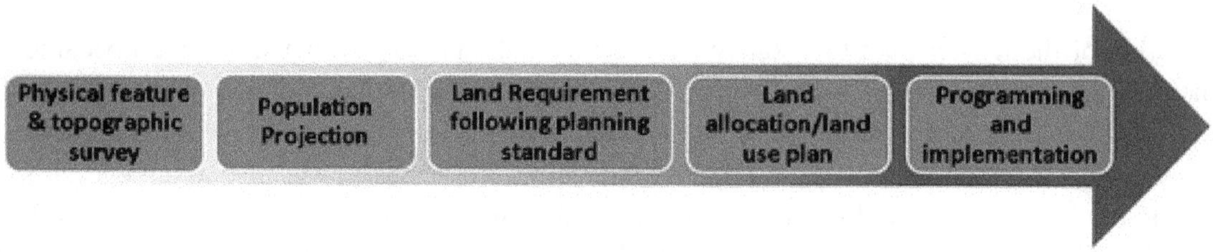

Procedurally, both planning packages followed the same path. However, master plans for Zilla Sahar emphasized more on the sectoral studies compared to the master plans of Upazila towns. It is to be noted here that the stakeholders for whom the plan was prepared had no contribution in developing the mental image of the planners.

Till this point in planning history, the planning approach of Bangladesh complied with the international practice. Even after being severely criticized internationally, the master planning approach did not lose its importance in Bangladesh. As of today, it is still used for the development management. However, instead of being a stand-alone tool for development management, it became a part of the multi-tire planning package "… as the age-long practices of urban planning and development control could not be replaced with a compatible system" (Rahman, 2016).

RATIONAL-COMPREHENSIVE PLANNING APPROACH: THEORETICAL UNDERPINNINGS

"The urban community is a system comprised of interrelated elements, but little is known about how the elements do, will, or should interrelate." (Davidoff, 1965)

The origins of the rational planning theory can be traced back almost indefinitely (Batty, 1976). Two major theories contributed to the development of the rational theory in the 1960s. The system view of planning (*sustentative theory*) describes what needs to be considered for planning and why. This theory was proposed by Brian McLoughlin in his seminal book "Urban and Regional Planning: A Systems Approach" at 1969. This theory considered the planning environment as a system of interconnected parts (cities and regions etc.) (McLoughlin, 1969). Enthusiastic readers are advised to go through Jacobs (1961) and Taylor (1998). The procedural theory introduces the rational process of decision-making. This theory was adopted in planning during the 1960s as a response to the international wave of adoption of quantitative scientific methods in policymaking and growing interest in corporate style management systems. Andreas Faludi's seminal book entitled "Planning Theory" (1973) contributed expressively to developing the process of this planning approach. Out of the criticisms of the master planning process (Survey > Analysis > Plan), a new generalized planning process comprised of five steps was proposed. These steps (i. Identification of problem, ii. Making alternative plans/policies, iii. Evaluation of the alternatives, iv. Implementation and v. Monitoring) are tied together by forward and feedback loop that makes the planning an ongoing and continuous process (Taylor, 1998; p. 68).

"Advocacy planning is about the development of an overall change strategy depicting long-term plan that embodies the vision of the community reflecting the gap between the existing and the expected situation as well as the course of accomplishing the expectation." (VeneKlasen, 2002, p. 99)

The advantages of the combination of these two theories (System and process) are very clear. Planning, which was previously considered as an act of art emerged as a scientific process through these theories. Although the rational-comprehensive plan is still the most popular planning approach not only in Bangladesh but also all over the world, criticisms and counter theories started surfacing almost simultaneously with the rationalist approach (Friedman, 1971; Flyvbjerg, 1996). The rationalist approach was confronted by mainly three major theories (among many)- Incremental planning (Lindblom, 1959 and 1979), Mixed-scanning (Etzioni, 1967) and Advocacy Planning (Davidoff & Reiner, 1962; Davidoff, 1965). Although none of these are beyond criticism, all these theories brought some new insight into the rationalist approach. A comparison of these planning theories can be found at Hudson, Galloway, and Kaufman (1979). As none of these theories are completely implementable for Bangladesh, in-depth discussion is beyond the scope of this chapter.

STRUCTURE PLANNING APPROACH IN BANGLADESH

Being a long-lasting British colony, Bangladeshi physical planning system is traditionally influenced by the British planning system, where rational planning approach emerged out of Structure Plan and Area Plan. "Structure Plan" was introduced through the Town and Country Planning Act in the United Kingdom (UK) in 1968 that explicitly proposed a rational planning process (under the heading "New Provisions as to Development Plans"). From 1975 till 2010, Bangladesh followed the British style of structure planning approach. Afterward, infusion of the North American Comprehensive Planning Approach (General Plan) is observed.

Following the same path of the master planning approach, structure plan (not only by name but also by process) was also adopted in Bangladesh at 1977 through a project entitled "The National Physical Planning Project". It was initiated by UDD to prepare Area Development Plan for 10 Districts. These plans were to be prepared for regional, urban, and local levels. Although the project remained incomplete, a few Structure Plans were prepared. While differentiating itself from its predecessor (master plan), it stated-

A structure plan differs from the older style master plan by concentrating only on broad structure of the city. It is not concerned, as are master plans, either with the details of physical layout or with detail of individual development areas, which cannot be implemented until the later stages of the plan period. It would not be possible to determine the detailed form of landuse and development implied by master plan over the period of the year 2000 bearing in mind the uncertainties associated with growth. (UDD, 1986, p. 4)

Each of the plans was prepared in two parts (i) the structure plan (that included strategy and structure) and (ii) the action plan. The process of preparation of structure plan (figure 02) becomes very clear from the content, which is the same as the steps proposed by the original procedural theory. Ignoring the environmental and ecological issues, these plans accentuated on employment creation and economic development.

Since the 1990s, a significant number of strategic plans have been prepared under different titles in Bangladesh by different agencies (e.g., Development Authorities, Local Government Engineering Department (LGED), City Corporations, Bangladesh Economic Zones Authority (BEZA), UDD etc.). Other than specialized plans (e.g., economic zone plans, corridor plans, transport plans, drainage plan, etc.), most of the plans followed the Rational Planning Approach. Substantive change led to procedural Change. Nevertheless, the environmental agenda remained neglected in most of these plans.

Figure 2. Simplified planning process adopted by Structure Plans of Bangladesh in the 1990s.

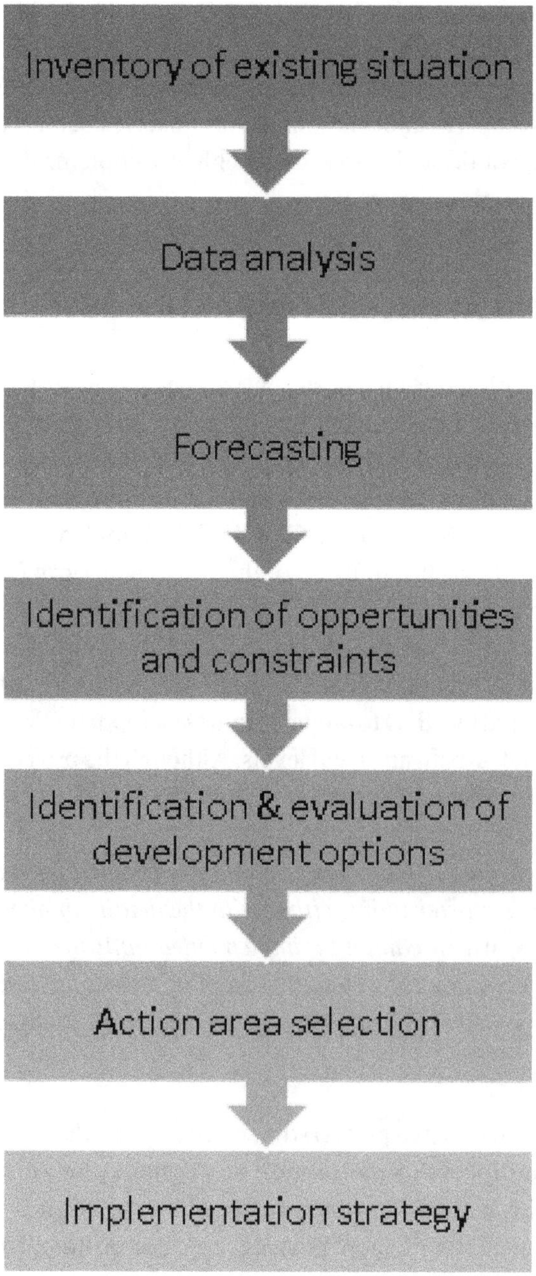

After about two decades of the second master plan of Dhaka, another project entitled "Dhaka Metropolitan Area Integrated Urban Development Planning Study" was initiated by the Government of Bangladesh at 1981 that detailed out some strategies for future development of Dhaka. It also emphasized the essence of a strategic plan for Dhaka. Following the recommendation of this project, the Capital Development Authority (RAJUK) initiated another project entitled "Preparation of Structure Plan, Master Plan, and Detailed Area Plan-Metropolitan Development and Plans Preparation and Management in Dhaka" in 1991. Outcome of this project is popularly known as Dhaka Metropolitan Development Plan (DMDP). Around this time, environmental agenda started getting its momentum. This plan incorporated hydrological issues systematically in all tiers of the plan. However, other environmental issues were still missing much-needed attention and scientific treatment. This project popularized a three-tier planning package (Structure plan, Master Plan / Urban Area Plan, Detailed Area Plan) in Bangladesh. MMP also adopted the same approach (KDA, 2011). The Structure Plan (Volume I) followed the typical path of rationalism in terms of the planning process. This structure plan fixated on 07 (seven) sectors (population, economic development, shelter and housing, social facilities and open space, transport, flood control and drainage, and utility services). Urban Area Plan (Volume II) followed the same path as the master planning approach. Detailed Area Plan (Volume III) was prepared later under another project.

Validity of the structure plan expired in 2015. To revise this plan RAJUK, initiated a new project called "Regional Development Planning (RDP)". Newly developed structure plan under this project is not yet (as of May 2021) approved by the government. Nonetheless, it followed the typical path of the rationalist approach. Urban Area Plan (second tier) was abolished in this package and it is infused in the Detailed Area Plan.

Like the structure plan in the UK, General Plan of USA also follows the rationalist approach of planning and includes multiple tires. Usually, it consists of Master Plans (also called community plan or area plan), Specific plans, Guidelines (design and development), and Zoning Code (State of California, 2017, p. 23; The Institute for Local Government, 2010, p. 17). General Plan is developed keeping a long-term vision and guiding principles in mind. This vision is then reflected through all the plan components. Hierarchical consistency among these plans is one of the major statutory requirements of this process.

Recently, Local Government Engineering Department (LGED) took an initiative to prepare master plans for 16 Pourashavas through a project, "Third Urban Governance and Infrastructure Improvement (Sector) Project (UGIIP-III)". If the Terms of Reference (ToR) and Draft Reports are evaluated, one can easily argue that the plan format (partially) followed the structure of the General Plan. No significant difference or innovation can be observed in terms of the planning process. Institutional and governance issues influencing the plan-making process remained as it is. Emphasis was given on the formulation of the vision, objectives, and principles of the plan. Following the policies of the structure plan, a network of plans is developed, e.g., Landuse Plan, Transportation and Traffic Management Plan, Drainage and Utility Services Plan, Environmental & Disaster Management Plan, Urban Services Plan, etc. These plans follow the traditional path of the master planning process. Several components of the general plan packages are missing in this plan, such as: zoning code, area-wise design guidelines, specific plan or any such plan, etc.

Most of the metropolitan cities of Bangladesh face a couple of common issues, like water logging, encroachment of water bodies, landslide (for hilly regions), loss of community spaces and vegetation, earthquake, very high population density (especially for Dhaka), lack of services and facilities etc. Although environmental issues have started getting their much-needed attention in the physical planning in recent years, policies related to the same often lacks scientific proof and are not translated spatially.

As an exception, hydrological model is used in some plans (e.g., DMDP, RDP and Jessore-Benapole Corridor Plan). To understand these issues clearly, MMP has been examined to know how it has treated point source air pollution (Sulfur Dioxide to be more specific) of a coal-fired power plant at Rampal and draw some conclusion upon it.

SPATIAL DISPERSION OF THE AIR POLLUTANTS OF THE COAL FIRED POWER PLANT AT RAMPAL

Air pollution has been one of the major considerations of landuse planning for quite a long time. Through the advancement of geospatial technology and atmospheric dispersion models, inclusion of air pollution in landuse planning has become increasingly popular. Theoretical underpinnings of air pollution sensitive landuse zoning have been developed through several academic researches during the 1950s to 1970s (Wronski, 1966; Frenkiel, 1956; Jarir, 1977; Linsky, 1954; Bosanquet, 1936; Henry, 1969; William, 1960). Since then, a long list of academic papers has been published to address this issue (Atkins, 2018; JRC, 2018; Macdonald, 2003; Bluett, J. et al., 2004; NOAA, 2015). However, air pollution is still neglected in Bangladeshi physical plans.

With a promise of ensuring electricity for every household, BDT 115.4 billion was allocated for Power and Energy sector in 2014-15 Budget. The same budget allocated resources to establish coal-based power plants with 1426 MW electricity generation capacity (GoB, 2014). Subsequently, Bangladesh started the construction of the Rampal thermal power plant in February 2015. Environmental activists protested vigorously against this project because of its proximity to the Sundarbans. The authenticity of the EIA report of the project has been questioned (Shakil & Ananya, 2015). Most of the literature has focused on the impact of the power plant on the Sundarbans. However, nobody has paid heed to the impact on the people living close to the project site

As per EIA, major gaseous emission from the proposed power plant would be SO_x, NO_x, H_2O, CO_2, CO, and particulate matters (CEGIS, 2013, p. 271). Spatial dispersion modeling of all the gases is beyond the scope of this chapter. Instead, a simple Gaussian Air Dispersion model for SO_2 is developed to identify areas under different threat levels. The impact zone with the increase of the stack heights is also simulated to justify the height. Estimated emission rate of SO_2 is 70.764 ton/day (or 819 g/s for each unit)). Total coal requirement for the power plant will be 4.715 million tons per year (ibid., p. 108).

Operationalizing Atmospheric Dispersion Model and Results

Detailed review of the atmospheric dispersion models is beyond the scope of this chapter. Atmospheric dispersion can be set up to estimate downwind concentrations of contaminants over varying averaging periods (Bluett, J. et al., 2004). Dispersion of air pollutants is largely dependent on characteristics of source and metrology. Source characteristics include (1) emission rate of pollutant, (2) stack height, (3) exit velocity of the gas, (4) exit temperature of the gas, and (5) stack diameter, etc. Metrological conditions include (1) wind velocity, (2) wind direction, (3) ambient temperature, (4) atmospheric stability, and (5) mixing height (Modi, M. et al., 2013). Modeling air dispersion in a geo-spatial environment based on all these parameters is often easier said than done.

Gaussian model assumes that (1) the plume spread has a normal distribution, (2) the emission rate is constant and continuous, (3) wind speed and direction is uniform (4) total reflection of the plume takes place at the surface, and (5) the terrain is relatively flat, i.e., no crosswind barriers (Bosanquet, 1936). As the model is not modified, describing the same is pointless. However, the enthusiastic readers can go through Hanna et al. (1982) to know about air dispersion models in details.

For operationalizing the model, the author specified a typical summer day (July, 2020). The models required (i) the properties of Sulfur dioxide (molecular weight 64.06 g/mol; ambient boiling point -10.0° C; >1 atm vapor pressure at ambient temperature; 100.0% ambient saturation concentration; (ii) Wind parameters (1.7 meters/second wind speed from north, plain land, 70% cloud cover, 30° C air temperature, moderately unstable air, normal atmospheric condition (No Inversion Height), 75% relative humidity; and (iii) parameters of the source (819 gallons/sec SO_2 emission, 275 meters stack height, 125° C temperature at the top of the stack, source pressure is equal to ambient, release rate is 367 kilograms/min).

Maximum ground level concentration of SO_2 has been estimated for an area of 10 km (06 miles) after one hour. The yearly average wind rose shows that wind prevails flowing from south to north direction in most of the time in a year (CEGIS, 2013, p. 130). Thus, the model is only developed for predicting ground level concentration for the same directional wind flow. During winter, the wind flows in the reverse direction. However, the situation is overlooked considering negligible wind speed.

There are numerous numbers of rural settlements, growth centers, towns located around the project site. Figure 03 shows the location of the power plant along with the surrounding human settlements (mainly growth centers). It is to be noted here availability of electricity will lead to increased industrialization; this will attract additional people to settle close to the power plant. This process will be further amplified by several nationally and internationally important development projects now under implementation within close proximity of the project site. Knowingly or unknowingly, these people will be exposed to the pollutants.

The model predicts that the ground level concentration of pollutants exceeds Acute Exposure Guideline Level (AEGL)-2 within 40 minutes at 4.27 kilometer (please see figure 04). Areas under AEGL-1 and AEGL-2 are shown in figure 05. AEGL estimates the concentrations at which most people will begin to experience health effects if they are exposed to a hazardous chemical for a certain duration (NOAA, 2015). Details about AEGL of SO_2 can be found at National Research Council (2010, pp. 393-447).

Within the confidence zone of AEGL-2, two major rural growth centers (Rajnagar and Gaurambha) are located. Additionally, four other growth centers are located within its confidence zone (AEGL-1). The zone of confidence is the zone to be under the same level of a threat if the wind moves. It is to be mentioned here that these growth centers are the primary locations of future urbanization and population concentrations.

SO_2 is colorless at ambient temperature and pressure. It is also odorless if the concentration is below the odor threshold of 0.67 - 4.75 ppm. Sulfur dioxide can reform as sulfuric acid. Health hazards of SO_2 and Sulfuric acid are well known. SO_2 causes irritation of the upper respiratory tract and eyes. Direct contact with SO_2 may cause Conjunctivitis, corneal burns, and corneal opacity. Prolonged exposure may kill people from respiratory arrest. Among the other side effects bronchitis, bronchopneumonia, and fibrosis obliterative bronchiolitis. Bronchoconstriction accompanied by increased pulmonary resistance are quite common. More details can be found on National Research Council (2010).

It could have been better if the total number of affected populations in the next 20 years and monetary value of health hazard could have been identified. However, this is also beyond the scope of this chapter.

Figure 3. Location of the thermal power plant at Rampal, Bangladesh

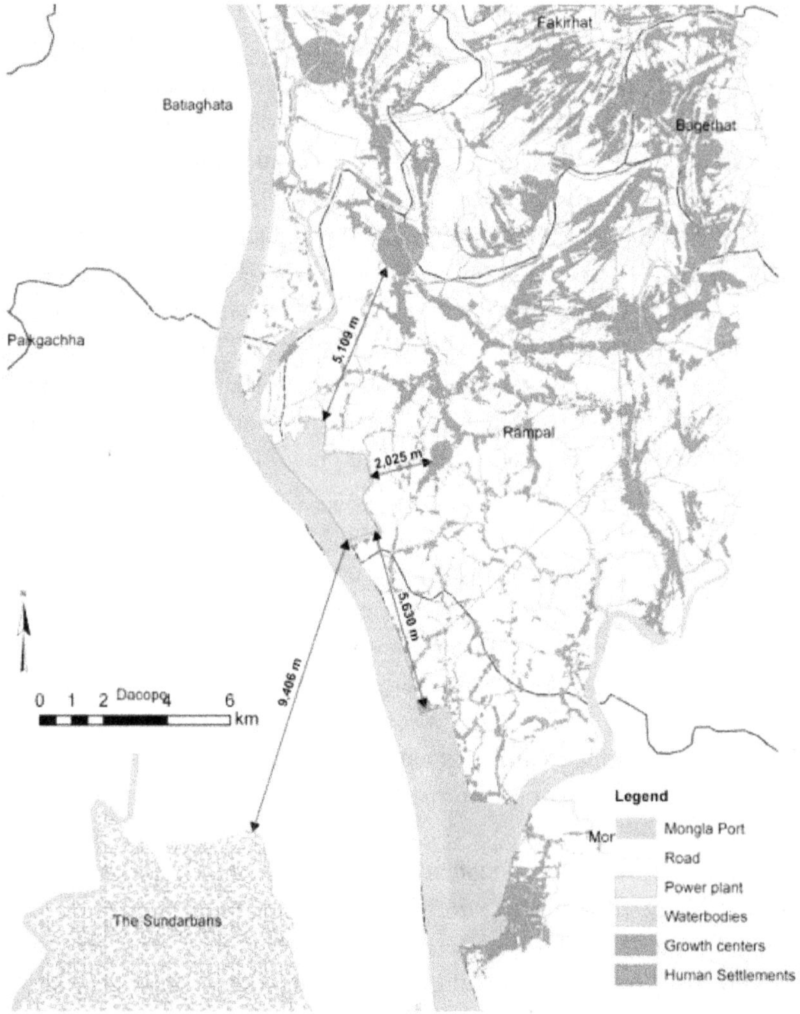

Figure 4. Threat at a point (4.27 km away from the source at downwind direction)

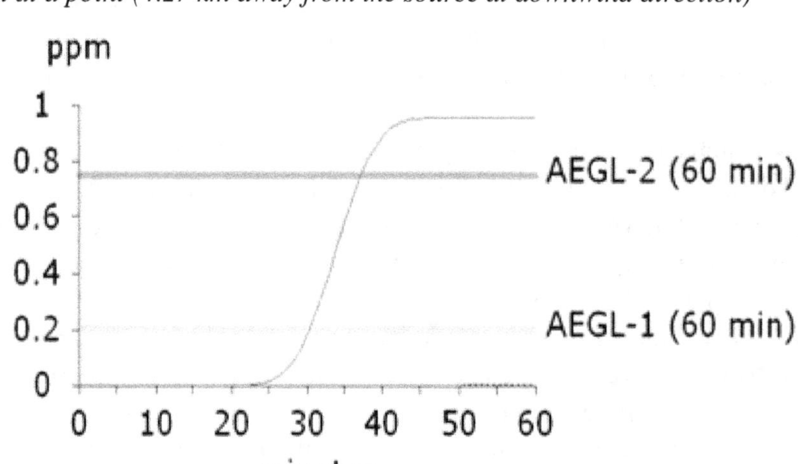

Figure 5. Output of Gaussian Dispersion model

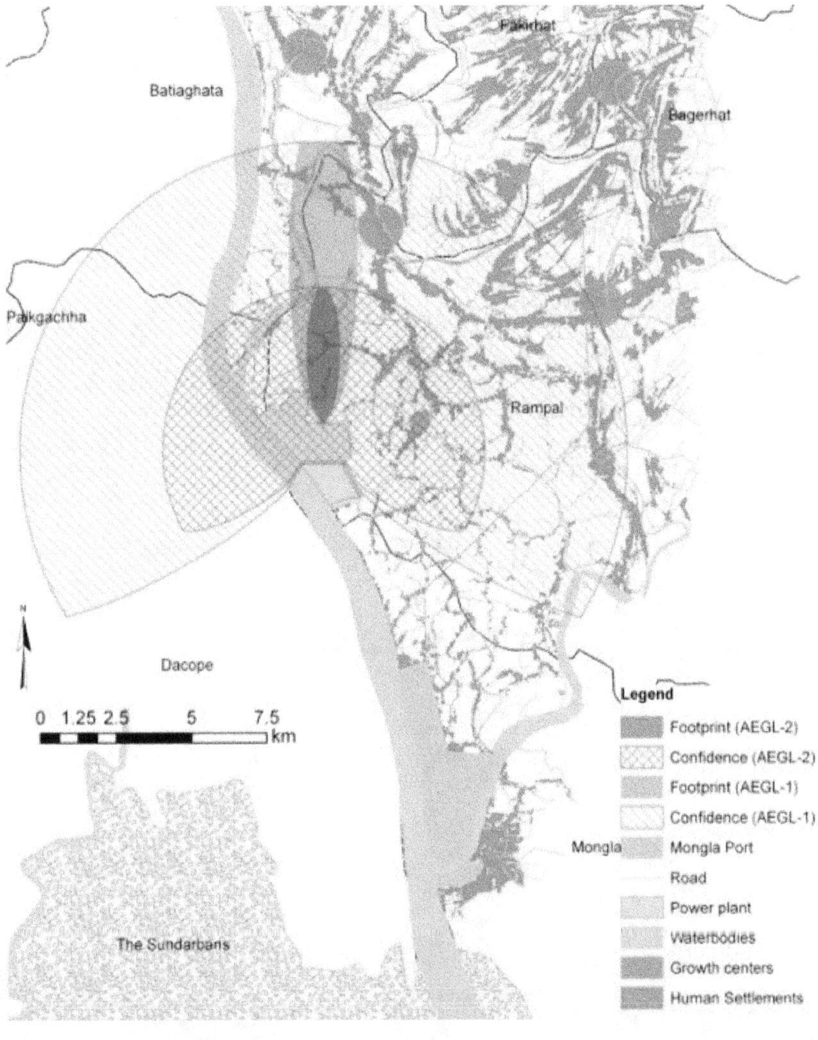

Evaluation of Stack Height

Primary intervention for reducing the impact is to adopt source control methods. There is nothing much spatial or physical planners can do in this regard. However, in this section, the author has attempted to see whether the increase in stack height helps to reduce the impact zone. For evaluation of stack height, the principle of sensitivity analysis has been adopted where the pre-specified model has been run with different stack heights. The result shows that with the increase of the stack height the threat zone shifts away from the source in the downwind direction and the area of the threat zone shrinks.

If the stack height is increased from 275 m (proposed height) to 300 m, the threat zone (AEGL-2) shifts to outward direction about 400 m (figure 06). If stack height is increased to 325 m, the footprint (AEGL-2) shifts more than 0.5 kilometers (figure 07). This shifting is 650m for 350m stack height (figure 08). The zone of confidence also shifts in variable scale in different directions. However, the proposed stack height is justifiable, and increasing the same will not incur beneficial results.

Figure 6. Output of Gaussian Dispersion Model for 300 m stack height

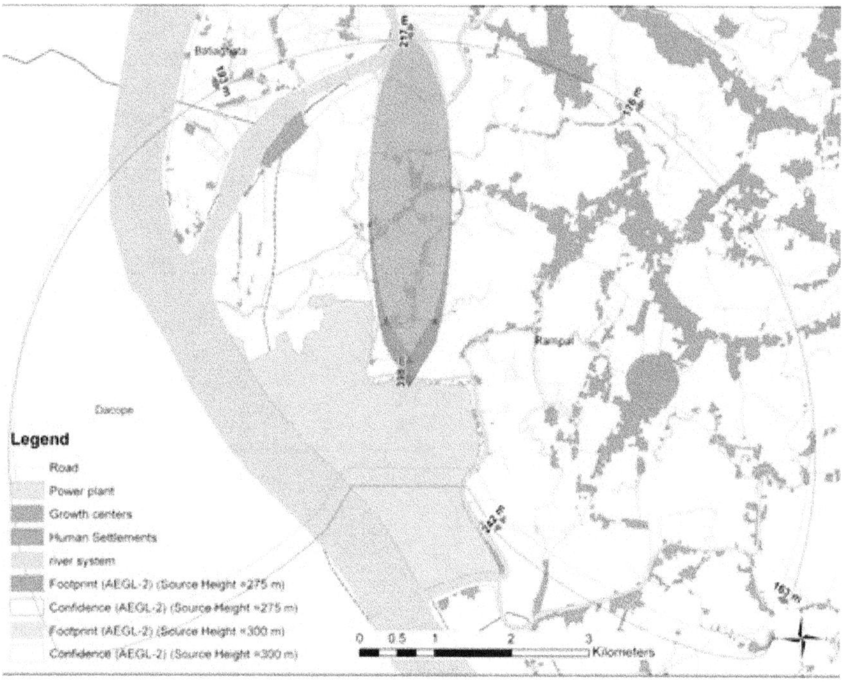

Figure 7. Output of Gaussian Dispersion Model for 325 m stack height

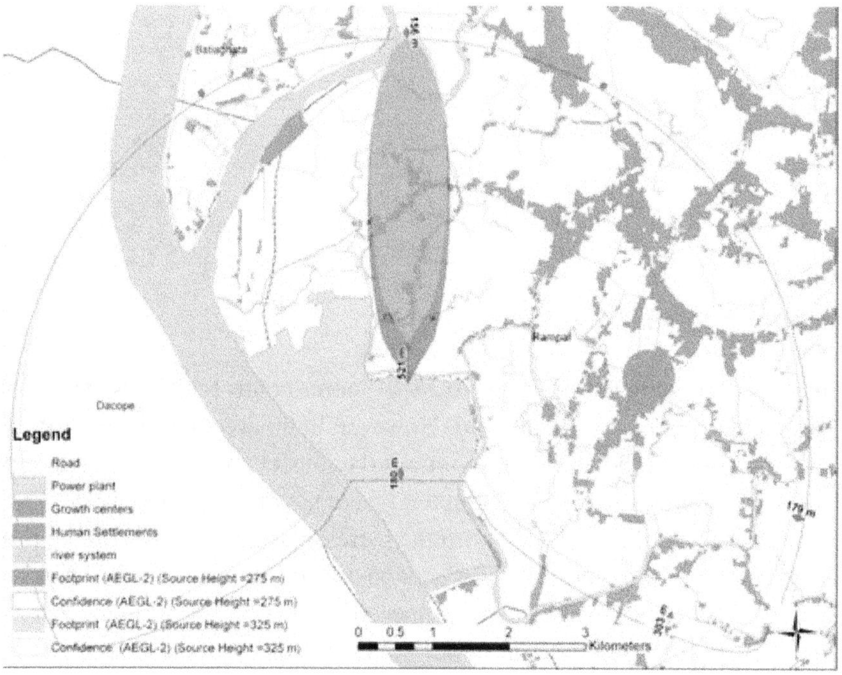

Figure 8. Output of Gaussian Dispersion Model for 350 m stack height

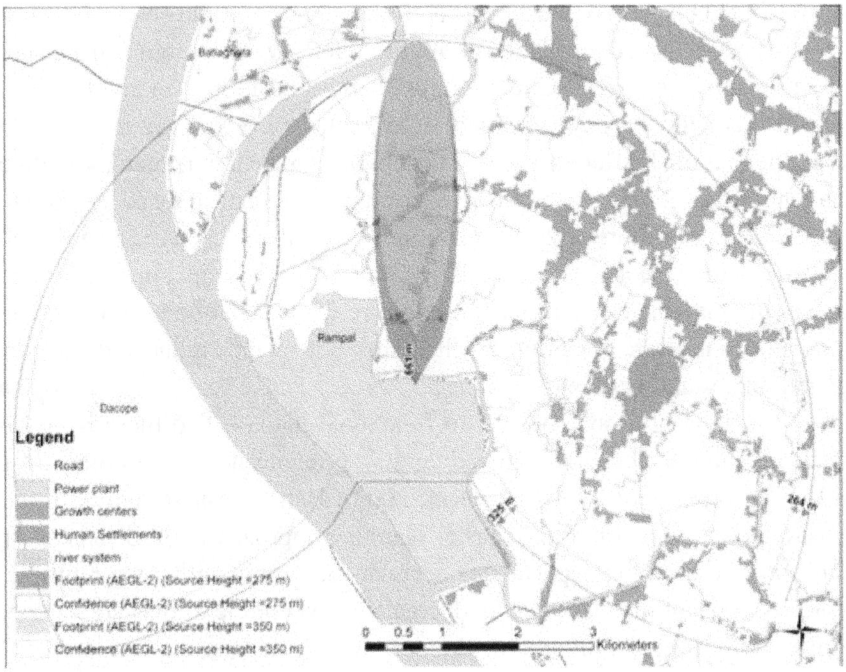

Air Pollution Sensitive Landuse Zoning

In the absence of any standardized process of planning in Bangladesh, often the planners and development practitioners attempt to resolve the issues that affect the people directly (e.g., drainage, circulation with waterlogging, traffic jam, etc.). Consequence of pollution on human health is a long-term process. People who are less educated, uneducated, and unaware often fail to recognize pollution as one of the major determinants affecting their life. As a result, planning organizations are reluctant to consider pollution both in the substantive and procedural parts of physical planning.

Landuse zoning for such ecologically volatile and economically potential areas is easier said than done. Several kinds of zoning have been proposed by scholars for protecting people from point source air pollution (e.g., air zoning, performance zoning). Nonetheless, to mitigate the impact of air pollution on the people living or will be living in the impact zone, the planners and policymakers need to accentuate regional planning rather than only local planning.

Integration of local plans with the regional plan would incur better results. Local plans should be prepared to ensure a safe, secure, economically viable (ample employment opportunity), and environmentally sustainable living environment. Several studies show that well-planned green zones (green corridors) help reducing the effect of pollution largely. The study area, being in the coastal region is dominated by many water bodies and wetlands. These water bodies need to be preserved as well. MMP has put forward appreciable effort in this regard both in terms of policy formulation in structure plan and its spatial application in Development Management Plan. Large scale blue and green network preservation program needs to be initiated after planning the same with a long-term vision. Public-private partnership is a must in this regard.

Regional landuse configuration is of utmost importance in this regard. Regional connectivity (road network) keeping the impact zone and intensity of pollution in mind would ensure targeted development. Planners also need to assign intensity of landuse (higher intensity away from confidence zone and vice versa) would serve the purpose largely. Combination of transportation plan and landuse intensity plan would be manifested as corridor type development. Nonetheless, no specific zoning system is applicable for such an area. Rather a mix of different zoning systems need to be incorporated for effective planning.

The most challenging part of such a plan is to encourage people to settle away from the power plant and restrict human settlement within the zone of confidence. Planners need to develop restrictive mechanisms for the zone of confidence and incentives for the outer corridors planned for human settlements. Restrictive mechanisms may include but not limited to restrictive public utilities and services policy, tightened development control, etc. Incentive mechanisms may include reduced tax to attract investment in the industrial and real estate sector etc.

The whole world is shifting away from mono-functional landuse and there is no disagreement in the scholarly literature about the benefit of mixed landuse (Greenwald and Boarnet, 2001; Cervero and Kockelman, 1997; Cervero, 1996; Ewing, R. et al. 2003). Bangladesh started with mono-functional landuse zoning with the introduction of the master planning approach in the 1960s. Responding to the global trend, planners also tried to adopt inclusionary landuse zoning allowing multiple activities in a single zone. Typically, most of the plans use landuse matrix where the usages are listed under different condition categories. Often the conditions for the permitted usages are not described. Usually, these conditions are described in the landuse regulation which is an integral part of the planning package. However, even after about seven decades of history of planning, landuse regulation has not been adopted in any of the plans of Bangladesh. MMP also lacks the same.

Some of landuses emit more pollutants compared to others. These landuses should be placed far away from residential and mixed landuse. Direction and velocity of airflow need to be considered carefully. Air pollution-sensitive services and facilities also need to be carefully planned.

CRITICISMS OF STRUCTURE PLANNING APPROACH OF BANGLADESH VIS-A VIS MONGLA MASTER PLAN (MMP)

Strategic plan still serves as the apex of the planning package. So, it is pertinent to conclude our discussion oriented more towards this plan. The key characteristics of a strategic plan are clarity, motivational impact, internal consistency, compatibility with the environment, appropriateness considering available resources, risk sensitivity, the extent to which it bridges the gap among the key actors, treatment of time, flexibility, workability, objectivity, and committed leadership (Albrechts, 2004). Vision is the key to achieve all these characteristics of a plan. A vision should reflect the in-depth understanding of the communities, their values, their preferred future, emerging trend and issues, and a road map showing the path of achieving the preferred future (Steiner & Butler, 2012, p. 39). True to visioning's roots in anticipatory democracy (ibid.). Meaning, the visioning process needs to be highly participatory and anonymous. Each plan should start by setting an agreed vision, some goals or principles that will determine which issues deserve higher priority compared to others, what process should be followed to achieve the desired outcome.

"A democratic plan may be recognizable not so much by its aims, or its policies, or its 'distributive justice'. Rather its identifying character is likely to be its tone, expressing the experience of inter discursive discussion." Healey, P. (1993).

Visions are translated into several principles. Without these principles, often plans lose their objectivity. Thimbu Structure Plan, for example, is developed based on 09 principles (RGoB, 2004). Sydney Strategic Plan is developed based on 04 goals carefully crafted to implement a concrete vision (Department of Planning & Environment, 2014). MMP had no vision or principles. Objectives are not at all rationalized and loosely coupled. Although, the plan took Coal-fired 1320 MW Power Station at Rampal under consideration. But its environmental impact on the people was completely ignored (KDA, 2011; Vol.-1, p. 72). The plan expects to achieve its objectives through 81 policies formulated for 19 sectors involving numerous numbers of implementing authorities. These numbers themselves open the door for some of the major criticisms. Some of the policies overlap with each other. The plan also lacks coordination and consistency among different sectoral policies. No policy was laid down related to the adverse impact of the coal-fired power plant.

"...the understanding of planning as a discipline that is shaping itself in and through practice and, in doing so, mirrors an image of the culture and environment that produced it." Sartorio, F. S. (2005)

Numerous numbers of development organizations, civil societies, NGOs, and local government organizations are working in any planning area. All these institutions need to share the responsibility of establishing, implementing, and ensuring the continued performance of collective spatial and environmental goals. In fact, the performance of any structure plan depends on how collective spatial policy objectives bridge the distance among these institutions and actors. I think it is unequivocally clear that the complexity of policymaking within multi-actor environment increases geometrically with the increase of the number of actors and spatial coverage. Salet and Faludi (2000) thus very correctly pointed out "*... goal-rational perspective of strategic planning could no longer flourish in a multi-actor context: one where numerous party all demand their own piece of the pie and do not automatically comply with the collective goals printed in a central plan.*" This is perhaps one of the major reasons (among many) behind the failure of the strategic plans of Bangladesh.

Lack of public participation is one of the most discussed issues of the Bangladeshi urban planning system. The rationalist approach of planning is inherently biased towards the socially affluent peoples (Etzioni, 1967). Public participation in the planning practice is further hindered by the Westminster style of democracy that is being practiced in Bangladesh where it is assumed that once elected, the political leaders will understand and speak for the underprivileged people (Khan, S., & Swapan, M. S. H. (2013). This assumption never remains valid.

For most of the planning projects, representatives from different public departments, local governments, professional groups, civil society organizations, business groups, media, elected political leaders, and academics are asked to attend different meetings in the plan preparation stage to develop a wish list as observed in Khulna (Rahman, M. A. U., 2016). Thus, almost always it is expert-led, rather than collective in nature (ibid). For MMP, the same approach of participation was adopted through three stages of consultations.

In Bangladesh, the structure plan's application is found lightly in landuse plan preparation. Almost none of the policies are spatially translated in the structure plan of MMP. In fact, it is difficult to spatially translate all the policies (totaling 81 in this case). Once prepared, landuse plan (master plan) is basically used for development management. The structure plan's role in this regard is almost unnoticeable. Usually, structure plan policies are not considered while preparation of new physical development projects. The increasing scarcity of resources, lack of people's participation (especially socially marginalized group), non-cooperation from different implementing agencies often contributes towards non-implementation of the plan.

CONCLUSION

Obviously, the planning system of Bangladesh is undergoing transition. One should not portray this transition negatively. Hopefully, urban professionals will learn from their mistakes and bring positive changes for the people. Every day new factors are added to the planning process and the professionals are trying hard to adapt to the change.

There are so many reasons behind the identity crisis of Bangladeshi Planning. Planning has not been institutionalized yet (often it is project-based) that makes incremental learning quite impossible. Bureaucracy hinders inter-ministerial cooperation at the top level that results in conflict and non-cooperation among the organizations directly or indirectly involved in implementing plans. Public fund is also allocated through ministries to respective organizations. Multi-sectoral funding is almost impossible which makes the place-making approach to planning impossible. Bangladesh does not have National Spatial Data Infrastructure (NSDI) and still lacks policies on inter-organizational data sharing. Meaning different organizations are spending resources for collecting the same data from the same area. These are making planning and its implementation extraordinarily challenging.

As of today, there is no national spatial planning framework available in Bangladesh. Regional Planning, although studied for quite a long time in the academia of Bangladesh, has never been practiced. Planning theory as a subject matter in the planning curriculum is relatively new addition in Bangladesh. Under such circumstances, the author tried to reveal the very elementary story of planning theory and relate it to Bangladeshi planning practices. At the same time, taking the coal-fired power plant at Rampal and MMP as examples, the author proved that Bangladeshi physical plans need to be more sensitive to environmental pollution. In doing so time, the author accentuated on scientific treatment of pollution in planning.

ACKNOWLEDGMENT

Air dispersion modeling in geospatial environment was performed using ALOHA (Areal Locations of Hazardous Atmospheres) program (a part of CAMEO® software suite). It is developed jointly by the National Oceanic and Atmospheric Administration (NOAA) and the U.S. Environmental Protection Agency (EPA).

REFERENCES

Albrechts, L. (2004). Strategic (spatial) planning reexamined. *Environment and Planning. B, Planning & Design*, *31*(5), 743–758. doi:10.1068/b3065

Steiner, F. R., & Butler, K. (Eds.). (2012). *Planning and Urban Design Standards* (2nd ed.). Wiley.

Atkins, N. (n.d.). Survey of meteorology - class notes - outline - Lyndon state college. https://apollo.nvu.vsc.edu/classes/met130/notes/

Batty, M. (1976). Models, Methods and Rationality in Urban and Regional Planning: Developments since 1960. *Area (23rd International Geographical Congress)*, *8*(2), 93-97.

Bluett, J., Gimson, N., Fisher, G., Heydenrych, C., Freeman, T., & Godfrey, J. (2004). *Good Practice Guide for Atmospheric Dispersion Modelling*. Ministry for the Environment. http://www.mfe.govt.nz/publications/air/atmospheric-dispersion-modelling-jun04/html/page5.html

Bosanquet, J. L., & Pearson, J. L. (1936). The spread of smoke and gases from chimney. *Transactions of the Faraday Society*, *32*, 1249. Advance online publication. doi:10.1039/tf9363201249

CEGIS. (2013). Final Report on Environmental Impact Assessment of 2x (500-660) MW Coal Based Thermal Power Plant to be Constructed at the Location of Khulna. Center for Environmental and Geographic Information Services (CEGIS).

Cervero, R., & Kockelman, K. (1997). Travel demand and the 3Ds: Density, diversity, and design. *Transportation Research Part D, Transport and Environment*, *2*(3), 199–219. doi:10.1016/S1361-9209(97)00009-6

Cervero, R. (1996). Mixed land-uses and commuting: Evidence from the American Housing Survey. *Transportation Research Part A, Policy and Practice*, *30*(5), 361–377. doi:10.1016/0965-8564(95)00033-X

Davidoff, P. (1965). Advocacy and Pluralism in Planning. *Journal of the American Institute of Planners*, *31*(4), 331–338. doi:10.1080/01944366508978187

Davidoff, P., & Reiner, T. A. (1962). A Choice Theory of Planning. *Journal of the American Institute of Planners*, *28*(2), 103–115. doi:10.1080/01944366208979427

Department of Planning & Environment. (2014). *A plan for growing Sydney: A strong global city, a great place to live.* Department of Planning & Environment. http://www.strategy.planning.nsw.gov.au/sydney/the-plan/

Etzioni, A. (1967). Mixed-Scanning: A "Third" Approach to Decision-Making. *Public Administration Review*, *27*(5), 385–392. doi:10.2307/973394

Ewing, R., Pendall, R., & Chen, D. (2003). Measuring Sprawl and Its Transportation Impacts. *Transportation Research Record: Journal of the Transportation Research Board*, *1831*(1), 175–183. doi:10.3141/1831-20

Faludi, A. (1973). *Planning theory*. Pergamon Press.

Flyvbjerg, B. (1996) The dark side of planning: Rationality and 'realrationalitat'. In Explorations in Planning Theory. New Brunswick, NJ: Rutgers.

Frenkiel, F. N. (1956). Atmospheric Pollution and Zoning in an Urban Area. *The Scientific Monthly*, *82*(4), 194–203.

GoB. (2014). *Budget in Brief 2014-15*. Finance Division, Ministry of Finance, Government of the People's Republic of Bangladesh. http://www.mof.gov.bd/en/index.php?option=com_content&view=article&id=269&Itemid=1

Greenwald, M. J., & Boarnet, M. G. (2001). The Built Environment as a Determinant of Walking Behavior: Analyzing Non-Work Pedestrian Travel in Portland, Oregon. *Transportation Research Record: Journal of the Transportation Research Board*, *1780*(1), 33–41. doi:10.3141/1780-05

Hanna, S. R., Briggs, G. A., & Hosker, R. P. (1982). *Handbook on Atmospheric Diffusion*. U.S. Department of Energy. doi:10.2172/5591108

Henry, W. H. (1969). The air diffusion model as an urban planning tool. *Socio-Economic Planning Sciences*, *3*(4), 329–349. doi:10.1016/0038-0121(69)90056-1

Hudson, B. M., Galloway, T. D., & Kaufman, J. L. (1979). Comparison of Current Planning Theories: Counterparts and Contradictions. *Journal of the American Planning Association*, *45*(4), 387–398. doi:10.1080/01944367908976980

Institute for Local Government. (2010) *Understanding the Basics of Land Use and Planning: Guide to Local Planning*. https://www.ca-ilg.org/sites/main/files/file-attachments/2010_-_landuseplanning.pdf?1387495993

Jacobs, J. (1961). *The death and life of great American cities*. Vintage Books.

Jarir, S., & Dajani, T. H. (1977). Air pollution control through land-use planning. *Urban Systems*, *2*(1), 35–43. doi:10.1016/0147-8001(77)90004-3

JRC. (2018). *Setting up the modelling exercise*. European Union Joint Research Centre. https://rem.jrc.ec.europa.eu/etex/37.htm

KDA. (2011). *Structure Plan, Master Plan and Detailed Area Plan for the Extension of the Khulna Master Plan Area up to Mongla Town. Khulna Development Authority (KDA)*. Ministry of Housing and Public Works, Government of the People's Republic of Bangladesh.

Khan, S., & Swapan, M. S. H. (2013). From blueprint master plans to democratic planning in South Asian cities: Pursuing good governance agenda against prevalent patron-client networks. *Habitat International*, *38*, 183–191.

Lindblom, C. E. (1959). The Science of "Muddling Through". *Public Administration Review*, *19*(2), 79–88. doi:10.2307/973677

Lindblom, C. E. (1979). Still Muddling, Not Yet Through. *Public Administration Review*, *39*(6), 517–526. doi:10.2307/976178

Linsky, B. (1954). Air Pollution Planning and Zoning. *Air Repair*, *3*(3), 155–156. doi:10.1080/00966665.1954.10467618

Macdonald, R. (2003). Theory and Objectives of Air Dispersion. MME 474A. *Wind Engineering*, 1–27.

McLoughlin, J. B. (1969). *Urban and regional planning; a systems approach*. Faber.

Modi, M., Ramachandra, V. P., Ahmed, L. S. K., & Hussain, Z. (2013). A review on theoretical air pollutants dispersion models. *International Journal of Pharmaceutical, Chemical and Biological Sciences, 2013*(3), 1224–1230.

National Research Council. (2010). *Acute Exposure Guideline Levels for Selected Airborne Chemicals* (Vol. 8). The National Academies Press. https://www.epa.gov/sites/production/files/2014-11/documents/sulfurdioxide_final_ volume8_2010.pdf

NOAA. (2015). *Acute Exposure Guideline Levels (AEGLs)*. National Oceanic and Atmospheric Administration (NOAA). https://response.restoration.noaa.gov/oil-and-chemical-spills/chemical-spills/resources/acute-exposure-guideline-levels-aegls.html

Rahman, M. A. U. (2016). Urban sustainability through strategic planning: A case of metropolitan planning in Khulna city, Bangladesh. *Journal of Urban Management, 5*(1), 16–22. doi:10.1016/j.jum.2016.06.001

RGoB. (2004). Thimpu Structure Plan. Department of Urban Development and Engineering Services, Ministry of Works and Human Settlements, RGoB, Bhutan.

Salet, W. G. M., & Faludi, A. (2000). *Three approaches to strategic spatial planning. In The revival of strategic spatial planning*. Royal Netherlands Academy of Arts and Sciences.

Sartorio, F. S. (2005). Strategic Spatial Planning. *disP - The Planning Review, 41*(162), 26-40.

Shakil, S. H., & Ananya, A. H. (2015). Effectiveness of Environmental Impact Assessment (EIA): Bangladesh Perspective. *Bangladesh e-Journal of Sociology, 12*(1), 115–125.

Shirazi, S., Baca, E., McCormick, M., & Litchney, S. (2017). State of California General Plan Guidelines. Governor's Office of Planning and Research, California, USA. https://opr.ca.gov/docs/OPR_COMPLETE_7.31.17.pdf

Swinburne, R. G. (1969). Physical Determinism. *Royal Institute of Philosophy Lectures, 3*, 155–168. doi:10.1017/S0080443600000662

Taylor, N. (1998). *Urban Planning Theory Since 1945*. SAGE Publications Ltd.

UDD. (1986). *Rangpur Structure Plan Bangladesh National Physical Planning Project, Urban Development Directorate*. UDD.

VeneKlasen, L., & Miller, V. (2002). *A new weave of power, people and politics: The action guide for advocacy and citizen participation.* World Neighbors.

Wronski, E. W., Anderson, E. W., Berry, A. E., Bernhart, A. P., & Belyea, H. A. (1966). Air Pollution Considerations in the Planning and Zoning of a Large Rapidly Growing Municipality. *Journal of the Air Pollution Control Association, 63*(3), 157–158. doi:10.1080/00022470.1966.10468458 PMID:5903516

KEY TERMS AND DEFINITIONS

Acute Exposure Guideline Level (AEGL): Acute exposure guideline levels (AEGLs) defines levels of chemical concentration that pose a defined level of risk to humans.

Atmospheric Dispersion Model: Downwind concentration of contaminants over varying averaging periods can be estimated considering the characteristics of source and metrology using Atmospheric Dispersion Model.

Landuse Plan: Landuse is defined as the function of human activities on land surface. Landuse management is essential for ensuring environmental sustainability, comfortability of living environment, and ability of a city to function smoothly. Policies of the structure plans are translated through landuse plan. It is mainly used for development management.

Master Plan: The term "master plan" is often used ambiguously. However, originally master plan of a city envisions the future a city in a rigid manner. It is a blueprint-based plan accompanied by an explanatory report. The blue-print depicts the end-stated of a city.

Structure Plan: To get rid of rigidity of master planning approach, strategic plan by the name "structure plan" was adopted on top of Master Pan. It provides strategic direction for its jurisdiction.

Chapter 5
Level of Urban Infrastructure Services and Coping Mechanisms of Households

Ankush Mishra
Swaniti Initiative, India

ABSTRACT

Basic infrastructure services like clean water, proper sewage and sanitation, etc. are important attributes in urbanization of a city to make it healthy, livable, and sustainable. Goal 6 of the Sustainable Development Goals (SDG) aims to ensure clean water and sanitation for all. Access to clean drinking water and basic sanitation facilities are crucial requirements for good health. Issues like intermittent supply, poor water quality, and low-pressure cause unreliable situations due to which community members resort to various coping strategies. Coping strategies are influenced by the household's level of education, income, and the level of unreliability of the service. The coping mechanisms used by the households are costly, and the poor people end up paying more because of the hidden cost. The poor water supply services result in induced costs by the residents in terms of the time costs for water collection and illness due to disease and water treatment costs. Because of these induced costs, the urban poor end up spending more for a clean and reliable water supply.

INTRODUCTION

As per the report of United Nations, Department of Economic and Social Affairs, in 2018 there are more than 2 billion people around the world who lacks access to safe and sustainable water services. Poor and children are at the main receiving end because of poor infrastructure facilities. More than 3,40,000 children are estimated to die from diarrheal diseases every year (United Nations Children Fund & World Health Organisation, 2017). India with more than 163 million people tops the list of people not having access to clean water followed by Ethiopia and Nigeria (WaterAid, 2018). Approximately of the total world population without access to clean water 19% lives in India. Travelling more than 30 minutes to get water is considered as without access to water. Unsafe water and the lack of hygiene practices are

DOI: 10.4018/978-1-7998-8331-9.ch005

linked with 21% of communicable diseases in India as per an estimate of the World Bank. Access to clean drinking water has a direct and indirect impact on the person's life in form of better health benefits because of disease control and improved living conditions. Author like Razzolini and Gunther has concluded that to promote adequate sanitary conditions and protect the health of the people supply of clean water in adequate quality, quantity, physical access and continuity is very important (Razzolini & Günther, 2008).

In developing countries and especially in slums or squatter settlements the access to drinking water varies from few minutes to few hours in a day. There are torrents of literature available which highlights a relation between poor and unreliable water supply and an impact on the health of the people, but not adequate study on coping strategies by households' members against unreliable water supply and its negative impact on the health and economic conditions. This study tries to understand how the society copes up with poor and rationed service and are their response efficient and sustainable keeping in mind environment, health and the financial condition of people.

Intermittent water supply affects water supply quality in the distribution system causing contamination affecting the welfare and health of households (Kumpel & Nelson, 2013). Various studies have shown the detrimental impact on health because of coping strategies adopted by the people against the unreliable water supplies. Case in point is urban Zimbabwe where in 2008-09 cholera outbreak happened because of the water contamination (Olayide & Ikpi, 2010). It was found that to tackle the problem of unreliable water supplies people resorted to digging shallow wells which was the main cause of water contamination in the country. In coming future, the problem of water scarcity will increase more and more because of which the water supply will become more infrequent and unreliable. So, it is very important to study the coping strategies used by the people and understand what can be the most effective way to ensure that the household gets safe and sufficient water at low cost with minimum adverse impact on the health of the people. So, this study estimates the total coping cost (direct and induced) of the households for unreliable water supply services and make comparison with the household water bill for piped water service.

In the first section of this paper the theoretical background was given regarding the status of unreliable water supply service in India and how it affects the health of the people. The second section contains the literature review on various coping strategies used and the coping cost incurred because of it. The third section provides detail on the study area and the research method used for this study. Households' profile and their coping strategies are described in fourth section while the coping costs of households are calculated in fifth section of this paper. The last section gives some concluding observations and way forward for future research.

LITERATURE REVIEW

Review of Water Supply Reliability in Developing Countries

In developing countries unreliable water supply remains a grave concern even in urban areas where the infrastructure exists. Goal 6 of sustainable development goal envisions provisions for safe water and sanitation for all but it falls short of any indicators on unreliable or infrequent water supply. Scope of all the indicators deals with physical infrastructure but it falls short in the provisions related to service delivery. Because of this reason despite having an acute problem of unreliable water supply in developing countries there is no unified definition on unreliable water supply or a way to measure it (Kleemeier,

2010). Many studies have been done in developing countries to understand actual behavior of households adopted by them to cope with the shortcomings in the services. This is an indirect method to assess the household demand for a service. Households resort to collective or individual strategies to fill the service gap, with the former one more common in rural areas whereas the later one is more prominent in urban areas (Zerah, 2000). In a study in urban Lebanon the authors tried to understand the impact on water quality and health because of intermittent water supply. A quantitative approach was followed in which before and after the storage of water in household tanks a sample was taken to test it. It was found that the water quality was degraded and in some locations the amount of chlorine was more than usual whereas in certain locations is was way below the desired level (Ayoub & Malaeb, 2015). Health cases were also reported because of the water contamination in such places where the water supply was intermittent and households resorted to various coping strategies.

In a study in urban Nepal the authors tried to understand how the coping cost and willingness to pay vary across different types of income groups and water users. They identified collecting, pumping, treating, storing and purchasing as five different kind of coping behaviors in households (Pattanayak et al., 2005). It was found that households incur significant induced (hidden) cost for coping with the inefficiency in services in terms of time and health cost. In case of Delhi, Dutta and Tiwari had conducted a study evaluating the true cost of services. They provided a framework to calculate in monetary terms the economic and environmental externalities caused because of unreliable water supply services (Dutta & Tiwari, 2003). The identification of full cost of water services are necessary to understand how much hidden cost is incurred by the households and can that additional cost be pivoted towards an improved service. Although there exist many studies on the unreliable water supply and coping mechanism employed by the households but they all fall short of accessing the hidden cost on health and environment which is an important attribute.

Coping Strategies

From literature review three main types of coping strategies were identified. First one deals with quantity of water available for which strategies to enhance and conserve it was practiced in many cases. Storing of water can be in large overhead tanks or in smaller vessel (Choe et al., 1996). The former one is easier to adopt whereas the latter one is a tedious process and has a hidden cost in terms of opportunity cost of time loss for filling the water and health cost because of transmission of diseases (Zerah, 2000). Buckets, drums and pots are the common vessels used for storing water, which are at times find at unhygienic conditions causing health harm to the people. Practices or strategies by households for conserving water was at times found to reduce washing of vegetables or fruits or less frequent bathing (Kudat et al., 1993) or restricting of use of flush in toilet to once per day (Chaminuka & Nyatsanza, 2013). Various literatures highlights households resorting to unhygienic and unsanitary practices to conserve water which in the longer run is unsustainable causing harm to members health and making the surrounding environment unclean.

Second and the most common coping strategy is improving the water quality by performing some kind of household water treatment method. Various studies have concluded that intermittent or unreliable water supply can lead to contamination of pipes (Kumpel & Nelson, 2013). Also, when the water is collected from some alternative source it has the possibility of either being unclean or of poor quality from the source itself or gets contaminated in the process of transporting and storing due to unhygienic practices by the households. So, it is very important to improve the water quality by treating it either by boiling or filtration or by use of some chemicals. The practice of water treatment differs from literature to literature based on socio-economic profile of the households which are studied. In a study in Kathmandu, Nepal the authors found the difference in behavior between rural and urban households with the proportion of water treatment higher in latter one (Katuwal & Bohara, 2011).

Other less common coping strategy is enhancing the flow rate which sometimes runs in parallel with the installation of overhead tank providing water at an increased pressure to the pipe lines (Choe et al., 1996). Some literatures have also identified installation of electric pumps to boost the pressure. Such coping mechanism have only economic impact on the households and there are no negative externalities to the human health. Although, such practices have a negative environmental externality attached to it as the people with high water pressure will use it more leading to water consumption. Factors like education level, income of households and land tenure plays a deciding role in the type of coping strategies employed by the households. Positive correlation was found between wealth and education level of the household and treatment strategies to get access to clean and safe drinking water (Caprara et al., 2009). Also, the coping behavior by households depend on the extent of unreliability or the perception towards it.

Coping Cost

From various literature installation and maintenance of storage facilities, buying of small vessels to store water, time spent in terms of opportunity cost when water was collected from an alternative sources, time and monetary loss due to diseases from unhygienic water, money spent on household water treatment are some of the identified coping costs due to unreliable water supply (Majuru et al., 2016). The concept of aversion behavior was highlighted by an Indonesian author who explains it as a behavior by the household to incur economic loss to avoid the negative environmental externality in form of pollution from poor water supply (Nastiti et al., 2017). In some cases the coping cost was equal to the amount paid in water bills (Choe et al., 1996) whereas in some other cases it was as high as approximately six times the bill amount paid (Zerah, 2000). It all depends on the availability of services and how good or bad it is. The coping cost in general was found to be higher in urban areas than in rural areas for almost all the literature because of the desire for better living condition and improved literacy on health impact due to unhygienic conditions. Many authors have related coping costs with the willingness to pay for improved services to find if there is a negative or positive difference between the two. If the willingness to pay comes out to be less than the coping costs of the households the natural course of action would be to use reliable and improved services at a lower charge. As in that case the individuals will end up saving the money which they are paying in terms of the hidden cost for improved services. Water supply has three main attributes namely the quantity, quality and pressure and when one or more of them are not available then the household's resort to coping strategies to minimize the risk associated with the unreliability.

STUDY AREA AND RESEARCH METHODS

Study Area Profile

For the study Bhopal city was selected because as per a report by KPMG in 2018 the city gets 90% of its water supply demand from surface reserves and only the remaining 10% is catered by ground water reserves (KPMG, 2018). So, it makes an interesting case to study how the municipality manages the demand for water supply and how reliable is it. The water supply network in the city has 100% coverage but no place gets the water supply for 24 hours. It varies from 3-4 hours in a day in some places to just 30 minutes in 2-3 days at other places. The extent of reliability of water supply is also not same for all the locations and varies based on the legality of the settlement. So, informal settlements like slums and squatter settlements gets poor services and the households have to resort to some kind of coping strategies to fill the gap between demand and supply. Two slums were selected for the detailed study, one having piped connection to majority of households and another not having it to compare and contrast on the coping strategies and coping cost among the two communities.

The two slums selected for the study are Damkheda Sector A slums and Damkheda Sector B slums in ward number 80 and 82 respectively as shown in Figure 1. As per census 2011 the number of households in both the slums are 619 and 299 respectively. The water supply system is relatively better in sector A slum, in sector B slum many households don't have in-house water connection and they collect the water from the communal tap.

Figure 1. Base map of Bhopal with slums locations
Source: Created by author, 2019

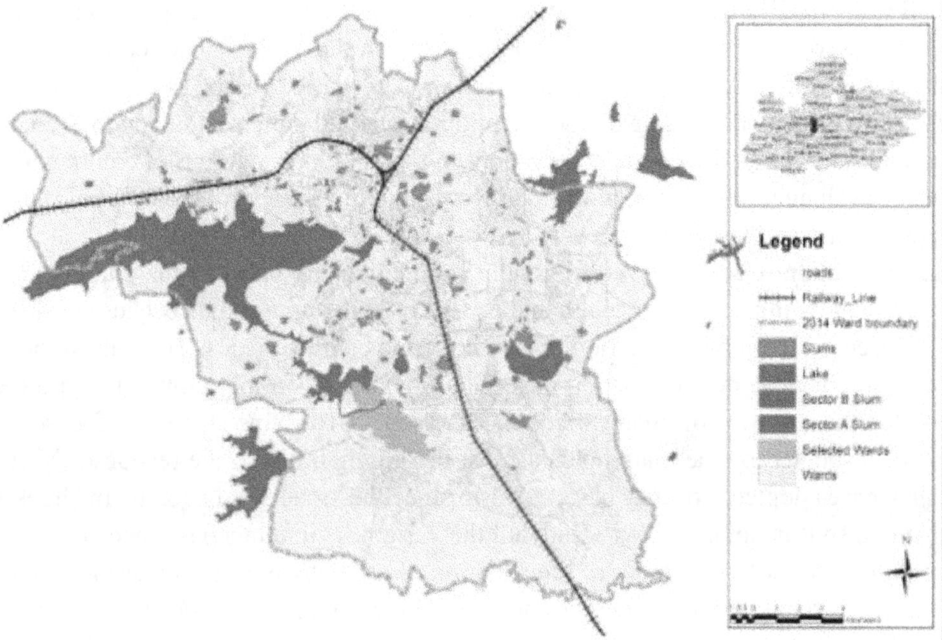

Sampling

The unit of analysis for this study was household because all the members will use the same infrastructure and hence the coping strategy and negative externalities associated with it will also be same. Sample size for the survey was calculated at 90% confidence level and 15% margin of error for Sector A slum & 95% confidence level and 20% margin of error for Sector B slum. Households surveyed from Sector A and Sector B slums were 35 & 22 respectively. Since the characteristics of both the slums were not same, with one having better in-house water connection facility than another so stratified random sampling of households was done to cover all the attributes available in the area. This method was followed because coping strategies depends on many factors like education, income etc. as identified by many literatures (Majuru et al., 2016; Pattanayak et al., 2005; Zerah, 2000). So, the random stratification sampling will help in covering all the different attributes which affects the coping behavior of the household and can be used in quantitative research analysis to understand a better association between two variables.

Data Collection and Analysis

The data collection technique was based on qualitative research method focusing on getting information related to household demographics characteristics and coping strategies adopted by households and calculate the coping costs accordingly. Questionnaire consists of structured and semi-structured questions depending on the type of information needed for analysis. The questionnaire focused on understanding the socio-economic characteristics of the respondents, reliability of the water supply service and how the household perceive such risk and adopt coping mechanisms. The primary survey focused on collecting information regarding the household's compensatory strategies to cope with the unreliable water supply. The questions asked in this regard were on household's storing strategies, collecting strategies, adaptation strategies and water quality strategies. While the questions on first three strategies focused on quantity and accessibility of water supply the last strategy focused on the quality of water supply and how the households cope up with it. Answer on more than one strategy was permitted as households are found to use more than one strategy depending on the extent of unreliability and socio-economic characteristics of the household. The author also made observations in field on the conditions of communal taps to see how it is managed along with the collection and storing containers used by the households to see if they follow proper sanitary and hygiene.

After data collection it was analyzed to establish a relationship between the unreliability of water supply services and its impact on the health and economy of the households because of different coping strategies used by them based on their socio-economic characteristics. To analyze the household choice of strategy linear regression models was found inadequate from various literature because the variables used in the study are discrete variables (Morrison, 2016). Since the study focuses on various coping strategies adopted by the households against the unreliability so the response will be discrete i.e., either they have adopted it or they haven't. Also, since the strategy adopted by the households can be more than two so it is not possible to study all the variables simultaneously because it would need information on the chronological order of the adoption of each strategy. So, binary logit models were used keeping one variable at a time for study to make things simple. The model was used to find the probability of household adopting that particular strategy keeping other factors unchanged. The next step was to calculate the coping cost for the strategies adopted by the household by summing up all the cost components (visible or hidden cost) incurred by the households. For estimating the hidden cost like loss of economic value due to health and opportunity cost of time for collecting water from alternative sources the proxy indicators were used. For capital expenditure the costs were annuitized to convert it in terms of monthly coping costs. For calculating the capital costs from the formula given in (Majuru et al., 2016) assumptions were made on the discount rate, scrap value rate and the life span of the capital. Similarly the coping costs in terms of health and time was calculated by annuitizing it (Dutta & Tiwari, 2003; Nastiti et al., 2017).

HOUSEHOLD'S PROFILE AND COPING STRATEGIES

Household's Profile of Slums Selected for Study

The sector A slum has maximum percentage of working population engaged in unorganized sector whereas in sector B slum the most common occupation is self-employed especially street vendors. The per capita income earning of households are higher in sector A slums in comparison to sector B slum as per a report prepared by BMC (Bhopal Municipal Corporation, 2015). The household's profile is an important attribute in decision-making regarding the coping strategies to be used by the household. The household profile is shown in Figure 2, Figure 3 and Figure 4.

Figure 2. Percentage of occupation of the total population
Source: BMC, 2015

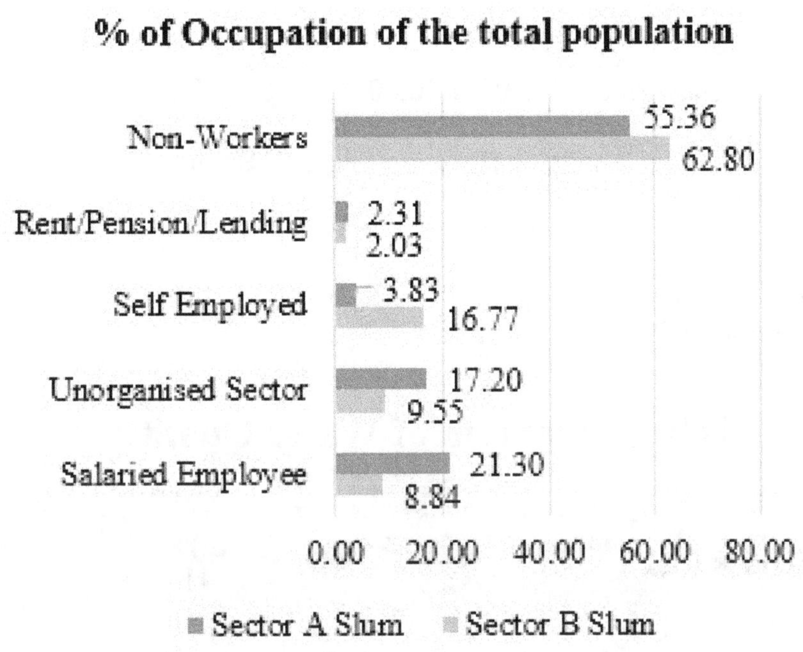

Water Collection Behaviour

From primary survey the different source of water for different purpose was found out for both the slums. It was needed to find out the primary source of water for different activities based on varied water quality. Because the quality of water from the public tap would not be that safe and secure for drinking purpose. But opting different source for different water use may increase the total coping cost. In the study area the treated tap water was at a distance of 800 meters from Sector A slum and 1.3 km from Sector B slum. So, for Sector B slum households the time coping cost will be more as shown in Figure 5 if they prefer treated water as their primary water source for drinking.

Figure 3. Percentage of HHs monthly income in INR
Source: Primary Survey by author, 2019

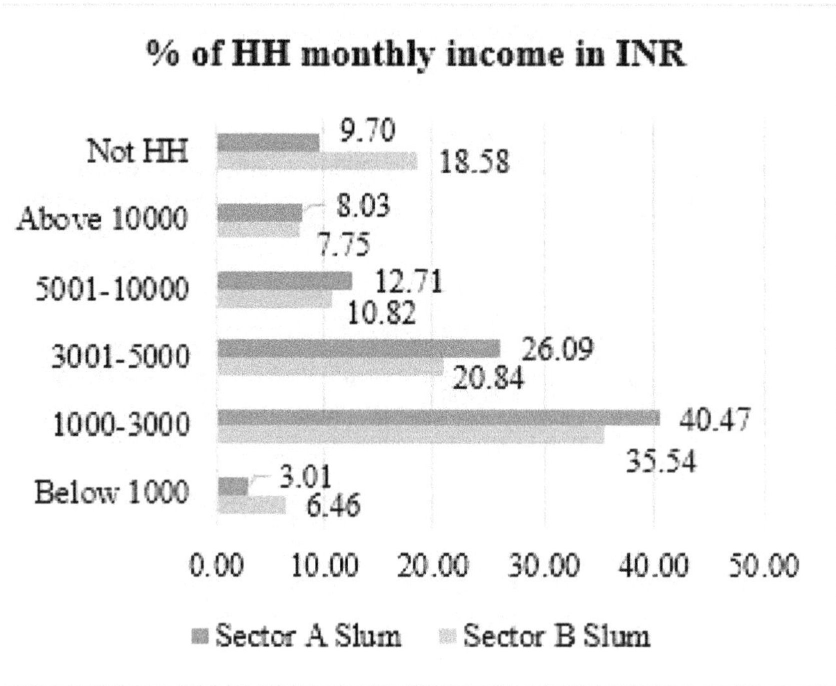

Figure 4. % of HHs perception of water quality
Source: Primary Survey by author, 2019

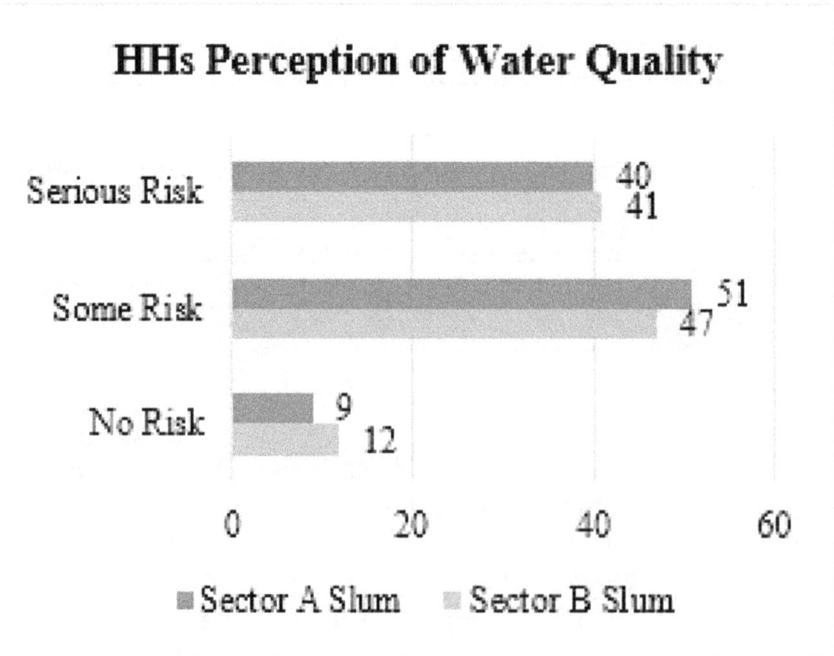

Figure 5. % of HHs using different source for different purpose
Source: Primary Survey by author, 2019

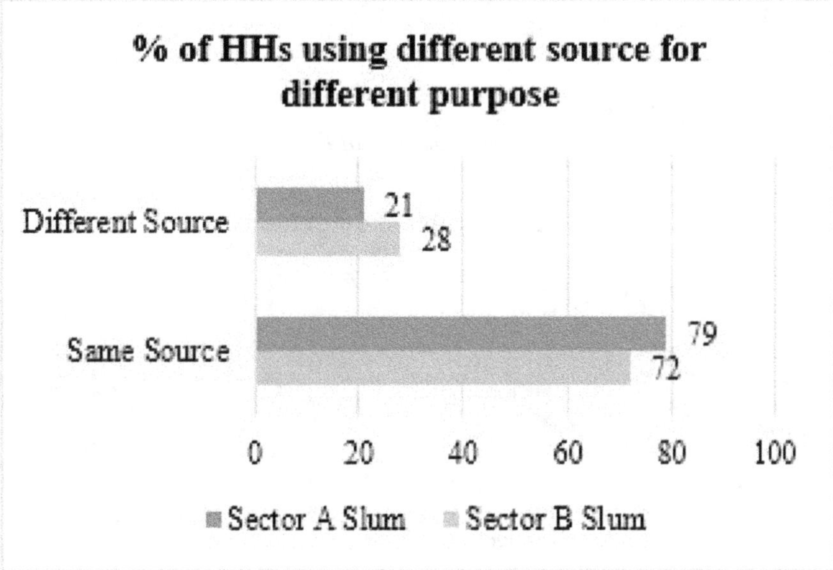

Coping Strategies of Households

Here strategy means the behavioral pattern of the households to cope with the unreliable water supply issues. The various coping strategies by households to get an improved water supply of good quantity and quality is not an end in itself but a means to achieve good health and other amenities associated with the water supply. So, the coping strategies are a function of households' utility which they try to maximize as shown in equation 1,2 and 3 (Cook et al., 2015).

$$U = (Z, T1, W(G), S (C (Tc, M)), \theta) \quad (1)$$

$$T = TW + TC + TL - TS \quad (2)$$

$$I \geq N + wTw - pM - Z \quad (3)$$

The total time available with the households is distributed among earning wages, in water collection, leisure and the time loss due to illness. The above utility functions show how the households try to maximize it by different coping strategies. The results derived from the primary survey of household's members on the type of strategy is shown in the Figure 6.

The above figure shows that households use more than one strategy to cope with the water supply issue. In both the slums the collecting strategy i.e., the collection of water from an alternative source which is not the in-house connection is very high. Despite large number of households collecting water from an unclean source the percentage of households performing water treatment at home is very low. Even some households having in-house water connection storage facility don't treat the water before using it for drinking purpose. Some households also practice rescheduling of activities like skipping cleaning of surroundings or toilet because of unreliable and intermittent water supply. It is clear from the result that the type and number of coping strategies used by the households is directly related to the income, education and number of members in the households.

Figure 6. Coping strategies of households
Source: Created by author, 2019; Data: Primary Survey by author, 2019

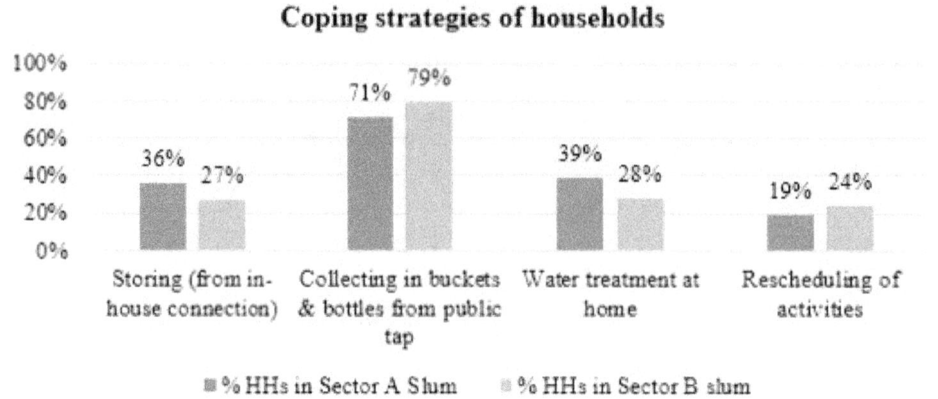

Probability Estimate of Different Coping Strategies Based on Unreliability

The result from the descriptive analysis of the households shows that the qualitative strategy to treat the water for drinking purpose increases to one in every three households when the household income is more than 8000 INR. On the other hand, the, every 4th household practice rescheduling of activities due to intermittent water supply when the household income is less than 5000 INR & the water supply is for less than an hour in two days. The binomial logistic regression model was done in SPSS to measure the probability of water treatment at home with respect to factors like household income, household size & education level of household. Here water treatment at home is a dependent variable to be measured on a dichotomous scale i.e., either the house treats the water or they don't treat the drinking water. And the three independent variables on which the probability of treatment depends in this study is the household income, household size and education level as shown in Table 1.

In this case the income level between 5001-8000 INR per month was taken as a reference point along with household size of 5 & above and secondary education level as reference point to conduct the test as shown in equation 4. The formula used for calculating the probability was taken from (Zerah, 2000).

$$P\left(Y_i = 1\right) = (1/1 + e^{-\left(x_0 p_0 + x_1 p_1 + x_2 p_2\right)} \tag{4}$$

Here Y is the household and x are the independent variables. For the reference point the probability of household using water treatment coping strategy was 14.8% as shown in Figure 7.

Table 1. Influence of income, household size and education level on household water treatment choice

Independent Variables	Coefficient	T^2 Statistics Test
Income (INR/Month)		
Below 5000	-1.37	11.8
5001-8000	Ref.	Ref.
Above 8001	1.19	12.2
Household Size		
4 or less than 4	-0.7	7.6
5 & above	Ref.	Ref.
Education Level		
Primary	-0.25	0.38
Secondary	Ref.	Ref.
Graduate	1.02	9.8

Source: Created by author, 2019

Figure 7. Probability estimate in % for water treatment at home
Source: Created by author, 2019; Data: Primary Survey by author, 2019

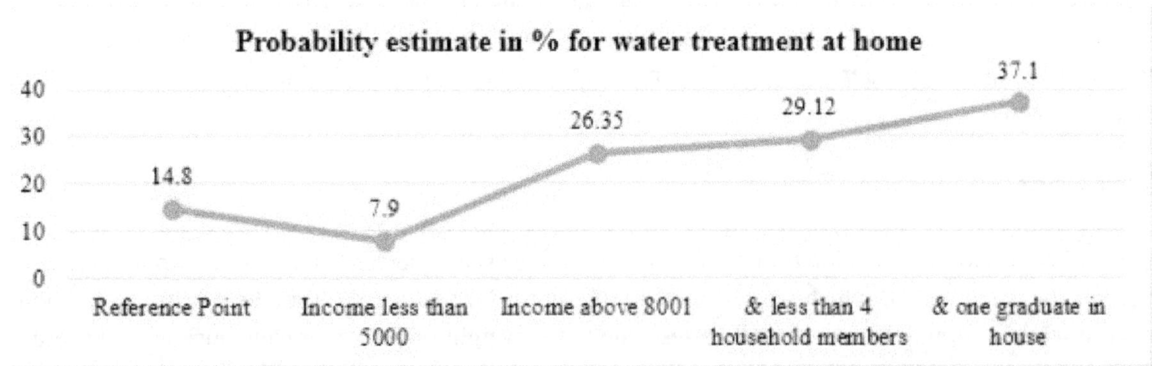

So, the test shows in Figure 8 that with the increase in household income or decrease in household size or the increased education level of the members the probability to choose water treatment strategy for coping with poor water quality increases.

For the water treatment coping strategy adopted by households the descriptive statistics result for different income category shows that with the increase in income the coping choice of households become more favorable towards a water treatment strategy. And when the education level factor is also considered the choice becomes even more favorable for a safe and clean drinking water. So, the decision for a healthy and hygienic life depends on the education and awareness about the disease and how it is transmitted and also on the economic status of the households.

Figure 8. Income-wise water treatment and diarrohea cases
Source: Created by author, 2019; Data: Primary Survey by author, 2019

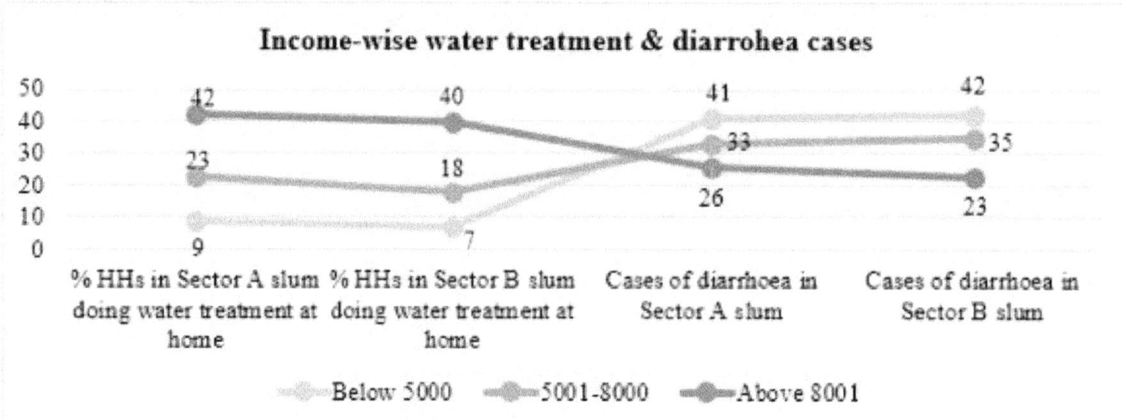

COPING COST COMPONENTS

This section deals with the estimation of total coping costs which the households end up paying for improved water service. It consists of four components namely the time cost for collecting water and loss of labour due to illness, water treatment cost, diarrhoea treatment cost and the capital cost which includes measures like water storage or buying from water tankers. The capital costs are amortized to bring it on same scale.

Time Cost

To calculate the water collection time the households were asked question on the minimum time it takes to reach the water source which was then multiplied with 1.75 to get the roundtrip time. It was done because while returning the speed will be less due to the weight of the water cans/buckets. An average waiting time was added to it based on the response from the primary survey. Then the mean time cost for collecting the water was calculated. Similarly, the households were asked on the number of days lost due to illness which was added to the mean time cost and was then multiplied with the minimum wage of India (40 INR/hour) to convert it into economic terms.

Water Treatment Cost

Through primary survey the households were asked on the water treatment practice performed at home like boiling, chlorination, filter etc. They were also asked on the frequency of water treatment done based on which the economic cost of treating water at home was calculated. Factors like the predominant fuel used for cooking in house were used as proxy to calculate the treatment cost.

Diarrhoea Treatment Cost

The households were asked about the incidence of diarrhoea cases in last three months and how much money was spent on it. The response was cross verified based on the market rate for treatment and medicine. Then the average treatment cost was calculated from the responses of all the households. It is to be noted that water and sanitation are not the only cause of diarrhoea but for this study the assumption was made.

Capital Cost

The capital cost consists of money spent on containers for storing the water or long-term measures taken by the household like a big water tank in the house. The cost of cleaning and the interval of cleaning was also asked from the household which was added in the total capital cost. All the cost was annuitized to convert it into monthly expenditure at a discount rate of 10%. The total capital cost also consists of financial cost of buying water from vendor for drinking or from water tankers for other purposes. All those costs were added to get the final total capital cost as shown in Figure 9.

Figure 9. Mean total coping costs of households
Source: Created by author, 2019; Data: Primary Survey by author, 2019

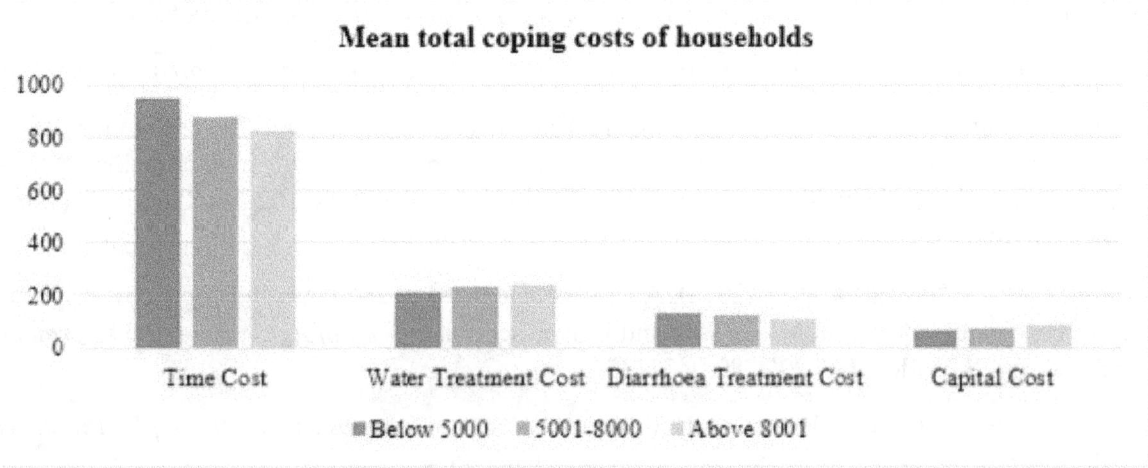

The time cost constitutes the maximum share of the total coping costs of the households. It is seen from the figure above that the mean total coping cost is higher for lower income group households due to less awareness and other alternatives available.

FINDINGS AND CONCLUSION

The paper shows that the households' resort to various coping strategies because of the unreliable water supply. Some of those strategies have an impact on the health of the households like the rescheduling of activities which is unhygienic. Also, the coping strategies have an economic impact on the households because of the time loss in collecting water which could have been used for other purpose. From the probability estimate the author find out that the probability of water treatment at home increases with the increase in households' income and the decrease in household size. From the reference point in which the probability estimate of water treatment at home was 14.8%, the increase in income and a smaller number of households and one graduate member at home resulted in 37.1% probability of doing a water treatment at home.

The total coping costs calculated shows that the households end up paying on an average 14% of the monthly income which is very high expenditure on something like water which is supposed to be a free good. Also, in comparison to the well serviced area the urban poor end up paying more in terms of the induced coping costs. From the last section of the analysis, it is evident that the mean total coping cost for a month is more than the water bill. This can be one source of revenue generation for the municipality from which they can provide infrastructural service to the households. A future study on this topic can be done on the willingness to pay of the slum residents for the improved service which will result in better livelihood and health condition along with financial savings from reduced monthly expenditure on water.

REFERENCES

Ayoub, G. M., & Malaeb, L. (2015). Impact of intermittent water supply on water quality in Lebanon. *International Journal of Environment and Pollution*, *26*(4), 379. doi:10.1504/IJEP.2006.009328

Bhopal Municipal Corporation. (2015). *Slum free city plan for Bhopal metropolitan area*. Author.

Caprara, A., Lima, J. W. de O., Marinho, A. C. P., Calvasina, P. G., Landim, L. P., & Sommerfeld, J. (2009). Irregular water supply, household usage and dengue : A bio-social study in the Brazilian Northeast. *Cadernos de Saude Publica*, *25*(suppl 1), 125–136. doi:10.1590/S0102-311X2009001300012 PMID:19287857

Chaminuka, L., & Nyatsanza, T. D. (2013). An assessment of water shortages and coping mechanisms of Harare residents : A case of Msasa Park and Dzivaresekwa. *Journal of Agriculture and Veterinary Science*, *4*(3), 21–35. doi:10.9790/2380-0432135

Choe, K., Varley, R. C. G., & Bijlani, H. U. (1996). *Coping with Intermittent Water Supply: Problems and Prospects, Dehra Dun, Uttar Pradesh, India*. Academic Press.

Cook, J., Kimuyu, P., & Whittington, D. (2015). The cost of coping with poor water supply in rural Kenya. *Water Resources Research*, *51*, 9127–9140. doi:10.1002/2014WR016259

Dutta, V., & Tiwari, A. P. (2003). Cost of services and willingness to pay for reliable urban water supply : A study from Delhi, India. *Water Science and Technology: Water Supply*, *5*(6), 135–144. doi:10.2166/ws.2005.0058

Katuwal, H., & Bohara, A. K. (2011). Coping with poor water supplies : Empirical evidence from Kathmandu Nepal. *Journal of Water and Health*, *9*(1), 143–158. doi:10.2166/wh.2010.151 PMID:21301123

Kleemeier, E. L. (2010). *Private Operators and Rural Water Supplies: A Desk Review of Experience*. World Bank. http://water.worldbank.org/publications/private-operators-and-rural-water-supplies-desk-review-experience

KPMG. (2018). *Green and Blue Master Plan for Bhopal Smart City*. Author.

Kudat, A., Bell, M. E., Boland, J., Humplick, F., Madanat, S., & Mukherjee, N. (1993). *Reliability of urban water supply in developing countries: The emperor has no clothes*. Author.

Kumpel, E., & Nelson, K. L. (2013). Intermittent Water Supply: Prevalence, Practice, and Microbial Water Quality. *Environmental Science & Technology*, *50*(2), 542–553. doi:10.1021/acs.est.5b03973 PMID:26670120

Majuru, B., Suhrcke, M., & Hunter, P. R. (2016). How Do Households Respond to Unreliable Water Supplies? A Systematic Review. *Environmental Research and Public Health*, *13*(12), 1222. Advance online publication. doi:10.3390/ijerph13121222 PMID:27941695

Morrison, G. (2016). Regressions with The Discrete Dependent Variables : Effect. *JMR, Journal of Marketing Research*, *9*(3), 338–340.

Nastiti, A., Muntalif, B. S., Roosmini, D., Sudradjat, A., Meijerink, S. V., & Smits, A. J. M. (2017). Coping with poor water supply in peri- urban Bandung, Indonesia : Towards a framework for understanding risks and aversion behaviours. *Environment and Urbanization*, *29*(1), 69–88. doi:10.1177/0956247816686485

Olayide, O. E., & Ikpi, A. E. (2010). Urban domestic water crisis in Zimbabwe: The case of Kadoma City. *Sustainable Development*, *12*(8), 90–99.

Pattanayak, S. K., Yang, J., Whittington, D., & Kumar, K. C. B. (2005). *Coping with unreliable public water supplies : Averting expenditures by households in Kathmandu, Nepal*. doi:10.1029/2003WR002443

Razzolini, M. T. P., & Günther, W. M. R. (2008). Health impacts due to deficient water access. *Saúde e Sociedade*, *17*(1), 21–32. doi:10.1590/S0104-12902008000100003

United Nations Children Fund & World Health Organisation. (2017). *Progress on Drinking Water, Sanitation and Hygiene - Joint Monitoring Programme 2017 Update and SDG Baselines*. WHO. doi:10.1111/tmi.12329

WaterAid. (2018). The water gap. *The State of the World's Water*. Advance online publication. doi:10.1029/2010GL044571

Zerah, M.-H. (2000). Household strategies for coping with unreliable water supplies: The case of Delhi. *Habitat International*, *24*(3), 295–307. doi:10.1016/S0197-3975(99)00045-4

Chapter 6
Ecosystem Valuation of Agricultural Land Use Change

Dimple Behal
SPA Bhopal, India

ABSTRACT

With the rapid pace of urbanization, land-use change is essential for economic and social progress; however, it does not come without costs. With such rapid urbanization, there comes pressure on the land and its resources, like that of food and timber production with a significant impact on the livelihood of millions of people. With the loss of agricultural land due to developmental activities, future agriculture would be very intensive. Therefore, it is likely with the existing pattern of allocating land uses for future development that we may lose the ecosystem services and highly productive agricultural lands. The value of these ecosystem services to agriculture is enormous and often underappreciated. The study focuses on identifying underlying causes of the land-use change, ecosystem services affected due to land-use change in peri-urban areas of Chandigarh using spatial mapping of affected ecosystem services and suggesting proposals for promoting agricultural ecosystem values using economically-informed policy instruments.

INTRODUCTION

Humanity on earth has been constantly altering the land for various reasons from the past, which may include resource extraction for its needs, farming and herding, and infrastructure. Since the 18th century, there has been a drastic pace and degree of land transformation induced as a result of population growth due to various anthropogenic activities. Urbanization is a continuing megatrend, with just 30% of the global population being urban in 1950. In 2014, 54% of the world's population live in urban areas and in 2050, two-thirds of the world 's population is expected to be urban, with 95 percent of that growth taking place in developing countries (United Nations, 2018). Currently, urban areas represent less than 2% of the Earth's surface, and are expected to rise by 4-5% in 2050 (Uwe R. Fritsche, Ulrike Eppler, Leire Iriarte, Sabine Laaks, Stepanie Wunder, Timo Kaphengst, Franziska Wolff, Dirk Heyen, 2015). As compared to the year 2000, the global urban land cover will increase by 200% in 2030 and 70% would be the global population growth in the same year.

DOI: 10.4018/978-1-7998-8331-9.ch006

India envisions rapid expansion of infrastructure across the nation (Department of Land Resources, 2013). Industrialization will boom with urbanization growing dramatically in the manufacturing sector, but agricultural land is essential, as a whole, because of its reliance on the population of rural areas. The 2011 census revealed an increase of 2774 towns over the decade, consisting of 242 Statutory and 2532 census cities. In 2001, there were 35 million plus cities / urban agglomeration shot up to 53 in the 2011 Census. Land is one of three major factors of production along with capital and labour in classical economics and an important necessity for the basics like that of housing and food production. Changes in land use are inevitable and are vital to economic development and social change. Pressure on land and its resources, such as food and forestry, has a huge effect on millions of people as a result of such a rapid urbanization.

CONCEPT OF AGRICULTURE ECOSYSTEM

Approximately 40% of the earth's surface is covered with the agricultural ecosystems, is a dominant mode of global land management (Power, 2010) but the arable land has been decreasing since the last decade, from 66.7% in 1950–1951 to 59.7% in 2008–2009 (Wan Guanghua, 2013). With the rapid urbanization by the end of 2050, there will be huge need of food production globally by 70% and in developing countries up to 100% more compared to 2009. Agriculture today is striving against the unusual challenge of growing production so as to meet the demand for food, whilst preserving natural ecosystems. The change in land cover and degradation of the agricultural ecosystem have major effects on agriculture's long-term productivity. Biodiversity, soil quality, deposition rates, as well as carbon sources and sinks, locally, regionally and globally, are affected.

Ecosystem services are the goods and services of the ecosystem are the utility of humans derive from ecosystem functions, indirectly or directly (Costanza et al., 1997). Agri-ecosystems therefore need to be sustainable because they contribute to providing numerous services to the ecosystems such as regulating water, mitigating risks from floods, biodiversity, cultural services and sequestration of carbon which benefit society. Core ecosystem assets and services – land, soil, water, pest control, and animal pollination – are the key to the production of agricultural products. More broadly, biodiversity at all levels plays an ever more dynamic and recognized role in the development of agriculture. The cost of biodiversity loss and ecosystem services is very important to assess and internalize in all development projects, particularly infrastructure. The importance to agriculture of these ecosystem services is huge and often undervalued. The future agricultural land would be intensively used expected to have a greater production (FAO, 2012). Agriculture can also be a source of various disservices according to management practices, including wildlife losses, fertilizer ruins, waterway sedimentation, greenhouse gas emissions and human and non-target species pesticide intoxication. The interactions between the provision of services and other ecosystem services and the disservices should be assessed in space, time and reversibility terms

India is among the top ten rich on the availability of highly endemic species. India accounts for 7-8% of the world's reported species, with 2.4% of the world's area. Due to soil, climate, topography and years of geological stability, the wide variety of ecosystems and habitats. India has four hot spots on global biodiversity (Eastern, Indo-Burma, West and Sundaland). More than 300 wild ancestors and close family members of cultivated plants, India is also one of the eight centres of the origin and variety of crops in Vaville.("Convention on Biological diversity-India," n.d.).

Since agriculture, like other ecosystems have significant ecosystem services, its services are still not been considered for the valuation. Rapid land-use changes and land parcels are converted into real estate at the regional and city level without undertaking the valuation of agricultural ecosystem services. The future agricultural land would be intensively used expected with a greater production (FAO, 2012) which means the existing agricultural ecosystems should be rich enough to cater the demand of the population and allocation of agricultural lands should be done critically. The current land use scheme in the country is inadequate and systematic, scientific land use planning is required (Department of Land Resources, 2013). Hence, there is a dire need to highlight the agriculture ecosystem services and its valuation with respect to other land uses for better development and sustenance of the settlements.

NEED FOR THE STUDY IN PERI-URBAN AREAS OF CHANDIGARH

The "Granary of India" title has always been associated with Punjab, as it produces 20% and 9% of wheat and rice respectively in India. At the international level, it signifies 3% of global production of these crops along with producing 2%, 2%, and 1% of the cotton, wheat, and rice of the World. India's agriculture share of total GDP in 1950-51 was 40% which has been decreased in 2014-15 to 18% while Punjab was having 40% in 1950-51 which has been reduced to a share of 19% in the year 2014-15. The share of agricultural production in State and national GDP can therefore be seen to decrease rapidly. Punjab has also experienced urbanization from 28% to 38% from 1981 to 2011 respectively. One out of two individuals is an urban migrate in Punjab, which makes it one of India's most urbanized states (Tripathi and Mahey, 2017). Thus, land-use changes are prevalent as the past is important for urbanization. The National Capital Territory of Delhi and the Chandigarh Union Territory are most urbanized among all States and Union Territories with 97.5% and 97.25% of the city's population (MoHUA, 2020).

Figure 1. Location of Chandigarh (Computed by author, 2020)
Source: Computed by Author, 2020

Chandigarh's UT was established in November 1966 and covers an area of 114 sq. Chandigarh has grown with a with a decadal growth rate of 17% as it has been planned for the population of half a million and present data show that its population has doubled that expected, i.e., 10,54,686, The area is experiencing higher growth and unplanned growth due to migratory burden and lack of an effective legal framework and enforcement in the peripheral area. In the absence of effective implementation of the periphery, the region has seen growth and unplanned growth almost contrary to Chandigarh's planning principles and therefore has been unable to take on the burden of immigration. Therefore, this paper will identify underlying causes of Agricultural Land-use change in the study area and its surroundings, identify ecosystem services affected due to shift from agriculture to other land-uses, assess the major ecosystem services affected through spatial mapping and will suggest proposals for promoting agricultural ecosystem values in study area using economically informed policy instruments

Topography and Physical Features

A large part of the Chandigarh area is under environmentally sensitive areas because of its strategic location on the foothills of the Shivalik region. Relatively the region is flat and has relatively mild paths south-west of the city, where all the rivers and rivers flow through the GCR. The height of the hills and the plains ranges from about 400m above MSL to about 200m MSL. The Ghaggar River with its tributaries is the principal hydrological surface of the region. During the inter-monsoon period, the water canals are dry, but swell in many places due to the flat topography around Chandigarh and south. Protection of floodable areas is extremely important for groundwater aquifers. With annual deposition of the river silt, the soils in this area are very fertile and therefore very productive for the increase of multiple crops through year. The Punjab Land Preservation Act (PLPA), as it is an eco-sensitive area, has locked a large number of protected and reserved forests and regions in the north-west of Chandigarh.

EVOLUTION AND DEVELOPMENT OF CHANDIGARH

Spatial Growth

According to the MoHUA report, "Delhi's national capital region and Chandigarh's Union territory are the most urbanized with 97.5% and 97.25% urban population among the Member States and the territories of the Union," according to 2020. Chandigarh is the only union territory of more than 90% and only about 2% in villages with the highest concentration of developed land.

The region has always been a growth driver, attracting people and promoting migration from neighbouring countries in particular, in search of better quality of life, jobs, better homes and therefore a magnet. The settlement pattern of Punjab is characterized by large towns most of which were located at the intersections of major roads and on the banks of seasonal waterways.

Indeed, the actual expansion has followed a combination of at least four of these conceptual forms. The six forms being:

1. Expansion of Chandigarh itself,
2. Building a "linear township" along the highways,
3. Building a new township around an existing small township,

4. Building a new town on a virgin site to function as a counter magnet,
5. Building a series of small satellite towns around Chandigarh and lastly,
6. Planned expansion of Chandigarh in the form of wedges and corridors".

Figure 2. Topographic map of the region
Source: Computed by Author, 2020

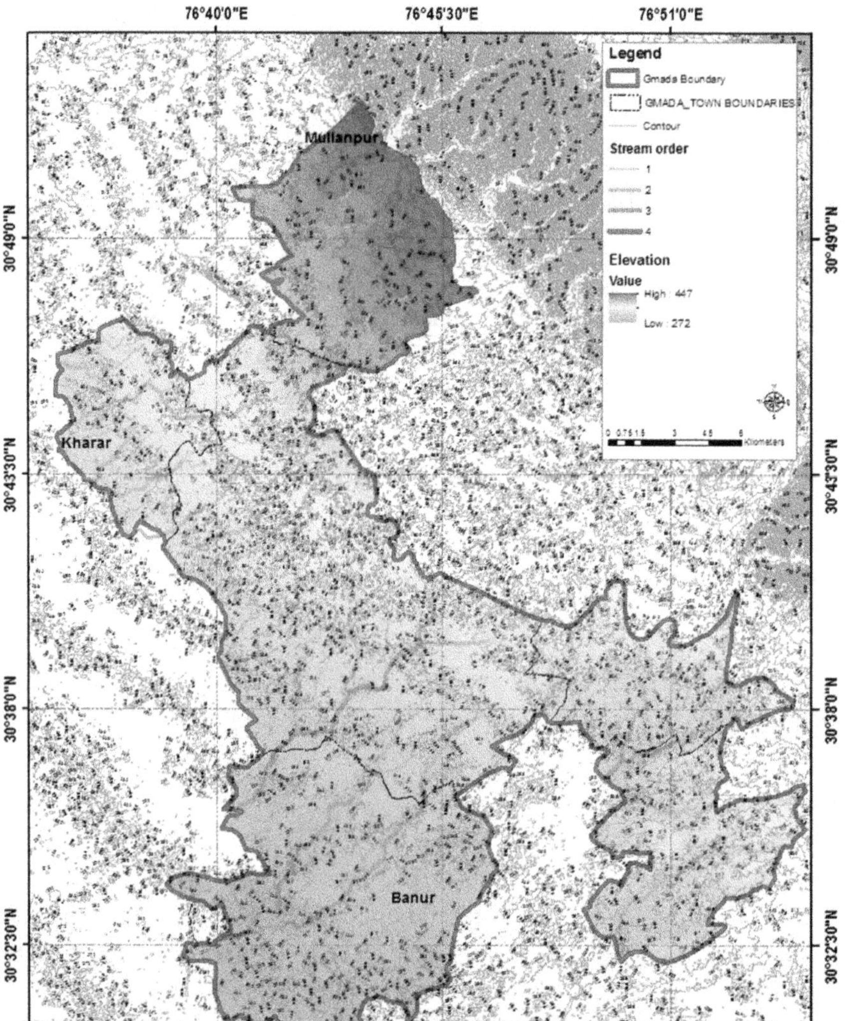

Peripheral Control Act

The Punjab administration enacted 'The Punjab New Capital Act of 1952' to direct and ensure the scheduled growth of Chandigarh. Initially, it was started with the peripheral control area, which had an 8 km radius and later expanded in 1962 to 16 kilometres. Le Corbusier, designed the city of Chandigarh for a permanent greenbelt. The Act was formulated to regulate the haphazard development in Chandigarh city, to protect the rural population against 'degenerate urban contact and urban dichotomy.

Four main goals of the act were:

- To control the rise of suburbs;
- To discourage land speculation";
- To conserve rural ways of life and traditional agricultural;
- "To protect and enhance the aesthetics of Chandigarh"
- Protect the plan's sacrosanctness and promote a peripheral view as a working rural landscape

Le Corbusier viewed the PCA as a critical feature protecting where he gave emphasis on rural activities in the periphery with no growth. Rural land within the Periphery was permitted to serve agriculture needs or the production of building materials for residents of the urban core. Such policies meant that Chandigarh had a permanent greenbelt around it. No more villages had to be developed in the periphery, as industrial or commercial construction was prohibited.

Land Use Land Cover Transformation (from 1990- 2000-2001- 2020)

The Land use Land cover transformation analysis has been done to understand the spatio-temporal changes in the GMADA region. Landsat imagery has been extracted from USGS of several years to perform the same. It can be observed, the built-up area has been increased rapidly since 1990 with the significant reduction in the forest area. The agricultural lands which were vacant in the year 1990 are now being used extensively for the plantation purpose although the agricultural lands which were not put into use has followed significant increase in 2000 with the introduction of agricultural policies but later on has been decreased.

The largest tract of the settlements is concentrated between the Junctions of NH 21-MDR A to Ropar & NH 95 to Ludhiana in the area directly bordering SAS Nagar. The other areas are found along the road to the Kharar Railway Station and the villages which are dispersed throughout Kharar LPA. The Villages within Kharar LPA are Khuni Majra, Bhokri, Rudiala, Daun Majra, Bhago Majra, Badali, Badala Niyashehar and Tirpari. The LULC of the year 2020, 2010 1990 and 2000 has been shown in **Error! Reference source not found.** with its area distribution shown in **Error! Reference source not found.**.

Table 1. Land use land cover distribution (in sq mts.)

Attribute	1990	2000	2010	2020
Water Body	10.25	28.5	105.39	348.7
Built-up	2579.2	5633.2	9179.1	12095.37
Barren Land	11687.4	7035.8	5534.4	7032.15
Agricultural Plantation	1764.3	3147.31	1762.9	1374.7
Forest	4880.4	981.29	110.18	285.45
Agricultural land: Vacant	20560.68	11079.3	8778.5	7988.65
Agricultural land	848.72	20770	21719.8	12816

Figure 3. (a) Land use land cover of GMADA region – 2020 (b) land use land cover of GMADA region – 2010 (c) land use land cover of GMADA region – 2001 (d) land use land cover of GMADA region – 1990
Source: Computed by Author, 2020

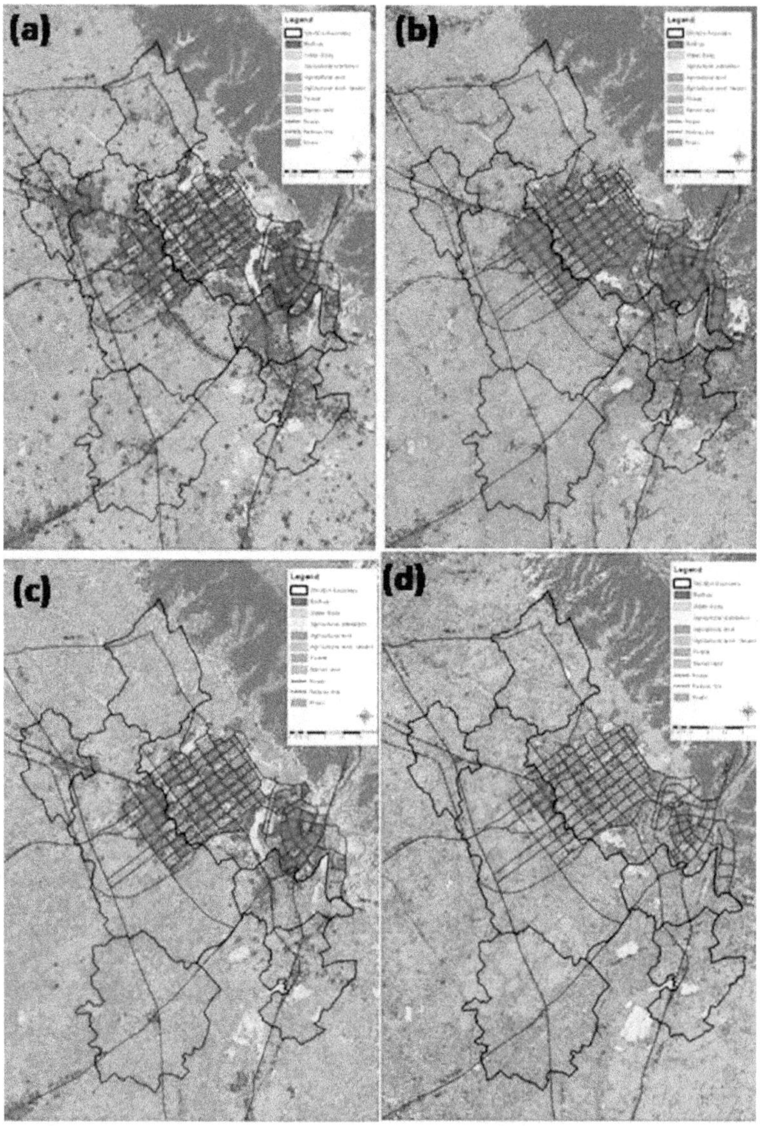

Settlement Hierarchy W.R.T Years (from 1991-2000-2011 census)

The census 1991, 2001 and 2011 population of villages and urban settlements were mapped on the basis of census town and village classification to map the concentration of high population villages and the change in the village population in the three decades. The urban and rural hierarchy of settlements is discussed in detail further.

Rural Settlements

The census 1991, 2001 and 2011 population of all the villages is mapped and categorised on the basis of total population into 6 categories as by the Census classification such as small hamlet, hamlet, small village, medium sized village, large village and very large village. Total number of rural settlements has been decreasing from 1991 to 2011 as more of the smaller order settlements are getting converted into higher order settlements. The large village in 1991, has been increased in 2001 and further the villages have been added into Municipal boundaries of urban settlements. The villages lying along the national and state highways grew at a faster rate.

Table 2. Classification of rural settlements

Settlement / Year	Small Hamlet (Less than 500)	Hamlet (500-999)	Small Village (1000- 1999)	Medium size Village (2000-4999)	Large Village (5000-9999)	Very Large Village (More than 10,000)	Total
1991	61	70	31	12	1	0	175
2001	49	52	40	24	2	0	168
2011	43	47	36	17	0	0	143

Source: (Generated by author based on Census of India)

Urban Settlements

There are 11 urban centres as per 2011 census and 9 as per 2001 census are mapped on the basis of population proportion and class, to understand the location of the different category towns based on census classification. The number of towns have been increasing from 1991 to 2011 as lower order settlements are converted into Class I towns. Chandigarh and SAS Nagar are the Class I towns of 2011.

Table 3. Classification of urban settlements

Settlement / Year	Class VI (0-5000)	Class V (5001 -9999)	Class IV (10000 -19999)	Class III (20,000 -49,999)	Class II (50,000 -99,999)	Class I (More than 1,00,000)	Total
1991	2	2	1	1	1	1	8
2001	0	2	2	2		2	9
2011	2	1	3	1	2	2	11

Source: (Computed by author)

Figure 4. (a) Settlements of GMADA 1991(b) Settlements of GMADA 2001 and 20119
(Source: Computed by Author, 2020)

Potential of Villages for Turning into Higher Order Settlements/Census Towns

On the basis of population and other criteria as defined by Registrar General of India, potential census towns for 2021 were identified and mapped. The methodology followed is as follows:

1. Mapping of Villages: The population, population density and male main working population of the villages were mapped for 2011 and villages were identified falling under each category based on which Potential census towns has been mapped.
   a. Identification of village
      i. Villages with population more than 5000
      ii. Villages with population 2000- 5000
   b. Mapping population density of villages
      i. Density 400 person per sq. km and more
   c. Male main working population
      i. 75% should be involved in non- agriculture sector

It has been assumed that the villages having population of 2000- 5000 will increase to more than 5000 in next decade. Also, to understand the extent of urbanization, the population projection has been done using Geometric Increase method for the year 2021. The method has been used considering the existing growth rate of the settlements around Chandigarh. These factors are overlapped to identify villages which may turn into potential census towns which is shown in Figure 5. It is observed after the analysis stated above, there are 26 villages (2000-4999) which are expected to be census towns in 2021. There are 9 Villages which are having the population more than 5000 fulfil the criterion of being higher order settlements which are in the surrounding of existing urban areas.

Figure 5. (a) Rural Population, 2011(b) male main working population in non-agriculture activities of rural area – 2011 (c) rural density, 2011 (d) potential census towns
(Source: Computed by Author, 2020)

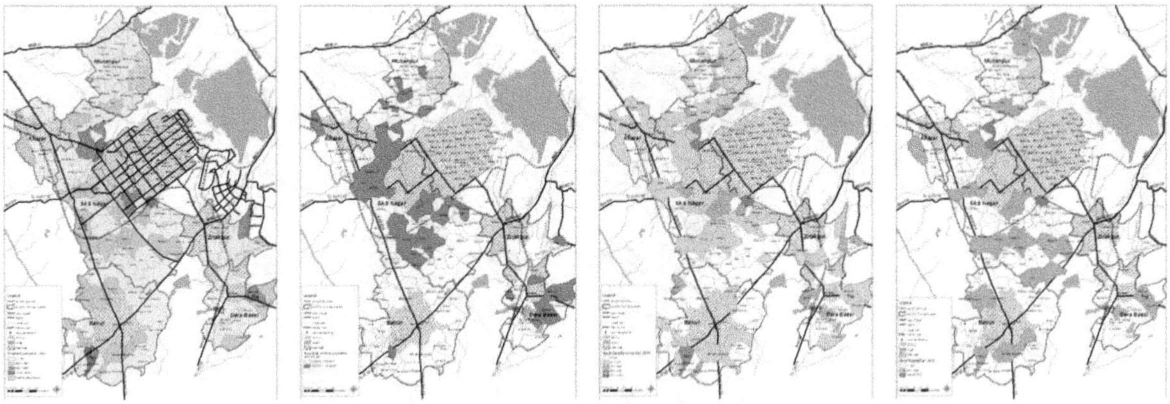

GMADA Regional Plan

To ensure the planned development, currently, GMADA takes care of peripheral areas of Chandigarh and has prepared a Regional Plan in the year 2008 with the total area of 119,236 hac. The 2001 population of GMADA totaled 711,210 people, with 38.9% of the urban population and concentrated mainly in S.A.S Nagar (44.6%) and remaining 55.4% spread to the remaining 9 municipalities. In Figure 6, GMADA Regional Plan has been showed with updated settlements by 2020.

In view of the wide variety of products, nutrients and high soil fertility of peri-urban farms (Brinkley, 2014) it is very deplorable to promote development following the current situation. The form of the adjacent cities is decided by policies in peri-urban areas. Therefore, looking into the rapid development of the Chandigarh its Peri-urban areas has been selected.

The peri-urban areas of Chandigarh are having six major towns in the state of Punjab.

The population of these towns of the past years and projected is given below in table 6.

In, it is presented that the population growth rate of Mohali/SAS Nagar is reducing while population of Kharar at the incremental rate. In Zirakpur and Banur, the population growth rate is less and unpredictable. Kharar has been selected as the second most populous city settlement in GMADA and represents about 15.3 percent of GMADA's urban population. In the Punjab state of SAS Nagar district, Kharar is a municipal city and council.

Figure 6. GMADA regional plan
Source: Computed by Author, 2020

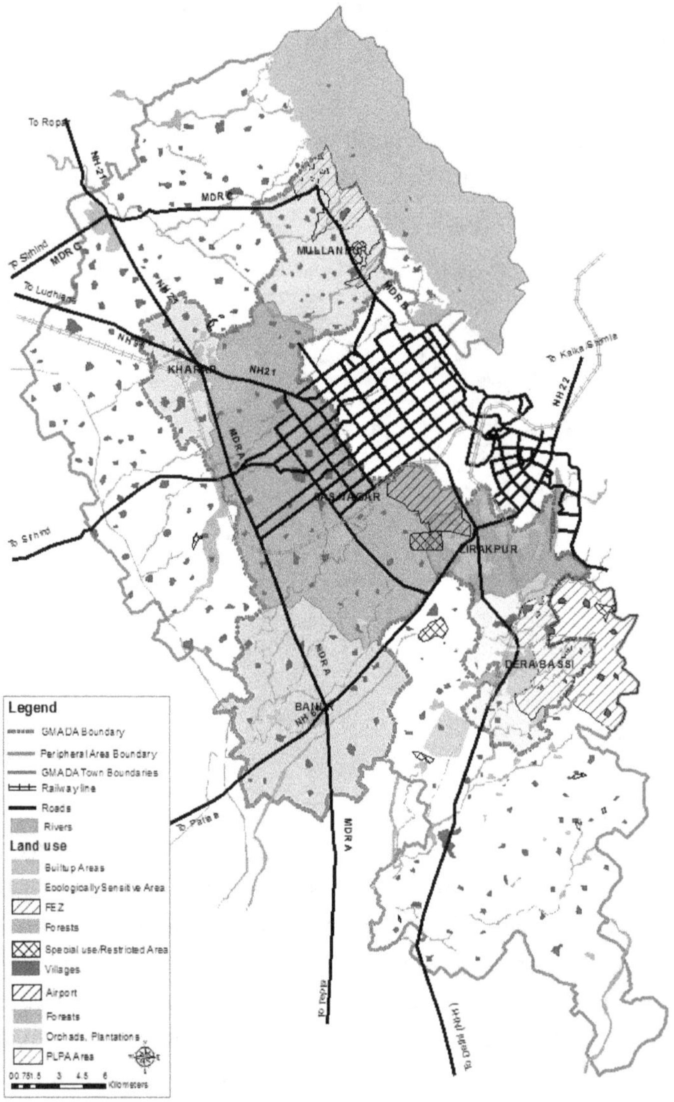

KHARAR LOCAL PLANNING AREA

Kharar Local Planning Area is located adjacent to SAS Nagar towards the north west of Chandigarh. It is approximately 10 km away from the town of Chandigarh. As in the 2008 regional GMADA Plan, 15.3% of the urban population of the GMADA Area has an increase of 4.9%. The city of SAS Nagar (44.6%), with the remaining 55.4% distributed over nine more towns, has a high concentration of urban communities within the GMADA district. In the GMADA area, Kharar makes up around 15.3% of the municipal population. The entire Kharar LPA is 3,324 hac. As of 2001 census, Kharar had a Urban population of 42,289 with 8,118 Households. As by 2011, population was 74460.

Table 4. Distribution of the urban population in greater mohali region in 2001

Urban Settlement	% of Urban Population
S.A.S Nagar	44.6
Kharar	15.3
Zirakpur	9.0
Kurali	8.3
Karoran	7.4
Dera Bassi	5.7
Bhankharpur	3.3
Mullanpur Garib Dass	2.2
Bhabat	2.1
Banur	2.0
Total	100

(Source: GMADA Regional Plan, 2008)

Table 5. Administrative jurisdiction and vision of GMADA towns

Town	Vision for 2050	Administrative jurisdiction	Class	Tehsil
Banur	Agro based	MCl	IV	SAS Nagar
Zirakpur	Agro based	MCl	II	Dera Bassi
Dera bassi	Industry	MCl	III	Dera Bassi
Kharar	Agro based	MCl	II	Kharar
Mullanpur	Eco Town	CT	V	Kharar
Mohali	Central SAS: CBD, Financial District, Arts and culture centre, Administrative Centre, Medical Hub.	MCl	I	SAS Nagar

Source: (GMADA Regional Plan, 2050)

Table 6. Population of GMADA towns

Town	Area in Sq. Kms	Population 1981	Population 1991	Population 2001	Population 2011	Density (Census 2011)	Projected pop 2058.
Banur	100.79	7453	10014	5426	18775	2209	1,19,558
Zirakpur	40.24	NA	NA	25022	95553	5307	1,29,038
Dera bassi	47.78	7421	9602	15841	26295	1315	70,222
Kharar	31.71	21208	25938	42289	74460	2349	1,20,000
Mullanpur	32.25	NA	4881	6147	6165	726	2,00,000
Mohali	165.25	32351	78457	123484	176170	6128	12,42,969

Source: (GMADA Regional Plan, 2050)

Figure 7. Location of Kharar
Source: GMADA

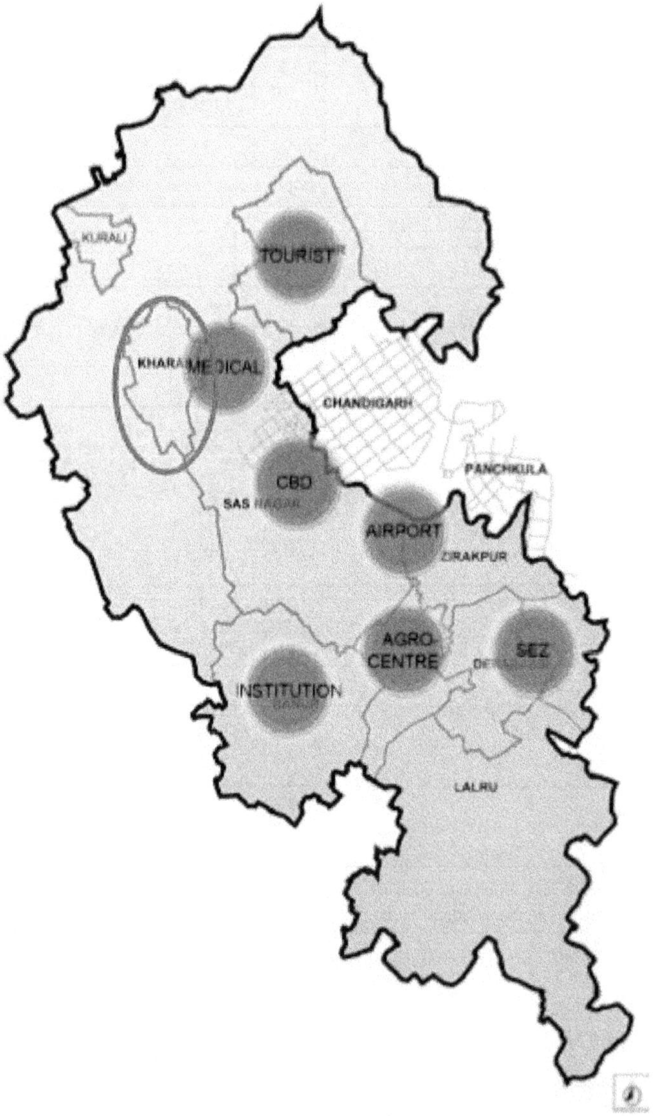

Climate and Geography

Kharar is geographically located at 30° 44' 4" N, 76° 38' 52" E. with an average elevation of 297 m. In the north of the region around Chandigarh lies the Shivalik hill which forms a fragile Himalayan ecosystem. Kandi (Bhabar) is located in the north east, and Sirowal (Tarai) and alluvial plains are located in the rest of the district. Two seasonal rivulets drain the area: in the east, Sukhna Choe and in the west, Patiala ki Rao. The central portion forms a division between surface water and has two small streams. The Jainti Devi River, a seasonal river, is bisecting Kharar LPA and is flowing from northwest to southwest.

Figure 8. Topography of Kharar
Source: Author, 2020

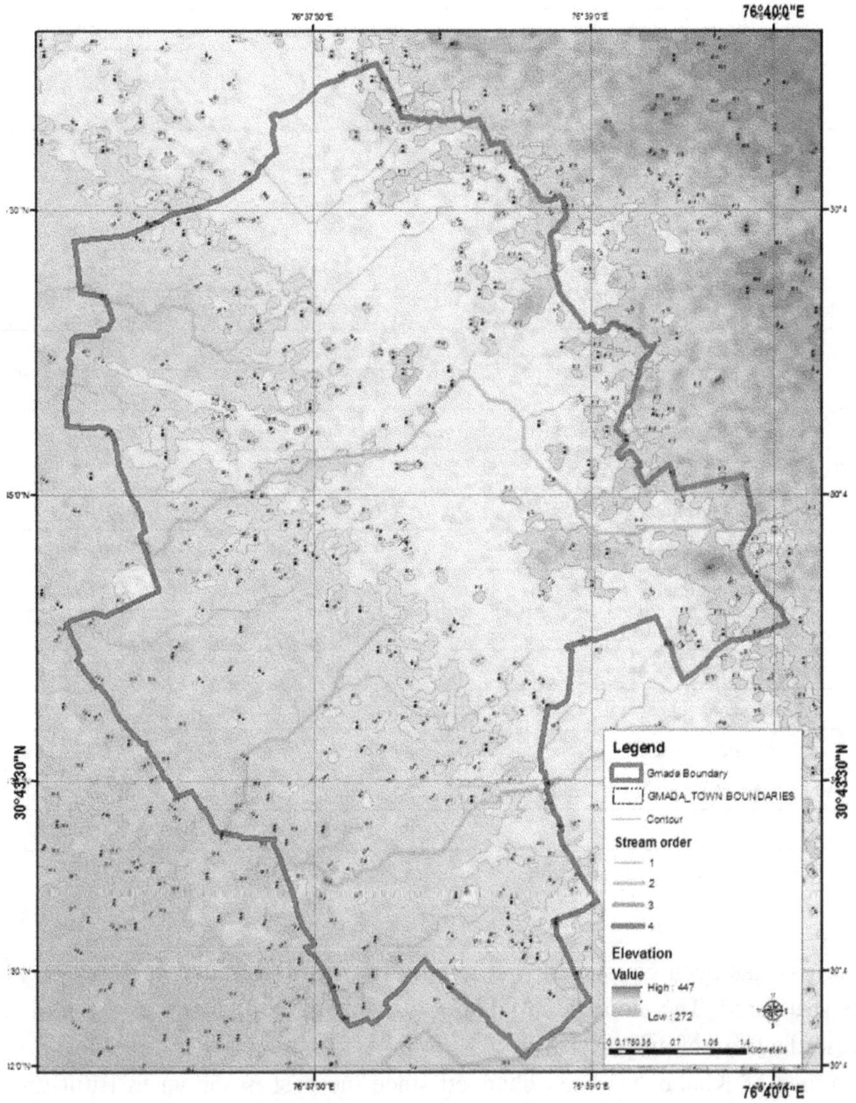

Existing Land Use 2020

A major part of the scheme, the Current Land Use Map 2020, is prepared based on a detailed land use assessment documenting existing land use in Kharar.

As stated above, Kharar LPA is having the area of 32.9 sq. km with pre-dominant land use as Agriculture with more than 61% of the land area. The existing land use reveals that only 29% of the total Planning area is developed, whereas 71% is undeveloped.

Residential area being the 28% of the total area and other land use distribution is shown in Land use distribution table.

Table 7. Land use distribution of Kharar -2020

Land use	Area (in sq mts)	Percentage
Agricultural land	20291385	61.5
Cremation and Burial Ground	4629	0.01
Drain	586985	1.78
Educational	249559	0.76
General Industry	184342	0.56
Government Land	28499	0.09
Government/Public Office	29747	0.09
Hospital and Health Institutions	16345	0.05
Local/Retail shopping	269211	0.82
Mixed Land use	402526	1.22
Parks/Open Spaces/Green belt	909530	2.76
Plant, Fruit Nursery and Orchads	6384	0.02
Play Ground	118260	0.36
Pond, Water Bodies, Lake	99044	0.3
Residential	8153818	24.71
Social and Cultural	5610	0.02
Stadium & Sports Complex	11551	0.04
Vacant Land	1432736	4.34
Water Works	58627	0.18
Whole sale market, Fruit Vegetable & Grain market	134139	0.41
Total	32992927	100

Source: (Computed by Author, 2020)

The development has been concluded that mainly through the transport network, along corridors continues to be structured. The old residential areas are mixed with cluster housing in nature and new development is in the form of property.

The built-up area of Kharar has been changed since the past is shown in Built-up area table and built-up map.

From the year 2000, 6.41 sq km agricultural land was lost in Kharar LPA even after the city had planned to be a farm-based economy. The projects for real estate are booming, and the development authority has allowed townships with a minimum area of 10-15 hectares for development purposes. It also means a wider growth in the future. Given the wide range of products and food supplied by peripheral urban farms and the high fertility of soil, it is highly deplorable to encourage development in the light of the current situation.

Figure 9. Built-up area and existing land use map, 2020
Source: Computed by Author, 2020

Table 8. Built-up area of Kharar

Year	2000	2010	2020
Area (in sq. mts.)	3066178	6349134	9473796

Source: (Computed by Author)

Administrative Setup and Agricultural Profile

Kharar Local Planning area falls in SAS Nagar district. The total area of district is 101.7 sq. km. with net sown area in the district is 750 sq.km, which constitute about 73% of the total area (Punjab and central ground water board, 2013). The district was created by the combination of certain changes in the profile by carving out 2 blocks (Sialba Majra and Kharar) from Ropar district and one block (Derabassi) from Patiala district (Punjab and Central Ground Water Board, 2013). The cropping intensity is 157%. Irrigation is mainly served by the use of tube wells in the district. Wheat, Paddy and Maize are the major crops of SAS Nagar. Paddy constitutes the main Kharif crop whereas wheat is the main Rabi crop.

Figure 10. (a) Agricultural lands in Kharar (b)Development along transit nodes
Source: Computed by Author, 2020

The district is divided into 3 tehsil viz., Derabassi, Mohali & Kharar comprising of 3 development blocks i.e, Kharar, Majri and DeraBassi as per the administrative setup. The District includes 414 villages. The important township in the district are Kharar, Kurali, Mohali, Zirakpur & Dera Bassi. The details about the blocks of SAS Nagar is given in Land Utilization Statistics (preceding 3 years average) (in hac)

Table 9. Land utilization statistics (preceding 3 years average) (in hac)

Block	Geographical Area	Forest Area	Land- under non-agricultural use	Cultivable	Permanent pasture	Current fallow	Net sown area	Gross cropped area	Cropping intensity (%)
Kharar	38326	13847	2858	3028	1100	546	28701	46179	161
Majri	26744	4000	735	-	-	277	13325	23892	180
Dera Bassi	36643	1228	636	1287	-	1360	32974	47936	146
Total	101713	19075	13374	4315	1100	2183	75010	118018	157

Source: (Development of Agriculture Sector, Punjab Agricultural University)

Table 10. Crops of Kharar block

Rabi Crops	Kharif Crops
Wheat	Paddy
Saron	Maize
Taramira	Sugarcane
Rabi oilseed	Mash
Massar	Arhar

Source: (Field survey, 2020)

It is evident the major kharif crop is paddy while major rabi crop is wheat in the block with most of its area coverage. Other significant kharif crops are sugarcane and maize.

Punjab being the exporter of rice at the national and global level, the block plays significant role in exporting the rice and wheat.

Figure 11. (a) Cropwise area of Kharif (in hac) (b) Cropwise area of Rabi (in hac)
Source: Economic and Statistical Department, Mohali

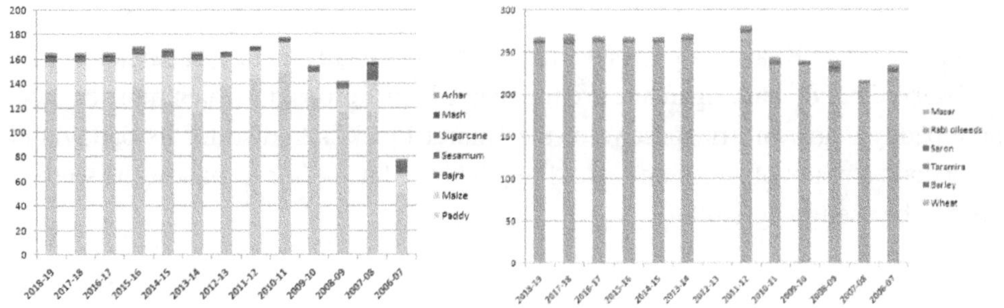

Figure 12. Average yield of wheat
Source: Economic and Statistical Department, Mohali

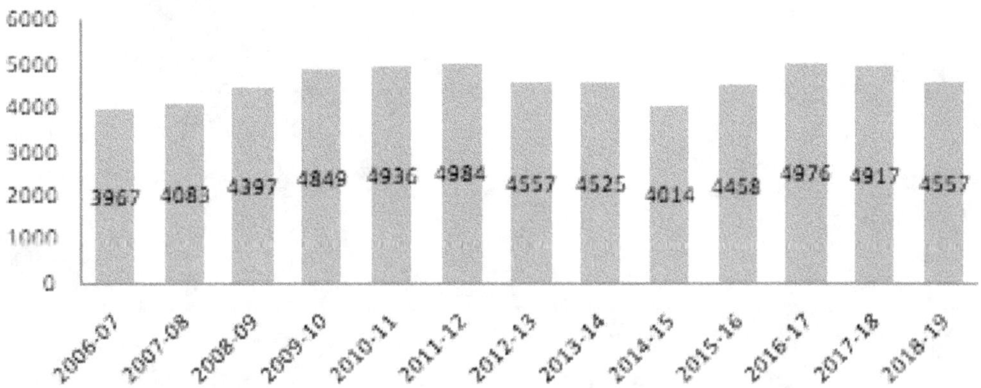

Figure 13.Average yield of paddy
Source: Economic and Statistical Department, Mohali

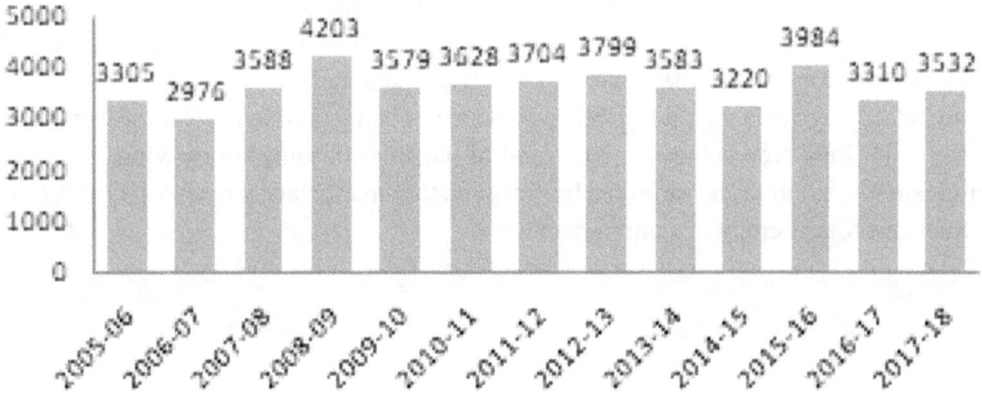

Agricultural Profile of Kharar Local Planning Area

Kharar Local Planning Area has the total agricultural land as the 20.2 sq. km which is more than 50% of the total land area. There are total 3458 number of agricultural land parcels. For the different seasons, there are different cropping patterns. While, there are also agricultural lands in which only one crop is sown in one year.

Kharif Cropping Pattern

Within the study area, the Kharif crops has the major crops as paddy, maize, sugarcane, mash and arhar. It has been observed paddy and sugarcane are the major kharif crops, in which most of the agricultural land is where paddy is grown. Irrespective of the Kharar block is known for its paddy and maize the study area has good amount of sugarcane also. Punjab being known for its paddy, the study area has the high potential to be used in agriculture.

Figure 14. Area distribution and number of land parcels (Kharif crops) (Computed by author as per the survey conducted)
Source: Computed by author, 2020

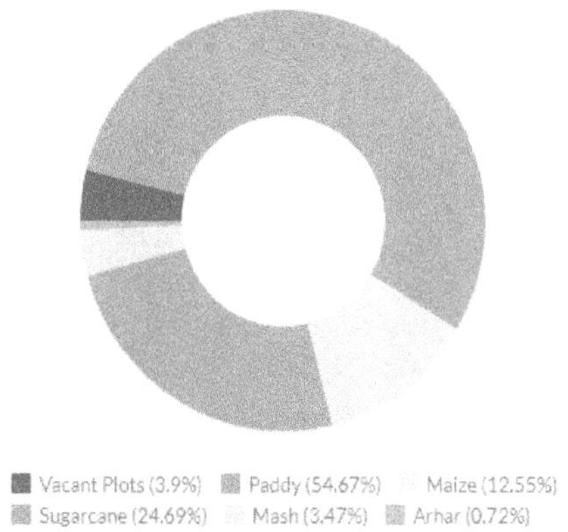

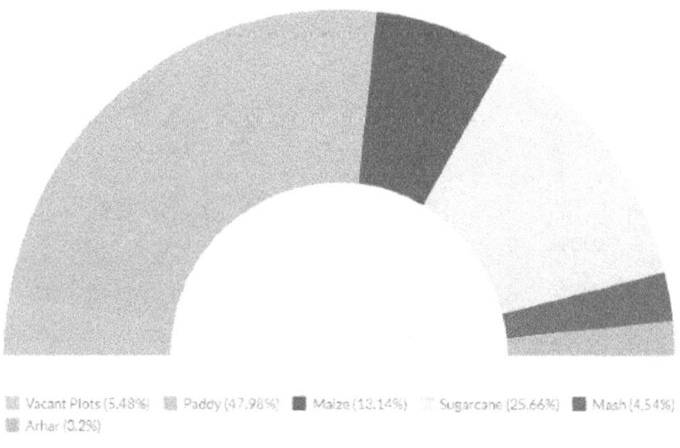

Rabi Cropping Pattern

The Rabi Crop is grown in almost all of the agricultural lands. This is because of the crops such as saron and taramira which is intensively used for personal gardening purpose and has been consumed widely in the region. Other crops being wheat, barley, rabi oilseed and masar.

There are total 35 agricultural land parcels which are not being used in the Rabi season for cropping. Wheat is the major Rabi crop in the area with most of the area consumed in growing.

The crops such as taramira and saron can be grown on the smaller land parcels which can be observed from the Rabi and Kharif cropping pattern map.

Figure 15. Area distribution and number of land parcels (Rabi crops) (Computed by author as per the survey conducted)
Source: Computed by author, 2020

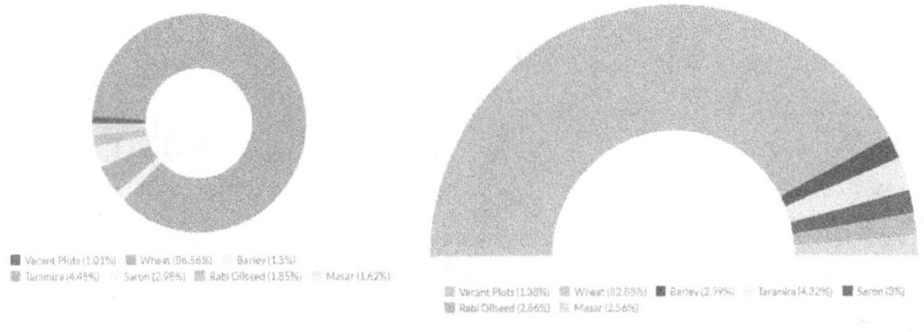

Figure 16. Rabi cropping pattern and Kharif cropping pattern (Computed by author, 2020)
Source: Computed by author, 2020

Land Holding Size

It can be observed, the land area with less than 0.8 hac (1.5 acre) are significantly more, i.e., 82% the land use changes would be rapid in that case. Considering land holding size less than 2.02 hac (5 acre), it is 97% of the total land holdings. Landholding size of the agricultural plots is shown in Figure 17

Figure 17. Land holding size distribution
Source: Computed by Author, 2020

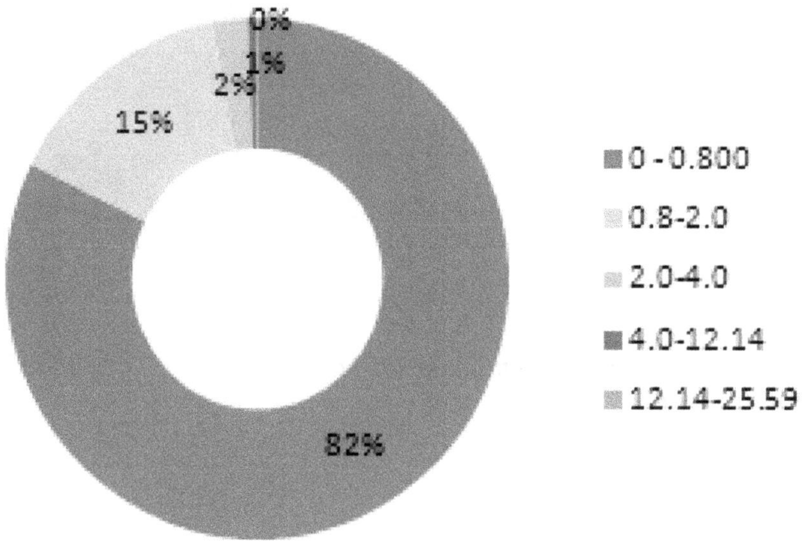

The high fixed investment in capital like mechanical machinery and high market prices of supplied inputs make this holding economically impossible. These farms' incomes are extremely small and can not meet their main demands, education, health and other family needs. Thus, giving up on farming and changing the occupation is another dimension. Any new generation would further break land pressure and ownership which means there will be higher probability of changing the land use from agriculture to developmental purposes.

Effect on Ecosystem Services due to Shift from Agriculture

Ecosystem Services Ranking by Stakeholders

To understand the value of each ecosystem service provided by agriculture several stakeholders has been surveyed during the data collection. The stakeholders involved were farmers, cultivators, local residents, real estate developers which are and will be directly affected from the change of agro-ecosystems. Experts from urban local authority and development authority were also contacted as they influence perceptions due to their significance in decision making process.

The interviews have been conducted in two parts. In the first part of the interview, stakeholders were asked to rank the ecosystem services on the basis of the benefits they provide to them considering their importance, if these benefits has been changed over the last 50 years and what were the causes of the change occurred. Existing ecosystem services were listed under the broad classification of ecosystem services; under the headings of provisioning, regulatory, cultural and supporting as stated. The interviews helped to provide information about the ecosystem services in the study area over the past 50 years, while the ranking of ecosystem services was based on the present condition. The ranking list has been circulated to various stakeholders to understand the stakeholder's relationship with the agricultural land and its ecosystem services. A table along with the sub-services listed under the above listed ecosystem services and a column for stakeholders scoring has been done.

Figure 18. Landholding size map
Source: Computed by Author, 2020

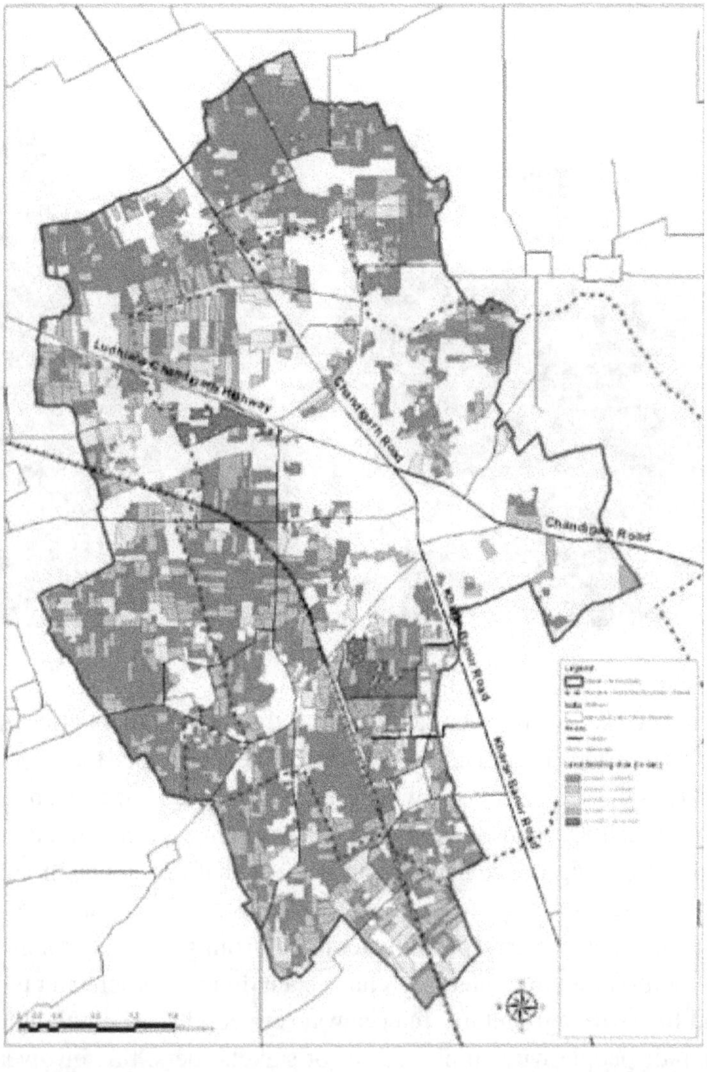

The ranking of 21 ecosystem services (Provisional, Supporting, Regulating and Cultural) has been done by stakeholders in which 1 indicated the ecosystem service which is least important to them while 21 being the most important.

The most beneficial ecosystem services as by the stakeholders came out to be Crop and livestock products, Soil protection/sediment retention, Carbon sequestration, water quality, Genetic diversity/habitat quality and biodiversity and pollination. As by the stakeholders perception,, crop and livestock products,, wildlife habitat, flood risk mitigation, genetic diversity/habitat quality, nutrient retention, soil protection and water quality were the most affected and altered since the past times. The major factor which has done the same is due to rapid urbanization and the extent of land use land cover changes have been drastic since the past times. The agricultural practices which are degrading the same like that of extensive use of insecticides/pesticides, stubble burning has been leading to agricultural disservices.

Figure 19. Interview with stakeholders

For each ecosystem service, people were asked about the Stakeholders affected due to change and stakeholders involved for Change in agro-ecosystems. The stakeholders had the opinion that farmers/cultivators (43%) and local residents (57%) are and will be directly affected due to change in agro-ecosystems.

Table 12 shows the stakeholders involved for each ecosystem service altered and the percentage of stakeholders involved in bringing the change. It can be seen that development authory's role is much wider in changing the agro-ecosystems. Also, as by stakeholders, ULBs play a key role in limiting the change. Farmers/cultivators have been responsible for poor farming practices for altering the ecosystem.

In the second part, semi-structured interviews have been done to understand their perception on the basis of their role and link with agriculture. Interviews were conducted at the individual level and occasionally in the group of people with similar sector of stakeholders, this involved special interaction with the farmers/cultivators to understand their perception and role in protecting the agricultural land. They have been asked about the cropping pattern, types of crops grown. Their opinion have been asked if they are willing to sell their land and their acceptance for the developmental purposes. It was hard to see as 43% of the agricultural cultivators/farmers were willing to sell their land if they are given the required amount more than the profit they could attain from selling agricultural products for several years while 57% people were not ready to give up on their land as they have some emotional values attached to it and also land is not just an asset for them but the source of income which can never be replaced if they would sell their land.

Residents have been asked about the significance of agricultural lands to them and if it is helping them directly or indirectly in their livelihood (may be through air quality enhancement, food products, sediment retention, landscaping etc). Every respondent responded positively as moreover they are dependent on it directly. One of the Residents responded as

"Eh dekho sarkar ki kar rahi ae. Har thaan jungle te kheti badi nu barbaad karke development leke aa rahe ne, dasso, insaan roti kitho khauga je khet hi na rahe ta, kissana da enna bura haal hoya peya, ohna da sarkaar kuch soch nhi rahi bas development de naam te road, flyover, ah real estate khade kitte ja rahe".

Table 11. Key ecosystem services and their proxy attributes of agricultural land

Benefits		Implied Ecosystem Services	Example of Measurement Attributes
Harvests	Avoided disposal cost	Pollinator populations	Wetlands/freshwater ecosystem, forests, grasslands, and conservation lands
		Soil quality	Soil quality
		Shade and shelter	Landscape metrics
	Avoided treatment cost	Water availability	Lakes, rivers, and wetlands
		Target fish	Fish populations and distributions
	Avoided pumping, transport cost	Crop populations	Croplands
		Target marine populations	Marine populations and distributions
	Birding	Biodiversity	Species richness index
Amenities and fulfilment	Angling	"Natural land cover in viewsheds"	Land covers
		"Wilderness, biodiversity, varied natural land cover"	Diversity of different land covers
	Swimming	Relevant species populations	Different species populations
Damage avoidance		Air quality	Air quality index
		Drinking water quality	Water quality index, total nutrient loadings
		"Land uses or predator populations hostile to disease transmission"	Threat assessment
		Wetlands Forests	Percentage of each land cover class in a parcel and surrounding areas
Waste assimilation		Surface and groundwater	Water yield
		Open land	Percentage of water body
Drinking water provision		Aquifer	Ground water level measurement
		Surface water quality	Water Quality Assessment
Recreation		Relevant species population	Species richness and distribution indices
		Surface water	Water availability
		Target population	Species richness and distribution indices
		Surface water	Water availability
		Beaches	Beach availability and accessibility

Source: (Boyd and Banzhaf, 2007 and Ma and Swinton, 2011)

The respondent directly emphasized on the pace of development (real estate, roads, flyovers) and loss of agricultural land. According to the respondent, if the same would be followed in future, we will give up on our food security and only government has to be blamed. People were not aware about the other ecosystem services of agricultural land, though they have been indicating the same through several factors. As one of the respondent responded as *"sheharan vich dekho kinni barbaadi ho rahi. Koi pashu-pakshi ni dissda, tuanu eh sab ethe hi disna jithe khet ne par eh sab v temporary a kyuki ethe v shehran vargi haalat hoi jandi, kheta nu loki convert karke ghar bnai jar he, jo bura haal sheharan da, jinve gandi hawa eh sab ethe houga."*

Figure 20. Ranking of ecosystem services by stakeholders -present condition
Source: Computed by author, 2020

The respondent has been indicating towards the poor air quality in urban areas due to unavailability of agricultural lands and open spaces. Also, he has been talking about loss of habitat and biodiversity in the urban areas and has been saying the same will happen here as the urbanization is coming here.

Real estate agents and developers have also been interviewed while for them more significance was of landscape amenities provided by agricultural land rather than any other ecosystem service as it would enhance the property's land value directly. They seek for more development and growth of the region as it will provide opportunities for the local people and will also provide connectivity. Real estate developers highly emphasized on "more the development, more will be our property price", therefore it can be seen the perspective of agents is to bring development no matter it is at the cost of losing our productive agricultural lands.

Figure 21. Ranking of ecosystem services by stakeholders - present condition and 50 years ago
Source: Computed by author, 2020

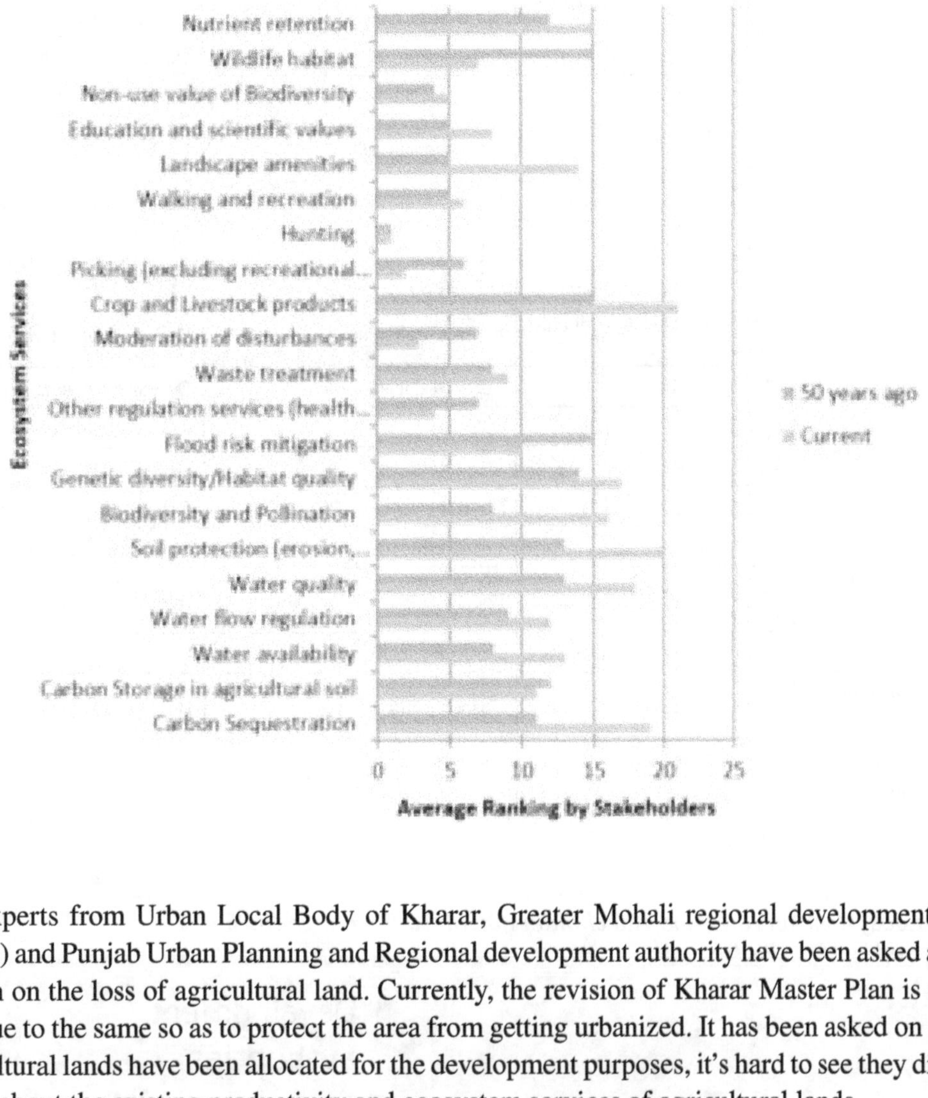

The experts from Urban Local Body of Kharar, Greater Mohali regional development authority (GMADA) and Punjab Urban Planning and Regional development authority have been asked about their perception on the loss of agricultural land. Currently, the revision of Kharar Master Plan is also under process due to the same so as to protect the area from getting urbanized. It has been asked on what basis the agricultural lands have been allocated for the development purposes, it's hard to see they did not gave a thought about the existing productivity and ecosystem services of agricultural lands.

Ecosystem Services Affected due to Developmental Pressure

For the purpose of mapping of ecosystem services, only two ecosystem services have been considered: Water Yield, and Habitat Quality. These ecosystem services are not exclusive but they have been selected on the basis of Stakeholders ranking of ecosystem services. Together, they highlight the abundance of ecosystem services that can be provided by agricultural lands, and are typical of the fundamental ecosystem services that agricultural lands should offer, regardless of the types of agricultural activities and products involved. Also, in the developing countries like India most of the data is not available to valuate the ecosystem services.

Table 12. Ecosystem services with stakeholders affected and responsible for change

S.No.	Ecosystem Service	Ecosystem service attributes	Major Stakeholder affected due to change in ecosystem service	Major Stakeholders responsible for change in ecosystem service
1.	**Regulation Services**	Carbon Sequestration	Local Residents	Development authority
		Water Quality	Farmers/Cultivators	ULB
		Genetic Diversity/Habitat Quality	Farmers/Cultivators	ULB
		Biodiversity and Pollination	Farmers/Cultivators	Development authority
		Water availability	Local Residents	ULB
		Water flow regulation	Farmers/Cultivators	Farmers/Cultivators
		Carbon Stored in agricultural soil	Farmers/Cultivators	Farmers/Cultivators
		Flood risk mitigation	Local Residents	Development authority
		Moderation of disturbances		Real estate agents
		Waste treatment	Local Residents	ULB
2.	**Cultural Services**	Landscape amenities	Local residents	Development authority
		Walking and recreation	Local residents	Development authority
		Education and scientific values	Local residents	Development authority
		Non-use value of biodiversity	Local residents	ULB
3.	**Provisioning Services**	Crop and livestock products	Farmers/Cultivators	Development authority
4.	**Supporting Services**	Nutrient retention	Farmers/Cultivators	Farmers/Cultivators
		Wildlife habitat	Local Residents	ULB

Source: (Computed by author, 2020)

Figure 22. Ecosystem services and stakeholders involved in change (in %)
Source: Computed by Author,2020

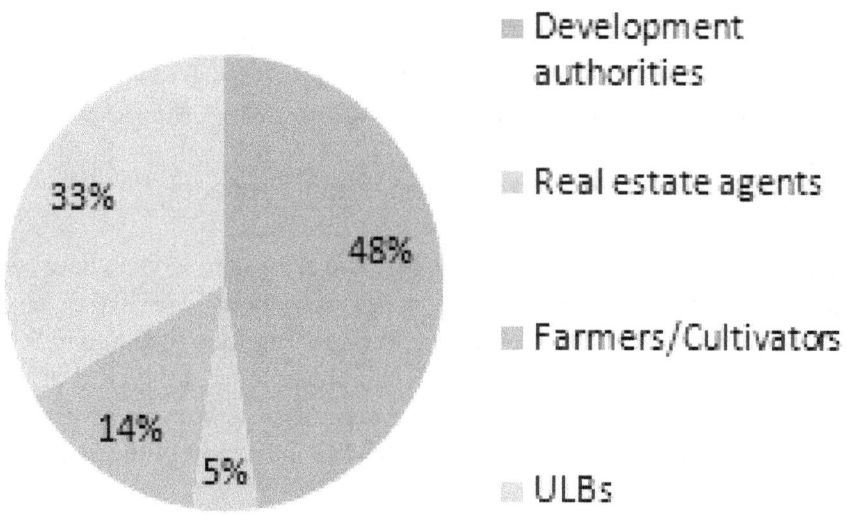

The water yield model simulates the annual average water yield by considering the biophysical data for the landscape of interest.To calculate the water yield following data inputs were used: LULC, rain events, precipitation, evapotranspiration, subwatersheds (stream flow), Crop/vegetation coefficient, NDVI etc.

Figure 23. Water yield and habitat quality (Computed by Author, 2020)
Source: Computed by Author, 2020

The habitat quality model generated the estimated states of habitat degradation and quality across the various study landscapes. Habitat quality and its extent are often regarded as proxies for use in measuring biodiversity. The model for this research employed data on LULC, sensitivities of the LULC types to threats, and the intensity and distribution of any specified threats. One assumption researchers made about this model was that larger areas with higher levels of habitat quality would support more flora and fauna species. Biodiversity is a vital indicator of the sustainability of an ecosystem. For decision makers, understanding the spatial distribution and richness of species available across a particular landscape can help them to make more informed decisions regarding land use and resource management, which can then bring about appropriate conservation strategies designed to maximize biodiversity. The model focused specifically on threats that could potentially affect the habitats within the study area. A current LULC map with a 1-kilometer buffer was generated during the preparation of this model for land uses. The built-up areas which are not supported by biodiversity while water bodies, parks can support the same. The buffer was used because there might have been potential threats at the greatest maximum threat distance that otherwise would not have been unaccounted for, resulting in an overestimated habitat quality score (Sharp et al., 2014).

It can be observed the areas near to the built-up areas like roads, railway lines etc. have low habitat quality. Also, the agricultural lands having multiple cropping have potential high habitat quality. The potential may also depend upon the agricultural disservices such as stubble burning, poor agricultural practices, extensive use of fertilizers etc.

LEGAL FRAMEWORK

There are some legal background which emphasizes on Ecosystem valuation. They are given as follows:

- The 1992 Convention on Biodiversity emphasized urgent efforts to stop the damage of biodiversity, to make sure that by 2020 ecosystems are resilient and provide essential services, ensuring the diversity of life on this planet and helping to achieve human wellbeing and eradicate poverty.
- MEA, 2001 promotes the true economic value of ecosystem services and provides financial instruments that take this value properly into account.
- TEEB (The economics of ecosystem and Biodiversity), widespread the values of biodiversity and ecosystem services into decision-making at all levels while focusing on i.e., Forests, Wetlands and freshwater ecosystems in India (TEEB, 2012).
- In India, the National Agriculture Board (1976) focused on the planning of scientific land uses for food health, autonomy and improved protection of living conditions.

Further in Table 13, the legal framework applicable in the study area is detailed along with associated gaps with it.

It can be attributed that there is no dedicated policy framework which is followed at the national, state or at the local level to incorporate the spatial mapping of ecosystem services so as to allocate the future land uses. It is evident that there are different policies for promoting agricultural resilience like National agro-forestry Policy, 2014 and National Urban Policy framework, 2018 for urban development while there is no comprehensive and integrated policy which addresses the issues of agricultural ecosystems into the development framework. Though peri-urban areas are given consideration but the agenda is to urbanize them rather than promoting the agricultural benefits.

Also, the Development Plans does not take into account the agricultural ecosystem potential while allocating the future land for the development purposes. Though the valuation is done at the national level (SEEA) but mapping of spatial attributes of ecosystem services is not done. Few studies have been conducted but in the isolation at the regional at state level but the comprehensive approach to prepare a dedicated plan is missing. Therefore, there is dire need to come up with a framework which has been followed in the research at the local level. Since development authorities are responsible for the preparation of the Plans at the local level comprising of peri-urban areas, the framework has to be adopted at the development Plan level. It is highly undesirable to see (evident in section and their responses) even after having the agro-based economy vision, no potential of agriculture lands has been taken into account. Land suitability analysis is done but the attention is not given on the areas having high and low potential for agricultural development. Therefore, it is recommended that the desired attention to be given on the ecosystem services of agriculture like every other ecosystem in the urban area by figuring out the spatial potential areas of it and then allocating the future land uses.

Table 13. Legal framework and associated gaps

S.No.	Level	Legislation	Provisions	Flaws/Things uncovered	Feature
1.	National	Draft National Land Utilization Policy, 2013	d. "The comprehensive and integrated land use planning in the country is still lacking which enables rationale and optimal utilization of land. Current land use planning in the country is inadequate and it does not cover all the levels of local, regional as well as at state levels. e. The per capita amount of agricultural land has reduced by 67% from 1951 (0.48 Ha) to 2007/08 (0.16 Ha)		Need for a systematic and scientifically based land use planning has been focused upon
2.		National Agroforestry Policy, 2014	"Promote resilient cropping and farming systems, to protect and stabilize ecosystems, and minimize the risk during extreme climatic events"	Does not take into account the land use practices to be taken care of	The policy is moreover oriented towards the affect of climate on agriculture
3.		National Urban Policy framework, 2018	The ten pillars are: 1. City Planning 2. Urban Economy 3. Physical Infrastructure 4. Social Infrastructure 5. Housing and Affordability 6. Transportation and Mobility 7. Urban Finance 8. Urban Governance 9. Urbanization and Information System 10. Environmental Sustainability.	"Does not take into account how agricultural lands in peri-urban areas have to be managed"	Natural ecosystems to be recognized systems within urban areas to enhance urban resilience. Peri-urban areas are emphasized to reduce social disparity, poverty and exclusion, providing better access to jobs and services. with urban agriculture
4.		SEEA	In September 2018, the CSO released India's first official environmental economic accounts containing asset accounts in physical terms of four natural resources – forest, land, minerals and water. Also, have done few studies on state-wise ecosystems- soil nutrient indices, water quality accounts etc.	Does not take into account the spatial aspects of ecosystem valuation at local level	
5.	UT	Inter State Regional Plan for Chandigarh Region (1984-2001)	The total area of the ISCR 2001 was 2421.3 sq km with an average radius of about 35 km from the core of Chandigarh (Chandigarh UT + Chandigarh Urban Complex + 16 Km radius Periphery control Area + Some portions beyond the 16 km radius in Punjab, Haryana and HP).	ISCR -2001 was advisory in nature. The Plan was not implemented as there was no enabling provision for notification and implementation of Regional Plan. Hence, the plan was a non-starter.	
6.		Chandigarh Inter State Metropolitan Regional Plan, 2000- 2021 (CISMeR)	Proposed plan was for a 50 Km radius around Chandigarh and for 60 Lac population.	Proposals not accepted by Punjab and Haryana Governments. The CISMeR Plan was submitted by Prof. Riberio in 2000, however, there was no further follow up.	

continues on following page

Table 13. Continued

S.No.	Level	Legislation	Provisions	Flaws/Things uncovered	Feature
7.	State	The Punjab New Capital (Periphery) Control Act, 1952	to control haphazard development around Chandigarh. The Act extended to the area within a distance of 5 miles (8 km) on all sides from the outer boundary of the land acquired for Chandigarh. Subsequently, the Act was amended in 1957 & 1962 to enlarge the periphery to cover an area upto 10 miles (16 km).	Haphazard development continued to grow till GMADA regional plan which authorized and regularized the developments later in the peri-urban areas No emphasis was given to save the agricultural land use	
8.		Punjab Regional and Town Planning and Development Act, 1995	"for preparation of Regional Plans and Master Plans and implementation thereof; for the constitution of a State Regional and Town Planning and Development Board, for guiding and directing the planning and development processes in the State; for Constitution of a State Urban Planning and Development Authority, Special Urban Planning and Development Authorities and New Town Planning and Development Authorities, for the effective and planned Development of planning areas; and for undertaking urban development and housing programmes and schemes for establishing new town"	The act does not take into account the significance to state's agriculture while focusing on development.	
9.		Agricultural Policy of Punjab	"Address various interlinked concerns of sustainability of the current cropping pattern and stagnating farm incomes through a simultaneous and multi-pronged action with an emphasis on improvement in production technology and infrastructure, pushing up capital formation, re-structuring the incentives and streamlining the institutions"	Does not take into account the measures to be taken for the land parcels which are getting sold by farmers for the purpose of development	Achieving agricultural sustainability by focusing on better livelihood and living conditions for farmers, particularly the small and marginal, and also the agricultural labourers. Area and wheat-paddy (water-intensive crops) has to be reduced
10.		Greater Mohali Region (GMR) - Regional Plan 2008-2058	The plan focused on the planned development in the peripheral areas of Chandigarh by delineating the regional boundary including few towns.	Though the plan focused on town based vision which included agriculture but the Plan does not take into consideration the agriculture based allocation of future land	
11.		Kharar Master Plan	The Plan envisions to be agro based economy	No specific framework have been used to allocate the agricultural lands All the un-authorized and un regulated development have been authorized. Extensive land use changes happening from agricultural to residential.	For approval of Real estate project, it should have minimum area of 10-15 acres

Source: (Computed by Author, Based on Agricultural and Development Policies)

Figure 24. Framework for how the management of ecosystem service flows through an agricultural landscape can improve the health, security and economic benefits to people
Source: Computed by Author, 2020

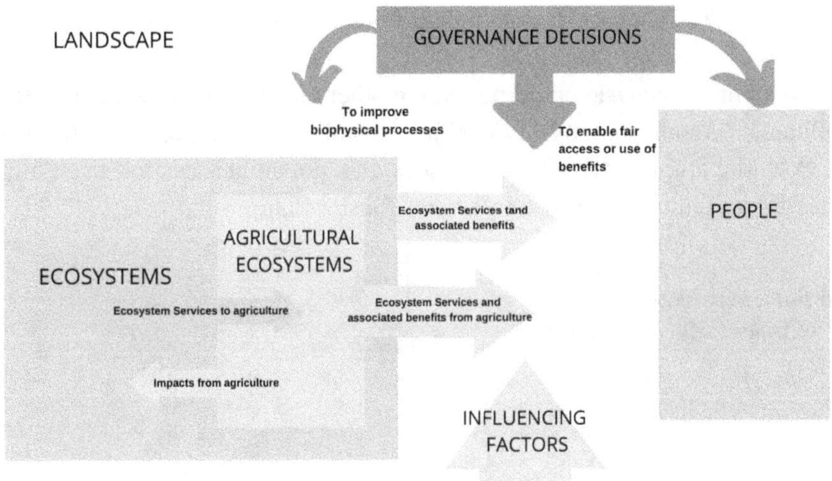

Figure 25. Potential areas of ecosystem services distribution
Source: Computed by Author, 2020

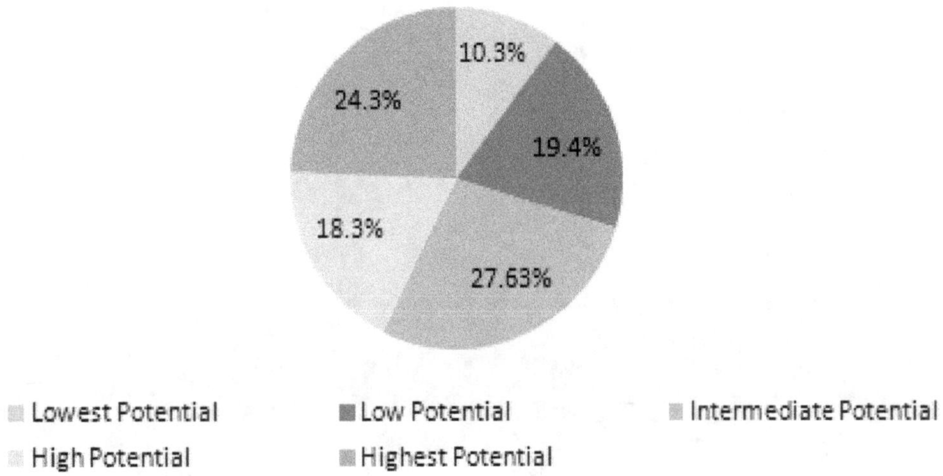

RECCOMENDATIONS

The potential areas of ecosystem services has been mapped spatially as shown in Water Yield and Habitat Quality map by weighted overlay the layers of water yield and habitat quality. Comparing the areas allocated to development in the Kharar Master Plan -2011 and the potential areas of ecosystem services, it can be seen the intermediate and high potential areas of ecosystem services have been allocated for the purpose of development in 2031. It can be observed the area along the corridors have less potential areas in terms of ecosystem services; the proposed Master Plan has allocated the development on it which is suitable. Considering the current development scenario of Kharar, the existing developed colonies are being regularized and authorized no matter it is coming in the agricultural zone allocated in the Master Plan. High potential areas has to be allocated for agricultural purposes only in future to preserve the agricultural ecosystem services and high productivity agricultural lands. Low potential areas can be converted into other land use such as Residential, commercial or any built-up area while Intermediate areas could be used for Parks/Open spaces.

The similar development scenario could be seen in other cities of India where unauthorized and un-regularized development is authorized by the development authority irrespective of the location attributes.

In figure 26, Potential areas of ecosystem services, low potential areas has been highlighted which has been allocated for agricultural purpose in Kharar Master Plan.

Figure 26. Potential areas of ecosystem services
Source: Computed by Author, 2020

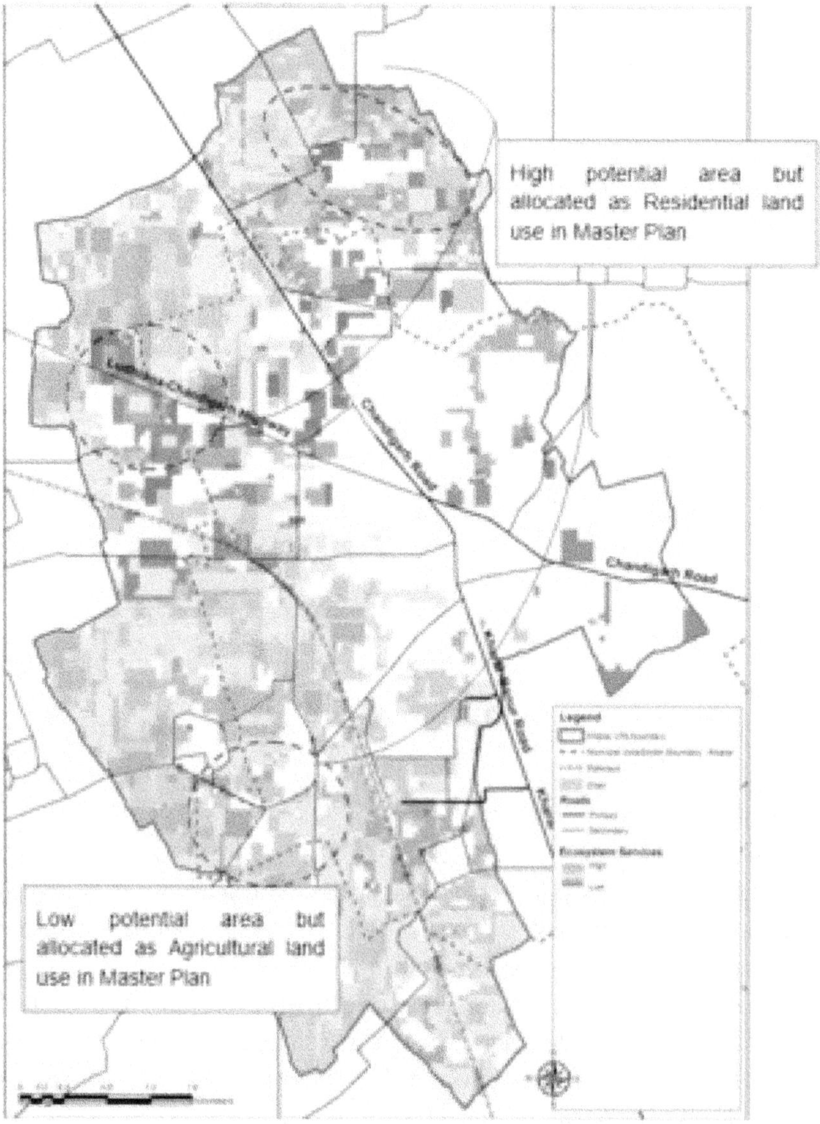

Hence, a framework has been carved out considering the legal framework, as emerging cities like Kharar face developmental challenges along with having the agricultural potential.

CONCLUSION

With the rapid urbanization, the agricultural lands are getting lost at the faster pace especially in the peri-urban areas where the city is expanding. The rich soil and diverse characteristics of agricultural lands in peri-urban areas, have high potential of agricultural productivity and ecosystem services which is often ignored. Punjab being the large contributor of agriculture at the national level is experiencing rapid urbanization due to which agriculture is getting compromised as urban sprawl of cities are converting the surrounding areas into built-up. The ecosystem services of agriculture are enormous and often underappreciated. The research is an opportunity to understand how the land use changes are affecting the agricultural ecosystems while focusing upon the drivers of land-use change in peri-urban areas. It is done followed by the spatial mapping of ecosystem services. Chandigarh is having high urbanization rate and settlements around it is experiencing extensive land use change from agricultural to built-up areas which has been analyzed with the help of spatio-temporal analysis. Kharar being one of them is envisioning 'agro-based economy' but with the passage of time the population is increasing at incremental rate. The built-up area is increasing at a faster pace with major share of Residential areas coming up in the form of real estate. The development could be seen majorly along the transit nodes due to connectivity.

The agriculture lands can regulate the capture, infiltration, retention and flow of water across the landscape. While the supply of surface water and groundwater inputs to agriculture through irrigation are indispensable but in case of Punjab the groundwater table is also depleting due to high amount of extraction of water and less water retention. While there are several other factors involved but one of the cause for that is urbanization and coming up of the built-up areas which has restricted the surface run-off and permeability of the surfaces. Due to this, water yield had been significantly changed and reduced. Habitat management within the agro-ecosystem may provide the necessary resources for pollinators or natural enemies. Many studies have identified the important role of perennial vegetation in promoting biodiversity in general and beneficial organisms in particular. But this is affected with the built-up areas coming up and restricting the pollination circles. Later, this impacts the agricultural productivity.

It is evident that current system of allocating land use in Master/development Plan is inadequate and the authorities have not incorporated agriculture ecosystem services into the decision-making process. The peripheral controlled area act which has been enacted to reduce urban sprawl around Chandigarh, is not proving to be the solution as unplanned development is booming in the peri-urban areas. Not only that, other policies at the national and state level of agriculture and urbanization is not integrated in the comprehensive manner. To fill the existing policy gaps between agriculture and urbanization are brought to the same level, and a comprehensive framework has to be used to allocate the land wisely based on agriculture productivity and its ecosystem services which could be adopted in the decision-making process. This framework will not only help to promote agricultural ecosystem services in the study area but can be promoted at each city level in India. There is a dire need to follow such framework as our food security, agricultural ecosystem services and productive lands would be lost on the name of development. The study provides a conceptual approach to developing a methodological framework for mapping biophysical flows and social values coming from agricultural ecosystems.

WAY FORWARD

Since India is a developing country and there is a great scope of new developments and infrastructure growth in coming future, the land allocation for future developments considering agricultural ecosystem services will not be difficult. All we need is technical expertise and availability of data so as to understand the developmental challenges. Developmental needs along with agricultural ecosystems of food production go hand in hand. The Research has helped to fill the gaps in the legal framework and decision-making process for allocating the future land uses. It has helped to bring out the challenges of peri-urban areas and the settlements which are emerging in nature. Considering the agricultural potential of the peri-urban areas, agricultural lands can't be lost at the cost of development as they have rich and fertile soils which is highly diverse in nature holding good productivity and ecosystem services. The research does not involve the agricultural disservices which may degrade the existing ecosystem services such as extensive use of fertilizers, stubble burning, use of insecticides and pesticides etc. which could be involved.

REFERENCES

Aljoufie, M., Zuidgeest, M., Brussel, M., & van Maarseveen, M. (2013). Spatial-temporal analysis of urban growth and transportation in Jeddah City, Saudi Arabia. *Cities (London, England)*, *31*, 57–68. doi:10.1016/j.cities.2012.04.008

Baral, H., Kccnan, R. J., Sharma, S. K., Stork, N. E., & Kasel, S. (2014). Spatial assessment and mapping of biodiversity and conservation priorities in a heavily modified and fragmented production landscape in north-central Victoria, Australia. *Ecological Indicators*, *36*, 552–562. doi:10.1016/j.ecolind.2013.09.022

Brinkley, C., (2014). *Evaluating the Benefits of Peri-Urban Agriculture*. doi:10.1038/jid.2014.371

Convention on Biological diversity-India. (n.d.). https://www.cbd.int/countries/profile/?country=in

Costanza, R., Costanza, R., Arge, R., Groot, R. De, Farber, S., Grasso, M., & Hannon, B. (1997). *The value of the world' s ecosystem services and natural capital 8009*. doi:10.1016/S0921-8009(98)00020-2

Dekolo, S., Oduwaye, L., & Nwokoro, I. (2015). Urban sprawl and loss of agricultural land in peri-urban areas of Lagos. *Reg. Stat.*, *5*(2), 20–33. doi:10.15196/RS05202

Department of Land Resources. (2013). National Land Utilisation Policy. Author.

FAO. (2012). The State of Food and Agriculture 2012. *The State of Food and Agriculture*. Advance online publication. doi:10.18356/3e6e633a-en

Fritsche, Eppler, Iriarte, Laaks, Wunder, Kaphengst, Wolff, & Heyen. (2015). *Resource-Efficient Land Use – Towards a Global Sustainable Land Use Standard (GLOBALANDS)*. Umweltbundesamt. doi:10.1017/CBO9781107415324.004

Gupta, R. (2014). The pattern of urban land-use changes: A case study of the indian cities. *Environment & Urbanization Asia*, *5*(1), 83–104. doi:10.1177/0975425314521539

Katyambo, M. M., & Ngigi, M. M. (2017). Spatial Monitoring of Urban Growth Using GIS and Remote Sensing : A Case Study of Nairobi Metropolitan Area, Kenya. *Am. J. Geogr. Inf. Syst.*, *6*, 64–82. doi:10.5923/j.ajgis.20170602.03

Ma, S., & Swinton, S. M. (2011). Valuation of ecosystem services from rural landscapes using agricultural land prices. *Ecological Economics*, *70*(9), 1649–1659. doi:10.1016/j.ecolecon.2011.04.004

Ministry of Statistics and Programme Implementation Government of India. (2013). *Green National Accounts in India A Framework*. Author.

MoEFCC. (2012). *The economics of ecosystems and biodiversity*. TEEB., doi:10.1007/978-90-481-9659-3_80

MoEfCC & GIZ. (2014). *The Economics of Ecosystems and Biodiversity TEEB - India Initative*. Work. Doc. 92.

Mundhe, N. N., & Jaybhaye, R. G. (2014). Impact of Urbanization on Land Use/Land Covers Change Using Geo-Spatial Techniques. *International Journal of Geomatics and Geosciences*, *5*, 50–60.

Power, A. G. (2010). Ecosystem services and agriculture: Tradeoffs and synergies. *Philosophical Transactions of the Royal Society of London. Series B, Biological Sciences*, *365*(1554), 2959–2971. doi:10.1098/rstb.2010.0143 PMID:20713396

Punjab Central Ground Water Board. (2013). Ground water information booklet S.A.S Nagar District, Punjab. Author.

Ray, D., Watts, K., Morris, J., & Handley, P. (2013). *Land Use and Ecosystem Services Project*. Academic Press.

Singh, D., & Vyas, A. (2014). Planning Strategies for the Development of Peri-Urban Area The Case of Indore Peri urban Area. *International Journal of Scientific and Engineering Research*, *5*, 27–32.

Tannier & Pumain. (2005). Fractals in urban geography: A theoretical outline and an empirical example. *Eur. J. Geogr.*

Team, P. (2015). *A Study on Punjab Export Potential and Strategy*. Rakesh Mohan Joshi Ram Singh Indian Institute of Foreign Trade.

TEEB. (2010). *Mainstreaming the economics of nature - A Synthesis of the approach, Conclusions and Recommendations*. TEEB.

TEEB. (2012). *The Economics of Ecosystem & Biodiversity- An Interim Report*. TEEB.

The National Research Council of National Academics. (2005). *Valuing ecosystem services: Toward better environmental decision-making*. The National Academies Press. doi:10.17226/11139

Tripathi, S., & Mahey, K. (2017). Urbanization and economic growth in Punjab (India): An empirical analysis. *Urban Research & Practice*, *10*(4), 379–402. doi:10.1080/17535069.2016.1227875

UN & UNEP. (2015). *SEEA Experimental Ecosystem Accounting: Technical Recommendations*. Final Draft.

UN, IIFM, & Ministry of Statistics and Programme Implementation. (2018). *Review of Existing Ecosystem Accounting Initiatives And Literature In India*. Author.

UNEP. (n.d.). Convention on Biological Diversity, in: Convention, The Diversity, Biological Cbd, The Protocol, Cartagena Plan, Strategic Secretariat, The Diversity, Biological Secretary, The Executive. UNEP.

United Nations. (n.d.). *System of Environmental Economic Accounting*. https://seea.un.org/

Wan Guanghua, K.S.S. (2013). *Urbanization in Asia: Governance, Infrastructure and the Environment*. Academic Press.

WWF. (2020). *About Agricultural Biodiversity*. http://www.environment.gen.tr/biodiversity/161-about-agricultural-biodiversity.html

Chapter 7
Impact of Land Use Land Cover Change on Sambhar Lake

Saloni Khandelwal
AIGGPA, India

ABSTRACT

The Aravalli range is one of the oldest mountain ranges in the world. Many lakes are sustained by the streams generated from the Aravalli range. In the last few years, hill systems are facing changes affecting the catchment of water bodies. Many waterbodies that were earlier able to provide water in peak summers are now dried up before the summer. The study aims to analyse land use land cover change in the Aravalli range on Sambhar Lake. Sambhar Lake is the largest inland saltwater lake in India and was selected as a "Ramsar Site." A catchment area of the lake is delineated, and micro-watersheds are selected for detailed analysis. A primary survey is done to know the impacts and causes of changes in the lake. The study indicates significant green cover changes in three seasons over the years. Because of the loss in green cover, the soil is losing its moisture, affecting the cropping pattern and agriculture production in the area. Factors causing the drying of the lake are affecting its ecosystem and the number of birds coming to it.

INTRODUCTION

The Aravalli range is one of the oldest hill systems in the world. They came into existence by an upheaval of organic nature at the close of the Dharwar era and dominated the geography of India in the Palaeozoic and Mesozoic periods. These are stretched in about 692 Km across Gujarat, Rajasthan, Haryana, and Delhi (Dhabriya, 1988). The Aravalli range plays a significant role in the area as they combat desertification, recharge groundwater, balance the ecology, check the drought occurrence, and provide minerals and other resources (Kohli & Arora, 2018). Many rives like Banas, Luni, Sakhi Sabarmati, Sahibi, Sota and others originated from it. Moreover, many streams originated from these hill system feeds various lakes like Pichola, Rajsamand, Kanota, Jaisamand, Anasagar, Siliserh, Man Sagar and many more (see Figure 1), one of them is a Salt Lake, i.e., the sambhar lake which is an important natural assets with

DOI: 10.4018/978-1-7998-8331-9.ch007

considerable aesthetic, recreational, scientific, cultural, economic, and ecological value (Singh, 2013). In the arid zone of the Thar desert, salt lakes are originated due to the formation of depression by neo-tectonic disturbances along lineaments (Kale, 2014).

For the last 30 years, these hill systems are facing changes because of mining, deforestation, population growth, and other activities. Because of all the above factors, the land use land cover (LULC) of the area surrounding the Aravalli range is changing rapidly. The land cover defines as the physical state of the land surface, including water bodies, forests, wetland, cropland, etc., whereas land use refers to biophysical resources used by humans. Therefore, land use land cover analysis is essential for environmental studies and natural resource management. This information also helps to understand the relationship between forest, wetland, cropland, settlement, etc. (Singh, Singh, Singh, & Mukherjee, 2016). Furthermore, the ecosystem functions of Aravalli's are changing due to changes in land use land cover, which is resulting in the drying of many rives and other water bodies. To understand this, the catchment of Sambhar Lake is taken as an example (see Figure 2 and Figure 3).

Figure 1. Important lakes around Aravalli range
Source: Created by author, 2019

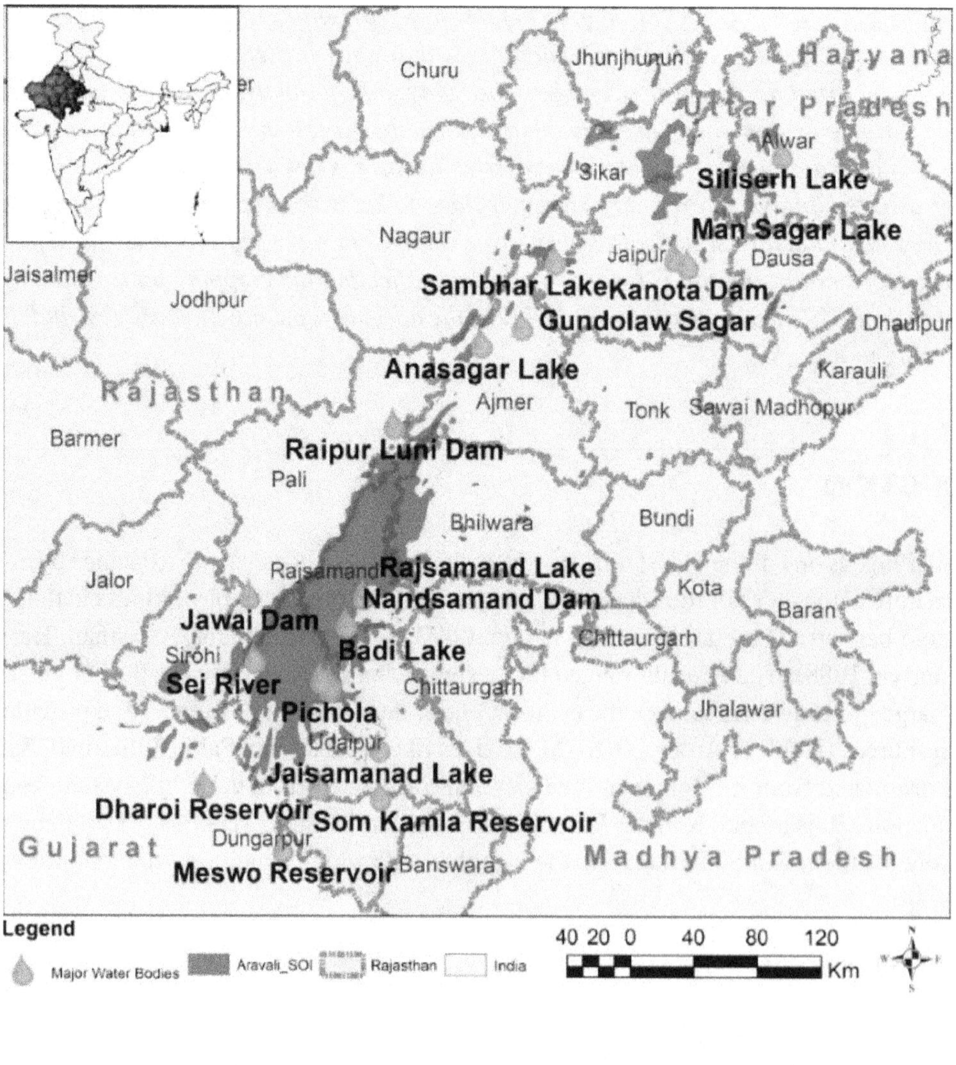

Figure 2. Google Earth image of December month- year 2001
Source: Google Earth

Figure 3. Google Earth image of December month- year 2018- increased salt mining area
Source: Google Earth

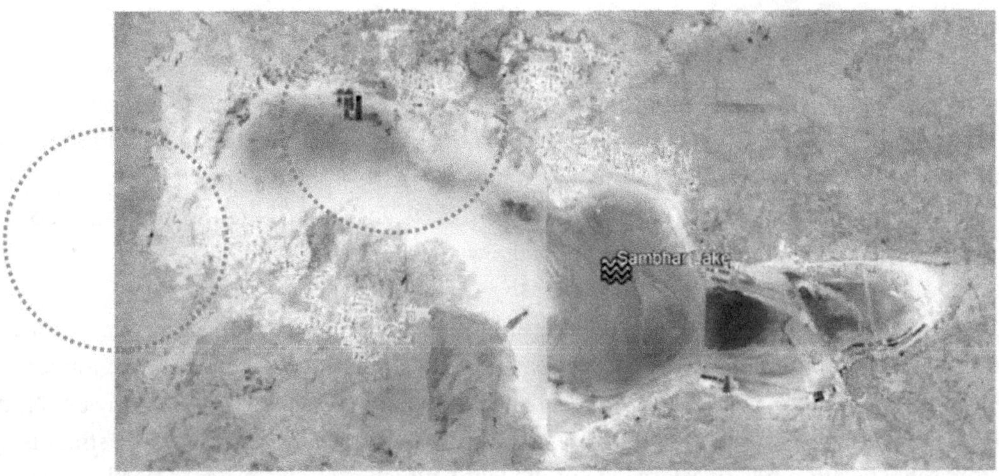

For achieving this study first catchment area is delineated, and watersheds are taken from the Soil and Land Use Survey of India (SLUSI), and then change in vegetation cover and land use land cover is assessed using Normalised Difference Vegetation Index (NDVI) and Geographic Information System (GIS). GIS technology and remote sensing have emerged as a decent tool for analysing the change in land use land cover of the area at spatial and temporal scales. (Singh, Singh, Singh, & Mukherjee, 2016) NDVI is used to calculate the value of vegetation index using multispectral remote sensing data (Zaitunah, Samsuri, Ahmad, & Safitri, 2018). Analysis of land use land cover is focused on the study, on three aspects- detection of changes in land use land cover using Landsat image between 1998-2008-2018 and Bhuvan geoportal image between 2006-2012-2018, identification of nature of changes, and assessment of the pattern of the changes. At the end of the study, some strategies are suggested for the sustainable management of the study area.

STUDY AREA - SAMBHAR LAKE

In the present study, the catchment area of sambhar lake has been selected as the study area. Sambhar Lake is India's largest inland Saltwater Lake, 70 km from Jaipur, located 26°51'56" to 27°03'19" N Latitude, 74°53'30" to 75°14'14" E Longitude(Space Applications Centre, ISRO Ahmedabad, 2013). It has a total catchment of about 6500 square kilometres (Figure 4 and Figure 5) which falls in Ajmer, Nagaur, and Jaipur district. The lake is also the largest playa in the Thar desert (Kale, 2014) (Playa means seasonal lake with flat bottom). Lake is periodically covered by water that slowly infiltrates into the groundwater system and evaporates into the atmosphere. This lake is an internationally important site as it is registered under Ramsar site (number 464)- in March 1990 and also an Important Bird Area (IBA site number In-RJ-16) (Zoological Survey of India, 2005). Lake is a wintering area for flamingos, and other birds migrated from Siberia and northern Asia. Areas in and around the Aravalli from where streams are generated are facing land conversion, due to which lake is getting affected. These changes are causing the deficiency of water in the lake, affecting the biodiversity dependent on it. Lake is internationally significant, but there is no specific lake boundary until now, and it does not fall under any category of Wildlife Sanctuary, National Park, or Reserves (Chakravartty, 2015). Various activities have already impacted the lake, especially salinization, surface inflow diversions, and other catchment activities like mining, pollution.

Geology and Geomorphology

Sambhar lake is at an elevation of 365 m above mean sea level (AMSL) located southeast of the Aravalli range. It is extending Northeast to Southwest for over 700 km. It is comprising of folded rocks of the early and middle Proterozoic age. These rocks contain primary minerals like felspar, amphibole, and biotite. These rocks weather to clay and iron oxyhydroxides and contribute the necessary constituents such as Ca2+, Mg2+, K+, Na+, to the river and lake water to eventually precipitate as salts in the playa during evaporation.

Sand ridges formed by the wind activities through the gap in the Aravalli range have bordered the Sambhar Lake. According to the groundwater board report, annual rainfall over this region ranges from 100 to 500 mm. Sambhar Lake is fed by two major rivers, Rupangarh, which flows Southwest-Northwest, and Medha, which flows Northeast-Southwest (see Figure 5). In addition, many other streams are generating from the Aravalli range (western side of the lake), which provides water to this lake. The surface area of the lake fluctuates from 200-300 square kilometres. The maximum depth of the lake is 3 meters, and the average depth is 0.61 cm. The maximum length of the lake is around 22.5 km, and its width ranges from 3 km to 11.2 km (Kale, 2014).

METHODOLOGY

In Figure 6 methodology of the study is explained in detailed. From the literature, the issue is identified on which research is required. After identifying the Aravalli range as the study area, different issues are listed down with the help of a literature study, and the need for study is formulated. In the study area, various important water bodies are studies and marked. After selecting the lake, slope, the contour is generated with the help of surface analysis in GIS. Basin, stream order, and catchment area are generated with the help of the GIS hydrology tool. Watersheds and micro-watershed are taken from SLUSI (Soil and Land Use Survey of India) and then digitized in Arc Map 10.3. Selected watersheds and micro-watersheds are given in Figure 7. Primary survey is done to know the impact of LULC change on the water body. Satellite images are downloaded, and then the analysis is performed, and impacts are identified.

Figure 4. Elevation profile of Sambhar Lake basin
Source: Created by author,2019

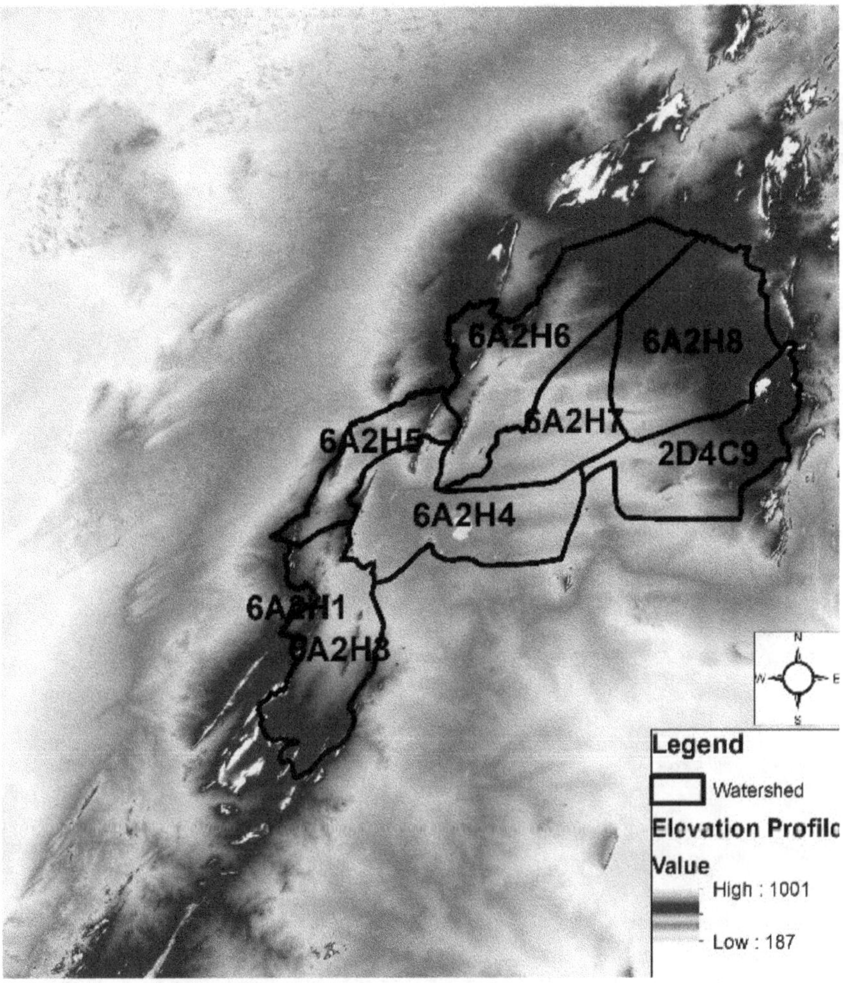

Five micro watersheds are selected as given in Figure 7. Primary survey is conducted in Sinodiya, Jhag and Gurna village as it falls in the selected watersheds for ground-truthing and to check land conversion is happening or not, and if yes, then what is the reason, how it is affecting the streams/waterbody. Population growth, number of survey details are listed in Table 1. The same questions were also asked in village Khakharkeri. Part of the Aravalli range falls in these villages. Streams generated from these hills are studied. Survey and focus group discussion (FGD) of various stakeholders (villagers, panch of the village, government officials of the agriculture department, forest department, town and country planning authority, and Non-Government Organisation working in that area) is conducted. Forty-four survey samples were collected from villages with a confidence level of 99% and a confidence interval of 20.73 percentage of farmers, five percentage of other workers, and 22 percentage of women were surveyed. Figure 8 shows the individual survey and focus group discussion conducted in various villages.

Image Acquisition

Satellite images of selected micro-watersheds have been acquired from The United States Geological Survey (USGS) Earth during the month of May-June, September-October, December-January for the years 1998, 2008, 2018 with minimum cloud cover (see Table 2).

Figure 5.Catchment area and watersheds of Sambhar lake
Source: Created by author,2019

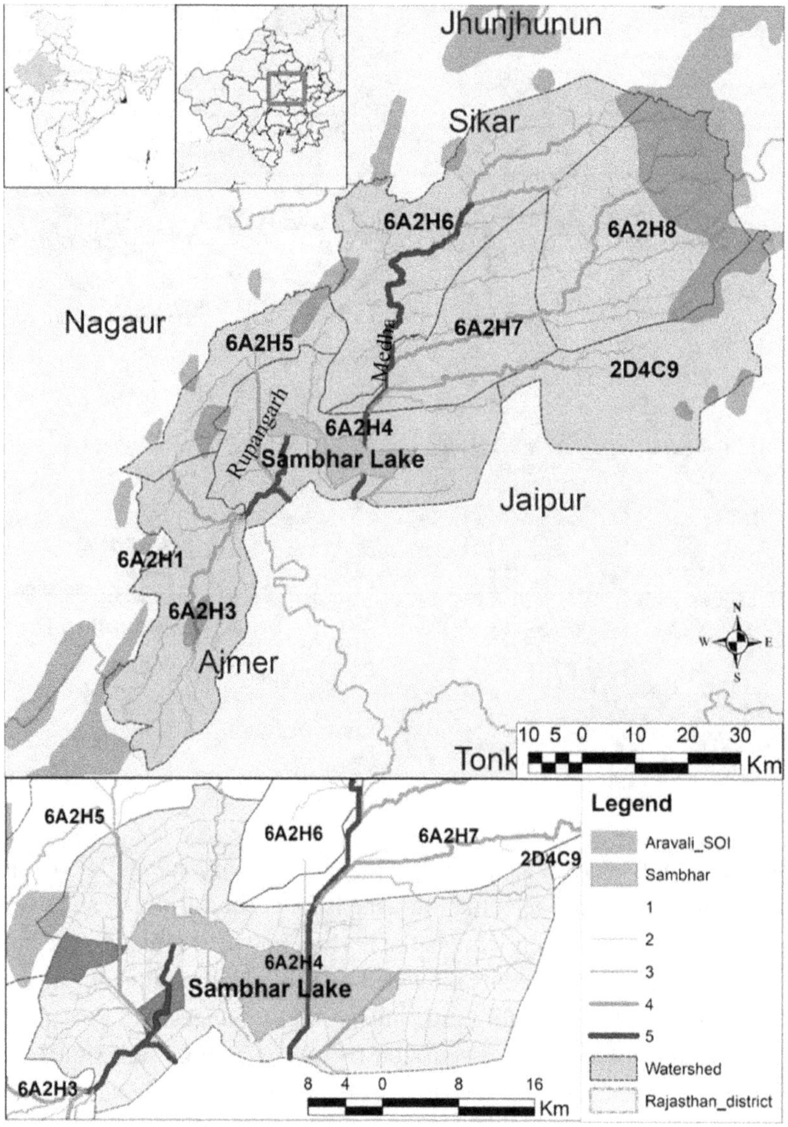

Figure 6. Methodology for study
Source: Created by author,2019

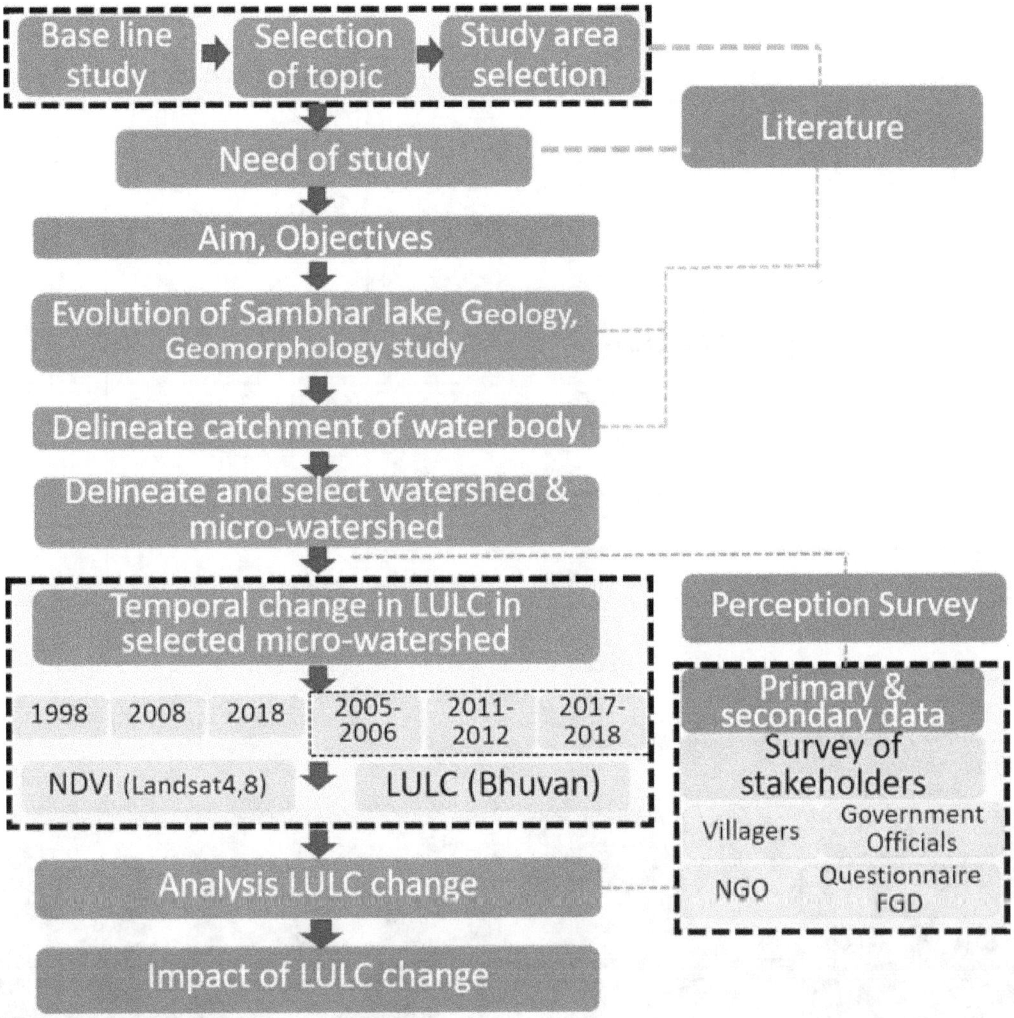

For detailed classification of land use land cover, data from Bhuvan is collected for the year 2005-2006, 2011-2012 and 2017-2018.

Image Processing

Landsat image consists of different bands, landsat8 has 11 bands, landsat4 has 7 bands, so layer stacking is essential to combine them into a single band. Layer stacking was performed on ERDAS IMAGINE 2013 software. First, stacked image tiles were clipped with the study area shapefile. Next, the image is opened in Arc Map 10.3 and then classified. The classified image is then reclassified by reclassify tool. Finally, the area is calculated for the reclassified image.

Land use land cover from Bhuvan is digitized in Arc Map 10.3 for three years, i.e., 2005-2006, 2011-2012, and 2017-2018.

Figure 7. Selected micro-watersheds and location of survey conducted
Source: (Minisry of Agriculture and Farmer Welfare,Government of India, 2018)

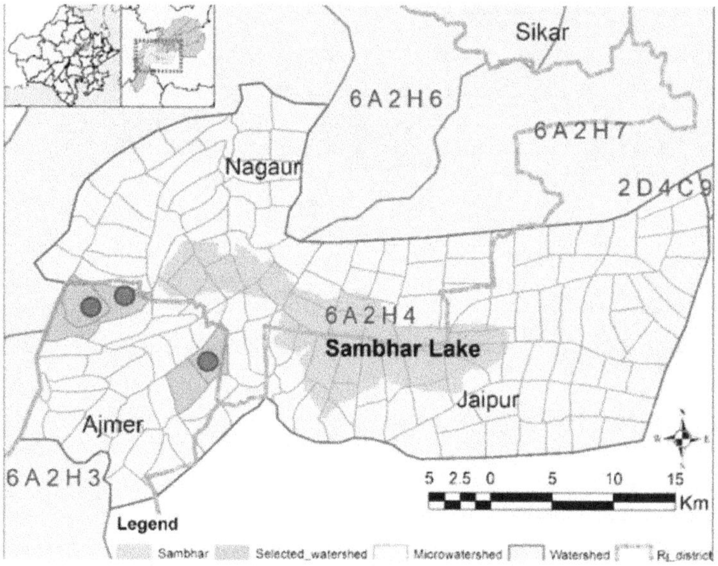

Number of micro-watershed	
6A2H4r8	17.41 Sq. Km.
6A2H4r6	
6A2H4w2	25.1 Sq. Km.
6A2H4w4	
2H4w3	

Figure 8. Survey in Khakharkeri village
Source: Created by author,2019

Table 1. Population growth in villages selected for survey

	Demography						
Villages	**2011**	**Growth (%)**	**Household**	**Household Survey conducted**	**2001**	**Growth (%)**	**1991**
Gurha	493	-14.41	87	6(7%)	576	140.0	240
Jhag	1067	-3.18	194	16(8%)	1102	64.7	669
Sinodiya	1566	-30.62	310	22(7%)	2257	20.6	1871

Source: (Government of India, 1991) (Government of India, 2001) (Government of India, 2011)

Table 2. Details of the image used in the study

Image	Season	Acquisition Date	Spatial Resolution	Cloud Cover
Landsat5	Summer	03/06/1998	30 meters	0
Landsat5	Monsoon	07/09/1998	30 meters	0
Landsat5	Winter	28/12/1998	30 meters	0
Landsat5	Summer	29/05/2008	30 meters	<5%
Landsat5	Monsoon	04/10/2008	30 meters	0
Landsat5	Winter	23/12/2008	30 meters	0
Landsat8	Summer	25/05/2018	30 meters	0
Landsat8	Monsoon	30/09/2018	30 meters	0
Landsat8	Winter	01/01/2019	30 meters	<5%

Source: (USGS, 2019)

Image Classification and Change Detection

Normalized difference vegetation index was calculated on ERDAS IMAGINE 2013 for the years 1998, 2008, and 2018 using Landsat images (4 and 8) for three seasons. The result of all the years was compared. Result of all the three years was also compared in Arc Map 10.3, which again show significant changes in land use land cover.

Vegetation Index

NDVI has a co-relation with the chlorophyll contents of plants. It uses band 3, i.e., red and band 4, i.e., near-infrared for landsat5, and band 4, i.e., red and band 5, i.e., near-infrared for landsat8. NDVI formula is given (see Eq. 1) as:

$$NDVI = \frac{IR - R}{IR + R}$$

Where IR is the reflection value of the infra-red band (band 4 and 5), R is the reflecting value of the red band (band 3 and 4). NDVI value ranges from -1 to +1, where -1 is no vegetation and +1 is highly dense vegetation. Water, barren land, stone, and other materials have a value less than 0. NDVI density classes are categorized into no vegetation: less than 0.1, very small vegetation- between 0.1 to 0.2, small vegetation-between 0.2 to 0.3, medium vegetation- between 0.3 to 0.4, and high vegetation, which is greater than 0.4 (Tilahun, 2015).

RESULT AND DISCUSSION

Primary Survey

In the primary survey, out of the total surveyed farmers, 90 percent of farmers have engaged in both farming and salt mining, and out of this, 47 percent of them shifted their occupation from agriculture in the last ten years. Tanker water supply is there in the villages. Eighty-four percent of the survey indicates that the livestock population is decreasing. Moreover, all farmers said that the cropping pattern is changing because of soil moisture loss and changing groundwater levels. Ninety-three percent of surveyed people said that the number of migratory birds has decreased over the years, and the salt-affected area is continuously increasing (see Table 3). Villages are stated that earlier hills around the villages were lush green. However, slowly population of villages started increasing, the frequency and intensity of eastern heat waves increase, people started over-utilizing the forest product, and slowly, these forests started decreasing. Different wild animals like leopards, tigers, and deer were spotted, but now not a single wild animal is seen. Although now most of the forest area in the Aravalli is under the forest department, people are still clearing the trees to get fuelwood.

Figure 9. Occupation in village and occupation shift
Source: Created by author,2019

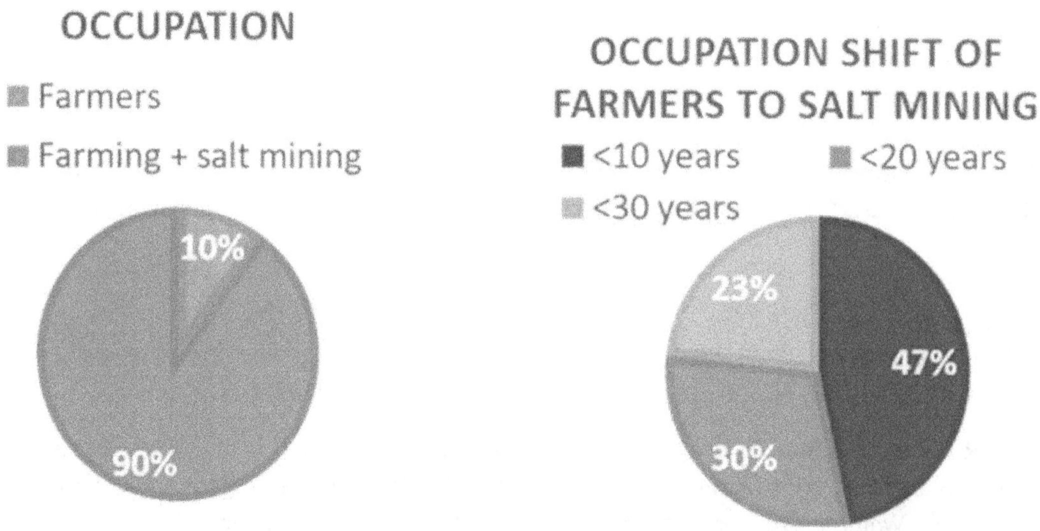

Table 3. Results of primary survey

	Livestock decrease	Shift in cropping pattern	Waterborne disease	Migratory birds decreased	Groundwater table decrease
Yes	37	44	44	41	44
No	5	0	0	0	0
Don't know	2	0	0	3	0

Source: Created by author,2019

Salt miners are using groundwater for salt manufacturing as most of the time lake is dry only. Most of the region is dry, so agriculture production is not good. Changing environmental and soil conditions have affected agriculture in the region very severely. Therefore, more and more people are shifting to the salt manufacturing industry. However, it affected their life expectancy. In all the surveyed cases, people are facing joint pain and skin problems. However, they do not have any other option but to continue this work to earn their living. Villages in the region have changed due to earnings from salt mining; most of the houses are pucca. People have different equipment and other facilities at their homes. However, people are not happy there.

NDVI of Micro-Watershed from 1998 to 2018

Based on the above NDVI classification, images of the selected micro-watersheds are generated as given in Figure 10, Figure 11 and Figure 12 for micro watershed1 and Figure 13, Figure 14 and Figure 15 for micro watershed2. Their area is calculated for summer, monsoon, and winter season for the year 1998, 2008 and 2018.

Micro-watershed:1- (6A2H4r8, 6A2H4r6) These watersheds are selected as Rupangarh river passes through these two micro-watersheds, and these are affected by salt mining.

In the year 1998, during the summer season, 12.58 sq. km area had NDVI value less than 0, which increased to 16.64 sq. km in the year 2018. In both monsoon and winter season areas under small vegetation, medium vegetation, and high vegetation had continuously decreased, as given in Table 4. Earlier in the year 1998, there was some parcel of land with medium and high vegetation, but with decade those areas were gone. Areas having NDVI values less than 0 are majorly salt-affected areas.

Micro-watershed:2-(6A2H4w2, 6A2H4w3, 6A2H4w4) These watersheds are selected as it has part of the Aravalli range, streams are generated from it, which provides water to the lake, especially during the monsoon season.

In the year 1998, during the summer season, 11.7 sq. km. of the area had NDVI value less than 0, which increased to 21.88 sq. km in the year 2018, as given in Table 5. Thus, in the year 1998, there were comparatively more medium and high vegetation areas in the watersheds, which are continuously decreased, especially in the monsoon and winter season.

Based on the analysis, it is clear that vegetation cover in the area is continuously decreasing. These micro-watersheds need to be maintained. Salt affected area is increasing. Based on the survey and site inspection it is evident that people are storing the river water from streams before it reaches the lake, resulting in a decrease in the lake's water. A decrease in green cover during the summer season also depicts that people live in these areas only practicing the Kharif crop.

Land Use Land Cover of Micro-Watershed from 2005 to 2018

Land use land cover data from BHUVAN is taken and then analysed in Arc Map 10.3 software for the year 2005-2006, 20011-2012, and 2017-2018. Figure 16, Figure 17 and Figure 18 show the LULC of micro-watershed 6A2H4r8 and 6A2H4r6. In the 2017-2018 classification, the government has merged the fallow and agricultural land.

In Table 6 it is shown that salt-affected areas and wetland areas have decreased to -6.52% and -15.16% in the year 2017-2018 from the year 2005-2006, but salt mining area has increased by 13.78 percent in the year 2017-2018.

Figure 10. NDVI-1998 summer season: micro-watershed-1
Source: Created by author,2019

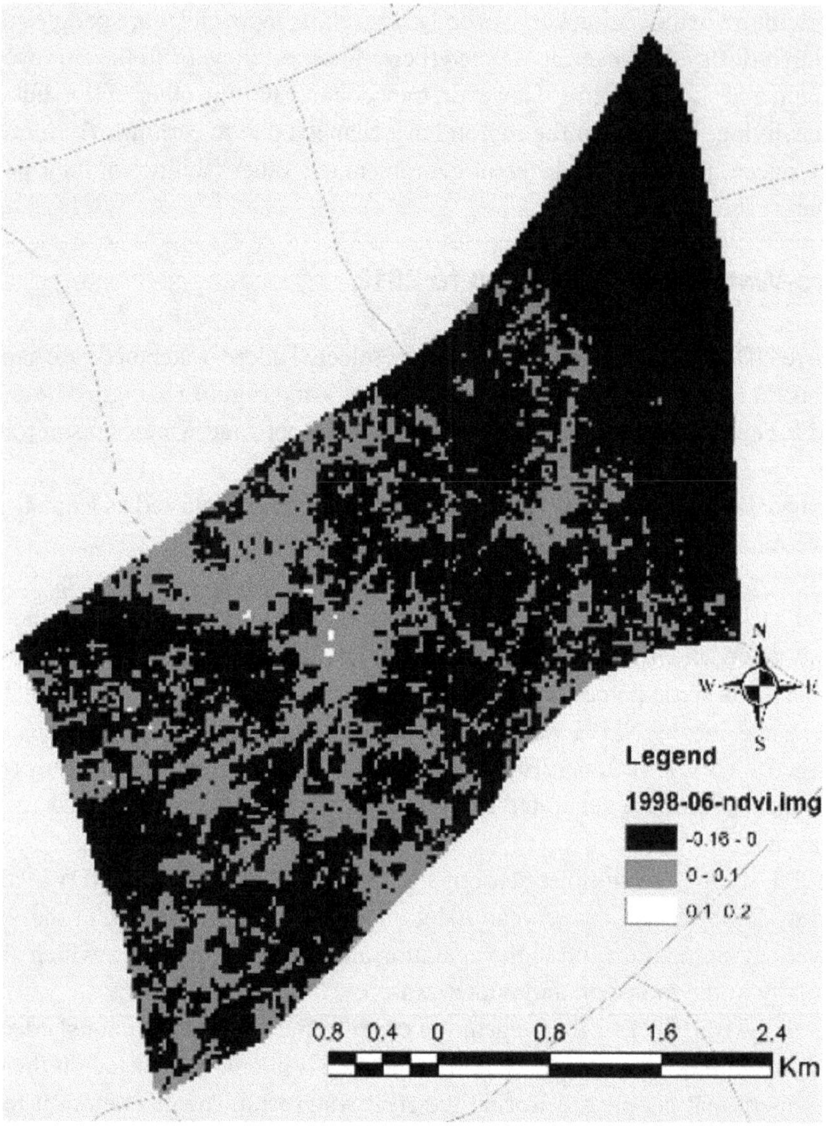

In the surveyed area, there is no scheme and groundwater for irrigation (primary survey); all agricultural land is rainfed (Kharif). Although agricultural land has increased, the production has decreased, and the cropping pattern has changed (Primary survey). Land parcels that were earlier wetlands are now used for salt mining, and most of them are illegal. In the Rabi season, there is no crop production; during this time, people are stopping the river water before it reaches the lake or using the groundwater and practices salt mining. As a result, the river area has also decreased in the selected area. Figure 19, Figure 20 and Figure 21 show the land use land cover of micro watershed 6A2H4w2, 6A2H4w3, and 6A2H4w4 from the year 2005-2006, 2011-2012, and 2017-2018. LULC data collected from Bhuvan and then digitized in Arc Map 10.3 software.

Figure 11. NDVI-1998 monsoon season: micro- watershed-1
Source: Created by author,2019

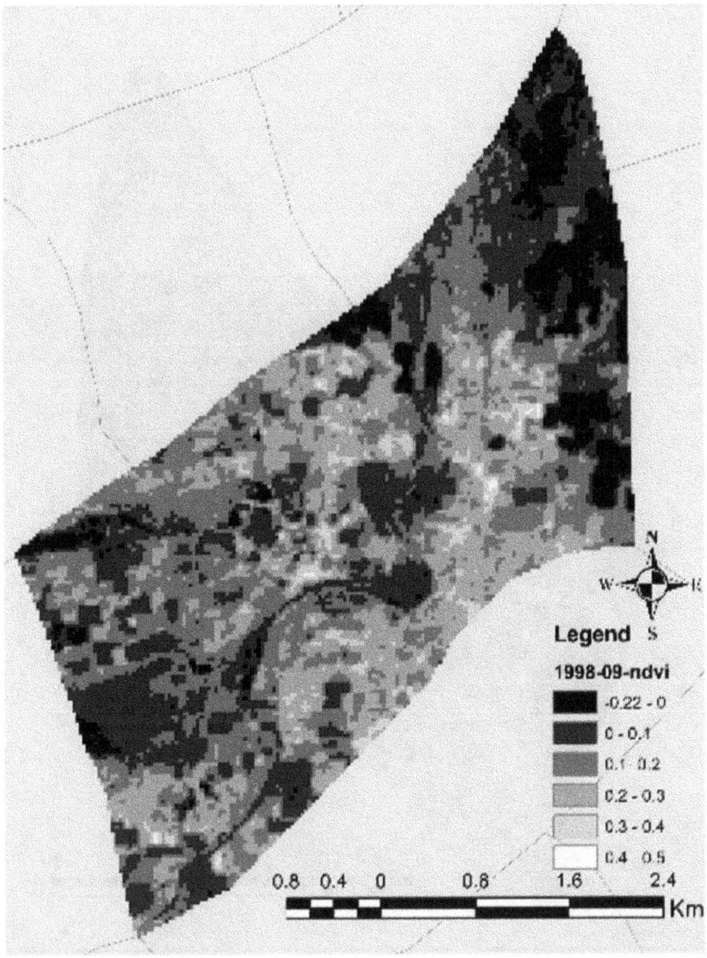

Table 4. NDVI results of watershed 1

Type of vegetation		Summers			Monsoon			Winters		
	NDVI	1998	2008	2018	1998	2008	2018	1998	2008	2018
No vegetation	<0	12.58	15.96	16.64	1.64	2.36	2.54	1.98	2.63	2.04
	0-0.1	5.15	1.77	1.09	5.04	4.36	4.25	11.46	12.42	12.86
Very small vegetation	0.1-0.2	0.01	0.01	0.01	6.22	8.35	8.64	2.81	2.44	2.46
Small vegetation	0.2-0.3				4.25	2.62	2.27	0.92	0.19	0.37
Medium vegetation	0.3-0.4				0.56	0.05	0.04	0.36	0.04	0.01
High vegetation	0.4-0.5				0.03			0.16	0.02	
	0.5-0.6							0.05		

Note- Area in Sq. km.
Source: Created by author,2019

Figure 12. NDVI-1998 winter season: micro-watershed-1
Source: Created by author,2019

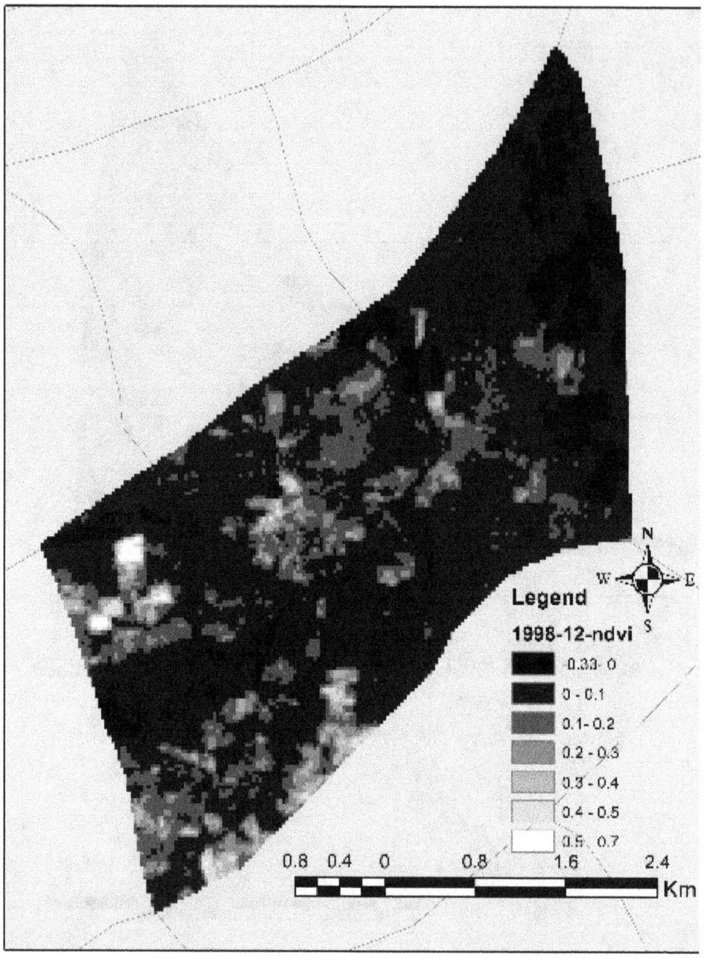

Table 5. NDVI results of watershed 2

Type of vegetation		Summers			Monsoon			Winters		
	NDVI	1998	2008	2018	1998	2008	2018	1998	2008	2018
No vegetation	<0	11.7	13.82	21.88	1.5	1.12	1.26	1.78	1.49	3.06
	0-0.1	13.74	11.63	3.56	7.6	4.3	2.18	15.83	18.12	22.31
Very small vegetation	0.1-0.2	0.02	0.01	0.02	8.96	15.17	18.41	4.09	5.55	0.07
Small vegetation	0.2-0.3				5.9	4.59	3.6	1.3	0.265	0.02
Medium vegetation	0.3-0.4				1.4	0.28	0.01	1.4	0.02	
High vegetation	0.4-0.5				0.1			0.99	0.01	
	0.5-0.6				0.001			0.07	0.005	

Note- Area in Sq. km.
Source: Created by author,2019

Figure 13. NDVI-1998 summer season: micro-watershed-2
Source: Created by author,2019

Figure 14. NDVI-1998 monsoon season: micro- watershed-2
Source: Created by author,2019

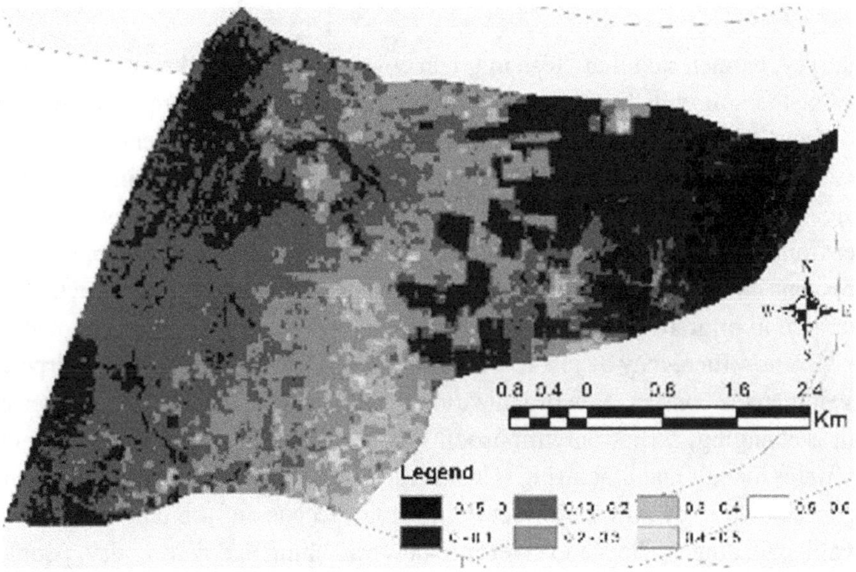

As depicted in Table 7, forest area has increased to 0.05 sq. km, but forest cover has decreased as per NDVI. Salt mining area has increased to 0.4 sq. km. which was under wetland in the year 2008-2009. Salt affected area has decreased to 16.09 percentage from 2.61% to 2.19%. The scrubland area has been increased by 417% from 0.58 sq. km to 3 sq. km. In the year 2017-2018, water bodies have also decreased. The built-up area in the selected villages has been increased.

Figure 15. NDVI-1998 winter season: micro-watershed-2
Source: Created by author,2019

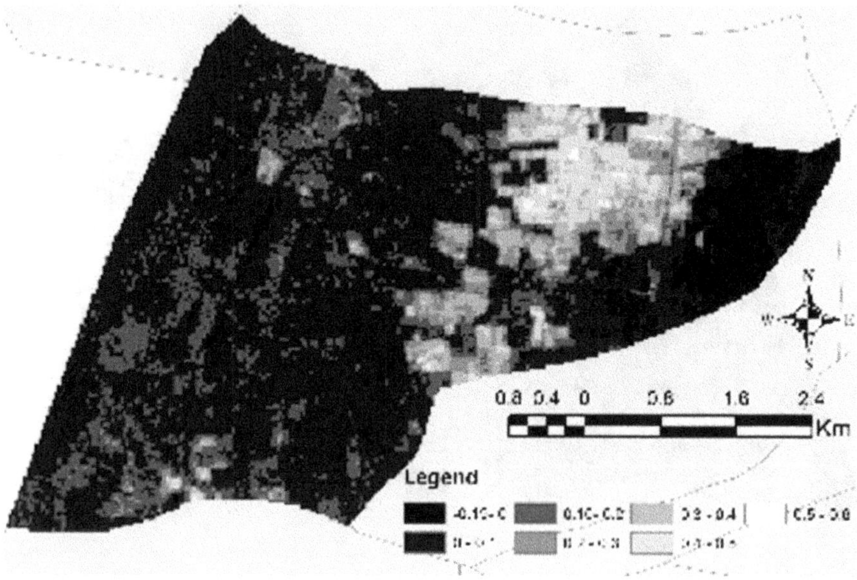

IMPACT OF LULC CHANGE

In the primary survey, women said that "loss in green cover in the Aravalli resulted in the loss of water holding capacity of the soil, which is affected grazing land and agriculture. Livestock population is also affected". According to the farmers, the water table is getting lower every year; that is why they are shifting to salt manufacturing. During Kharif season, they harvest, and during the rabi season, they go for salt mining. In every focus group discussion and survey, a common thing which came is "lake is not getting water from last 14-15 years". Lake is not getting enough water; it is supposed to occupy 230 square kilometres, but it has not had any water for the last 10 -15 years. The destruction of habitat has resulted in a decrease in migratory birds. In 1982-83, 5,00,000 flamingos were counted, their number fell to 20,000 in 2008 now there may be just few hundred says Kadyn (founder and chairperson of people for an animal in Haryana, a conservation organization). The illegal Salt pan area is increasing. The way of extracting salt is changing; earlier only monsoon water was used for salt extracting, now people are pumping groundwater for salt manufacturing (Chakravartty, 2015). This excess groundwater extraction has lowered the groundwater by almost 40 feet in the area. Because of the land use land cover change around the Aravalli's, feeding streams and rivers are now vanishing and getting dry; people are shifting from agriculture to salt manufacturing industries. Salt production units are hiring water tankers for salt manufacturing. Existing situations are creating acute drinking water shortages in villages surrounding the lake and causing migration. The age limit of people working in the salt industry is decreased to about 50 years which is a very horrible situation. Workers of the salt industry, if they get any injury, then they will never be able to recover. Because of the increasing salt pans (as seen in Figure 2 and Figure 3) the agriculture area and green cover are decreasing around the lake. There is an urgent need to address the changes happening in the lake's catchment area; otherwise, the time is not far when this lake will be a thing of the past (Table 8).

Figure 16. Landuse landcover 2005-2006
Source: Created by author,2019

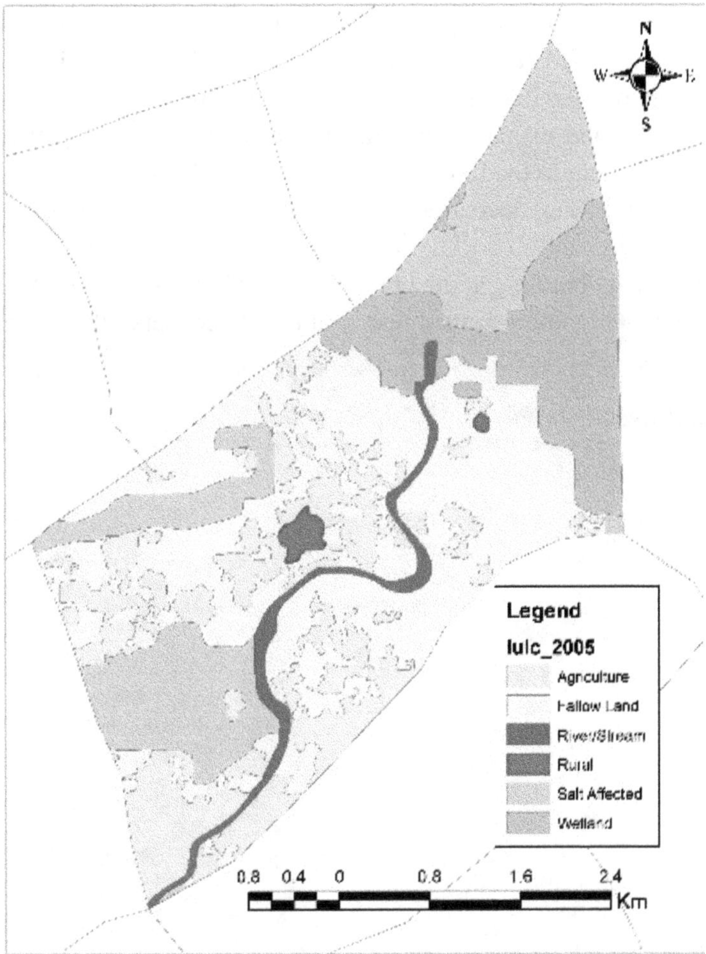

RECOMMENDATIONS

The success of the developmental activities depends upon the scientifically effective and sustainable utilization and management of Natural Resources. Unfortunately, the unsustainable use of valuable natural resources has resulted in acute detrimental effects upon our biosphere. Decentralized planning of the management process through public participation in the identified gram panchayat will help achieve the required goals. After analysing the extent of land degradation in the Aravali range region through different methods and change in the LULC in the selected villages, few recommendations have been made in the following section.

Water resource management strategies are required in the region. On the basis of land suitability analysis for the Ground Water Recharge, construction of rainwater harvesting structures like anicuts, ponds, and other embankments at the village level can put significant evaporation losses of water and recharge groundwater. This water can be used for drinking and plantation purposes. This will also reduce the practice of villagers stopping river water from reaching the lake and will increase water inflow in lake wetland. Joint forest management strategies with the help of the forest department and community participation can help in improving the green/forest cover of the area. Aware people to use LPG by explaining their benefits. Use of biogas systems and their advantages (manure and fuel). Plantation will reduce the effect of heat waves and will increase the water-holding capacity of the soil. Crop-based management strategies can be used to reclaim saline soils. There is a great increase in scrubland in both the micro-watersheds; wasteland management strategies can be used for effective land management. Cattles of the villagers are going to the forest areas in the Aravalli. Because of overgrazing, the soil is getting loose, and plants are getting decreased. If the land around the forest area is under government, it can be converted to grazing land. It will help in preserving forest biodiversity and provide food for cattle. Awareness and capacity building at the gram panchayat level is needed. Already people are raising voice, support from NGOs and the government will help to conserve the Aravalli by local people.

Figure 17.Landuse landcover 2011-2012
Source: Created by author,2019

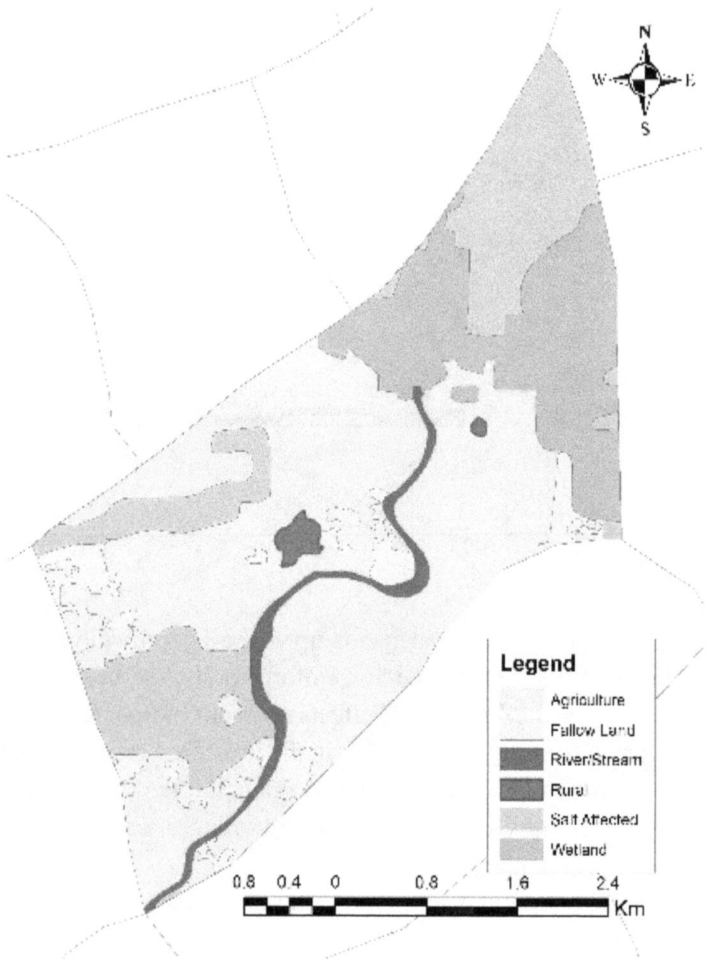

Figure 18.Landuse landcover 2017-2018
Source: Created by author,2019

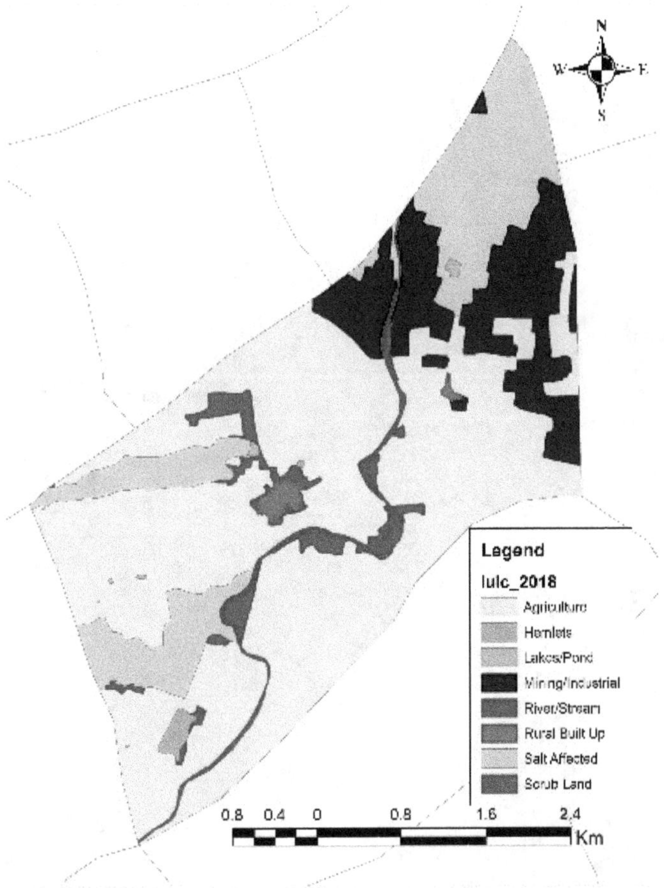

Table 6. Land use land cover change

Landuse 1	Landuse 2	Area Sq. Km 2006	Area (%) 2006	Area Sq. Km 2012	Area (%) 2012	%Change from 2006 to 2012	Area Sq. Km 2018	Area (%) 2018	% Change from 2006 to 2018
Agriculture	Cropland	3.5	20.10	8.84	50.76	30.66	10.76	61.81	41.72
	Fallow	6.4	36.75	1.07	6.14	-30.61	0	0.00	-36.75
Barren	Sal Affected	4.19	24.06	3.89	22.34	-1.72	3.05	17.54	-6.52
	Scrub Land	0	0.00		0.00	0.00	0.63	3.62	3.62
Builtup	Mining	0	0.00		0.00	0.00	2.4	13.78	13.78
	Rural	0.145	0.83	0.145	0.83	0.00	0.21	1.23	0.40
Waterbodies	Inland Wetland	2.64	15.16	2.98	17.11	1.95	0	0.00	-15.16
	River	0.54	3.10	0.49	2.81	-0.29	0.32	184	-1.26
	Waterbodies	0	0.00	0	0.00	0.00	0.03	0.17	0.17

Source: Created by author,2019

Figure 19.Land use landcover 2005-2006
Source: Created by author,2019

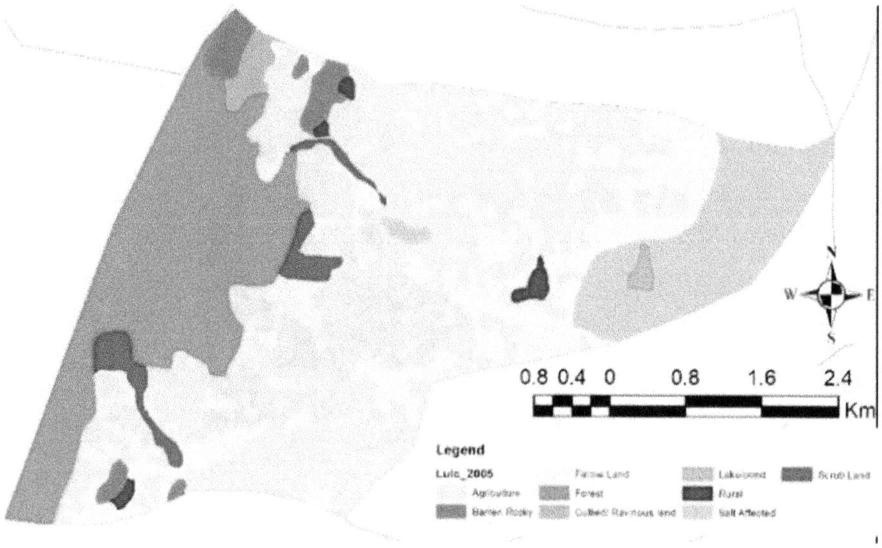

Figure 20.Land use landcover 2011-2012
Source: Created by author,2019

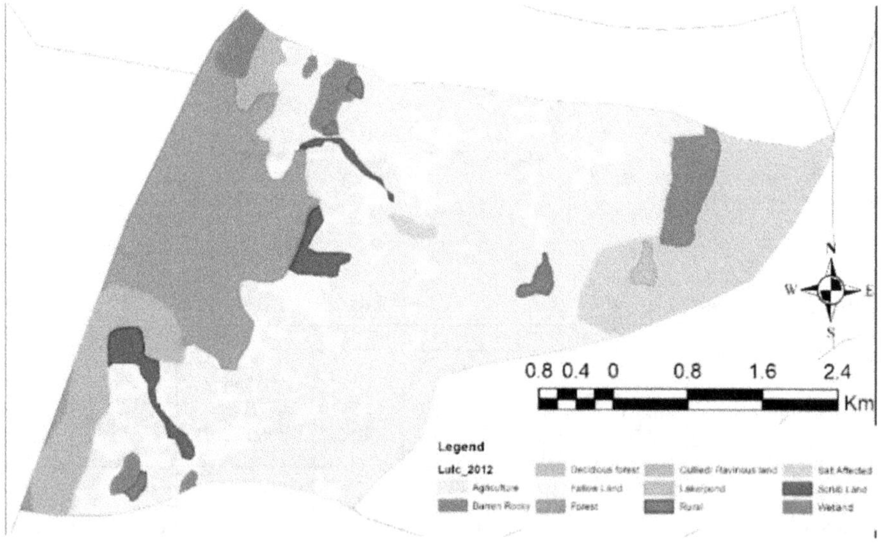

Figure 21.Land use landcover 2017-2018
Source: Created by author,2019

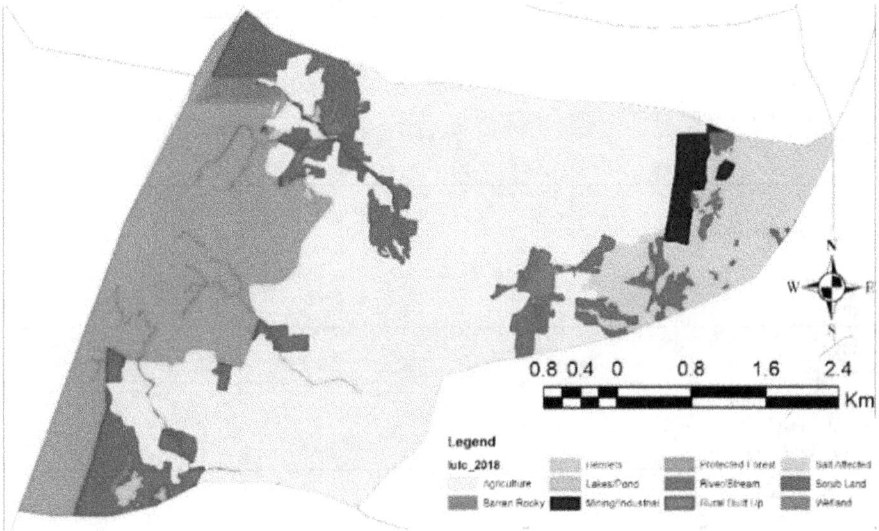

Table 7. Change in LULC

Land use 1	Land use 2	Area Sq. Km 2006	Area (%) 2006	Area Sq. Km 2012	Area (%) 2012	% Change from 2006 to 2012	Area Sq. Km 2018	Area (%) 2018	% Change from 2006 to 2018
Agriculture	**Cropland**	9.38	37.37	11.52	45.90	22.81	13.37	53.27	42.54
	Fallow	6.08	24.22	3.49	13.90	-42.60	0	0.00	-100.00
Barren	**Salt Affected**	2.61	10.40	2.55	10.16	-2.30	2.19	8.73	-16.09
	Scrub land	0.58	2.31	0.58	2.31	0.00	3	11.95	417.24
	Barren Rocky	0.58	2.31	0.58	2.31	0.00	0.25	1.00	-56.90
	Gullied/Ravenous	0.24	0.96	0.24	0.96	0.00	0	0.00	-100.00
Built-up	**Mining**	0	0.00	0	0.00	0.00	0.4	1.59	100.00
	Rural	0.19	0.76	0.19	0.76	0.00	0.25	1.00	31.58
Waterbodies	**Inland Wetland**	0	0.00	0.5	1.99	100.00	0.01	0.04	-
	River	0	0.00	0	0.00	0.00	0.19	0.76	100.00
	Waterbodies	0.09	0.36	0.09	0.36	0.00	0.04	0.16	-55.56
Forest	**Forest**	5.35	21.31	4.08	16.25	-23.74	5.4	21.51	0.93
	Deciduous forest	0	0.00	1.28	5.10	100.00	0	0.00	0

Source: Created by author,2019

Table 8. Series of impacts due to land use land cover change

Series of Impacts due to Land Use
• Green cover decreasing • Soil moisture decreasing • Water holding capacity decreasing • Shift in cropping pattern and agriculture production change • Decrease in water table (>400feet) • Increase in scrub land
• Livestock decreasing • Villagers building check dams on streams and rivers • Shift in job • Increase number of salt miners • Increasing salt pans
• Increasing salt pans • Water not reaching lake • Wetland converted to salt mining • Decrease in lake water level • Number of migratory birds decreasing

Source: Created by author,2019

Few methods for improving the green cover in and around the Aravalli range which will contribute to reviving the streams and water bodies are –

- Plantation for buffer areas of forest boundaries
    - National mission for sustainable agriculture
    - National Afforestation and Eco-Development Board
    - Mini Mission III of National Mission on Oil Seeds and Oil Palm (NMOOP)
- Plantation in physical gaps
    - Construct Green wall under National Action Programme to Combat Desertification (specifically using the local plant species)
- Water recharge structures
    - Help of NGO like Tarun Bharat Sangh (TBS): Learning from Thanagazi
    - Integrated Watershed Development
- Community strengthens and awareness
    - Help of NGO and SHG like Sambhav trust active in the region

FURTHER RESEARCH DIRECTIONS

This study is a very small step in relating change in LULC in Aravalli and its impact on the water bodies. Detailed research on each water body considering different aspects like social and environmental issues may be done so that specific issues and tentative solutions can be identified. These studies will increase the knowledge bank and contribute to giving a different perspective that can help conserve these natural resources. More and more such research should be promoted.

CONCLUSION

This study shows a considerable change in green cover from the year 1998 to 2018 in selected micro-watersheds. The same changes are happening in other areas around the Aravalli range. Aravalli hills are converting to stone deserts. Wetlands around the lake are fully converted into salt pans. Population growth around the Aravalli range has caused a change in LULC in the catchment area of the Sambhar Lake. The number of migratory birds has tremendously decreased. Like sambhar lake, other water bodies around the Aravalli range are drying. There is a high need to conserve the Aravalli range to save our water bodies and ecosystem. Immediate plans and programs are required to conserve the Aravalli and water bodies.

"Stop degrading natural systems to save ourselves"

REFERENCES

A salty sore: The dying Sambhar lake. (2019, March 25). Retrieved from India water portal: https://www.indiawaterportal.org/news/salty-sore-dying-sambhar-lake

Chakravartty, A. (2015, July 4). Choked on salt. *DownToEarth.*

Datta, D. S. (2017). *Sambhar Salt Lake- Why Conserve It?* academia.edu.

Dhabriya, S. (1988). *Eco-crisis in the Aravalli hill region.* Environmental Research Publication.

Ecology and Rural Development Society Jodhpur. (2014). *Status Survey of Important Bird Areas of Rajasthan.* Indian Bird Conservation Network.

Eghenter, C. (2000). *Mapping people's forests: The role of mapping in planning community-based management of conservation areas in Indonesia. Academic Press.*

Government of India. (1991). *Census Digital Library.* Retrieved from Primary Census Abstract: https://censusindia.gov.in/DigitalLibrary/TVDirectory.aspx

Government of India. (2001). *Census Digital Library.* Retrieved from Primary Census Abstract: https://censusindia.gov.in/DigitalLibrary/MFTableSeries.aspx

Government of India. (2011). *Census Digital Library.* Retrieved from Primary Census Abstract: https://censusindia.gov.in/DigitalLibrary/MFTableSeries.aspx

IUCN. (2000). *Communities and forest management in South Asia: A regional profile of the working group on community involvement in forest management.* The World Conservation UnioN.

Jellison, R., Williams, W. D., Timms, B., Alcocer, J., & Aladin, N. V. (2008). Salt lakes: values, threats and future. In N. V. Polunin (Ed.), *Aquatic Ecosystems* (pp. 94–110). Cambridge University Press. doi:10.1017/CBO9780511751790.010

Kale, V. S. (2014). *Landscapes and Landforms of India.* Springer. doi:10.1007/978-94-017-8029-2

Kohli, S., & Arora, S. (2018, September 20). *Degradation of the Aravalis Could Spell Ecological Disaster*. Retrieved from The Times of India: https://weather.com/en-IN/india/news/news/2018-09-20-degradation-aravalis-ecological-disaster

Kukla, G., Weiss, H., & Dalfes, H. (1997). *Third Millennium BC Climate Changes and Old World Collapse*. Springer.

Mathur. (2008). Geoscientific Studies and Managing Lakes of Arid and Semi-Arid Regions of Rajasthan. *Proceedings of Taal 2007*.

Ministry of Agriculture and Farmer Welfare, Government of India. (2018, February 4). *Soil and Land Use Survey of India*. Retrieved from https://slusi.dacnet.nic.in/

National Remote Sensing Center. (2019, February 3). *Bhuvan, Indian Geo Platform of ISRO*. Retrieved from Thematic Map: https://bhuvan.nrsc.gov.in/bhuvan_links.php

Qureshi, A. S., Qadir, M., Heydari, N., Turral, H., & Javadi, A. (2007). *A Review of Management Strategies for Salt-prone Land and Water Resources in Iran*. International Water Management Institute.

Singh, B. P. (2013). Modern salt (halite) deposits of the Sambhar Lake, Rajasthan and their formative conditions. *Current Science*, 1482–1484.

Singh, R. (2018, January 18). Sambhar: A Lake No More. *Times of India*.

Singh, R. P., Singh, N., Singh, S., & Mukherjee, S. (2016). Normalized Difference Vegetation Index (NDVI) Based Classification to Assess the Change in Land Use/Land Cover (LULC) in Lower Assam, India. *International Journal of Advanced Remote Sensing and GIS*, 1963-1970.

Space Applications Centre ISRO Ahmedabad. (2013). National wetland atlas: Wetlands of international importance under Ramsar convention. Ahmedabad: Ministry of Environment and Forests, Government of India.

Team, G. C. (n.d.). *Sambhar Lake Area Overview & Issues*. Retrieved from Gram Chetna Kendra: http://www.gck.org.in/aboutus/sambharlake.html

Tilahun, A. (2015). Application of GIS for Calculate Normalize Difference Vegetation Index (NDVI) using LANDSAT MSS, TM, ETM+ and OLI_TIRS in Kilite Awulalo, Tigray State, Ethiopia. *Journal of Environment and Earth Science*, 27–33.

USGS. (2019, March 15). *USGS Science for a Changing World*. Retrieved from Earth Explorer: https://earthexplorer.usgs.gov/

Zaitunah, A., Samsuri, A. G., & Safitri, R. A. (2018). Normalized difference vegetation index (ndvi) analysis for land cover types using landsat 8 oli in besitang watershed, Indonesia. *Earth and Environmental Science*, 1-9.

Zoological Survey of India. (2005). *Fauna of Sambhar Lake*. Zoological Survey of India.

KEY TERMS AND DEFINITIONS

Buffer Zone: A buffer zones provides barrier from human encroachment, enlarge the natural habitat, and reduce edge effects, enhance the environmental services provided by the reserves and help in conservation of species.

Deforestation: A deforestation is the decrease in the forest area and cover.

Ecology: The study of relationship between living organism and their physical environment.

Watershed Management: A watershed management is the process of guiding and organizing the use of land and other resources that are in a watershed to provide desired goods and services without adversely affecting soil and water resources.

Chapter 8
Climate Integration in Sustainable Urban Planning

Asia Lachir
National School of Architecture, Agadir, Morocco & Laboratory of Processes for Sustainable Energy and Environment (ProcEDE), Cadi Ayyad University, Marrakech, Morocco

ABSTRACT

Currently, cities are home to more than half of the world's population. The increasing urbanization rates create an unprecedented urban sprawl that worsens the urban climate situation. Urban areas modify their local climate and face the consequent urban climate impacts, which are particularly exacerbated by global climate change. This chapter shares scientific knowledge on how cities affect their climate and how urban spatial planning can mitigate the negative impacts of urban climate. Focus is given on the urban heat island, the most documented aspect of urban climate, directly linked to city spatial characteristics and functions. This phenomenon is explained, and tools and methods to assess it and mitigate its intensity are introduced in an attempt to help urban planners and designers to use climatic knowledge in urban planning to build more sustainable and climate-resilient cities.

INTRODUCTION

Urbanization and climate change are among the most defining issues of our time. Since 1950, the world has gone through a process of rapid urbanization. In 2007, for the first time in history, the urban population exceeded the rural population, and since then the number of the world's city dwellers has continued to grow; the population in cities reached 55.3% of the world's population in 2018 and it's projected to reach 68% by 2050 (United Nations, 2019). This makes cities in the front line of global challenges. Particularly, climate change is among the most threatening challenges facing urban areas. These are at the same time the major contributors to climate change as they account for a significant portion of total greenhouse gas emissions and home to the majority of the world's population that will be the victims of climate change (United Nations, 2020). Thus, cities are widely acknowledged as key actors in both mitigation and adaptation. The Second Urban Climate Change Research Network (UCCRN) Assessment Report

DOI: 10.4018/978-1-7998-8331-9.ch008

on Climate Change and Cities suggests that cities need to transform to achieve low emissions development and climate resilience to fulfill their climate change leadership potential (Rosenzweig et al., 2018).

Most cities are exposed to climate-related hazards such as flooding, rising sea levels, heat waves, and drought. It is estimated that the economic impacts of climate change on cities will range between 2.3% and 5.6% of their total gross domestic product by the end of the century (Long & Rice, 2018). Indeed, climate change impacts are often exacerbated in urban areas due to interactions with urban infrastructure, culture, economic activities and the growing population. For instance, impervious surfaces in cities increase peak flows associated with heavy rainfall events, which can lead to flash flooding when storm water exceeds the capacity of the drainage system (Miller & Hutchins, 2017). Also, urban forms and materials modify local climate in city centers. These develop specific microclimates and are often several degrees warmer than surrounding areas. This is known as the Urban Heat Island (UHI). It is the most documented effect of urban climate modification as it can yield a wide range of impacts on health and energy use (Lachir et al., 2016; Oke, 2011).

Most efficient measures to mitigate and to adapt to climate change in urban areas are often related to urban planning and design (Rosenzweig et al., 2018). The way we arrange our cities has a critical role to play in reducing greenhouse gas (GHG) emissions, the raised temperatures and the vulnerability to climate-related hazards. Urban planning has long been considered as an essential instrument to achieve sustainability in cities. According to UN Habitat (2010), "Sustainable urban development can be defined as the spatial transformation of the urban environment and the quality of life of its inhabitants - through participatory, equitable, accountable and effective planning, management and governance processes and systems – making neighborhoods, towns and cities more environmentally sound, economically productive and socially inclusive for their present and future generations". Sustainable urbanism recognize climate as a main component. It integrates disaster risk reduction and climate change adaptation and mitigation considerations and measures into urban and territorial development and planning processes (United Nations, 2017).

Recently, a new urban development paradigm has evolved as climate urbanism, a shift in urban planning that focuses on carbon as a central object of policy concern instead of the three pillars of sustainable development and the development of climate-friendly and climate resilient cities (Long & Rice, 2018). As such, climate integration in urban planning is a necessity to build our future cities better. In this regard, this chapter aims to extend the knowledge of urban climate, its characteristics and consequences. It particularly focuses on urban heat island and how it is affected by spatial planning in urban areas. An extensive literature review was achieved to provide general guidelines and support for urban designers and the professional planning community, to better understand and improve the urban climate situation by reducing heat stress, to improve urban life quality and create energy efficient cities.

ENERGETIC EFFECTS OF URBAN LANDSCAPES ON LOCAL CLIMATE AND ITS CONSEQUENCES

Urbanization is a major and a long-lasting form of land cover transformation. It affects land surface water and energy balance and alters local climate through several physical processes. According to Arnfield, (2003) ; Bounoua et al. (2009); Lachir et al. (2016); Masson (2006); Masson (2000) and Oke (1982), these processes can be summarized as follows:

- Natural land cover is replaced by impervious surfaces. This decreases photosynthesis and consequently evapo-transpiration and limits soil infiltration and water storage. Therefore, urban areas respond more quickly to precipitation, and water outflows by runoff are significant. This also affects the surface energy balance as sensible heat fluxes increase while latent heat fluxes decrease.
- Urban forms (e.g., building morphology and streets geometry) modify surface roughness and disrupt air flows. The increase in surface roughness in dense urban areas can cause a reduction in wind speed near the land surface. However, under certain conditions the wind speed may be locally higher in town than on the outskirts. This can be observed when the streets are parallel to wind direction, which will therefore channel it and accelerate it by the Venturi effect.
- Construction materials are characterized by a lower albedo. i.e., the ratio of reflected solar energy to incident solar energy. They create a radiative trap that absorbs more energy during the day and releases it at night. In addition, high vertical building fronts increase the surfaces exposed to solar energy and amplify the radiation exchanges.
- Anthropogenic heat is released with proportions closely related to the size and density of the city, industrialization, as well as its climate and air conditioning needs. It results mainly from building's heating and air conditioning, automobile traffic and industries.

Urban heat island is the most important and the most documented effect of urban climate modification. It was first studied by Luke Howard who provided evidence that cities are often warmer than their surrounding rural areas (Howard, 1833). Since then UHI have been extensively documented (Myrup, 1969). Figure 1 (a) illustrates UHI in a typical American city (US EPA, 2014). It shows the increased temperature difference in cities and map the temperature variation across the city, from the rural to the urban environment and back to the rural environment. The spatial field of air temperature isotherms generally mimics the outline of urban structures and forms and marks the urban-rural periphery (Oke, 1982). Figure 1 (b) shows both air and surface temperature profiles across the city. Air temperature is generally measured at two meters above the ground and reveals the urban heat island of the urban canopy layer ($UHI_{UCL)}$ and surface temperature variation describes surface urban heat island ($UHI_{surface}$). Surface temperature is highly sensitive to land surface characteristics. Every surface has a distinct thermal response and a unique energy balance. Thus, it results in a myriad of surface temperatures in a city, especially during the daytime when the land surface is exposed to solar energy. UHI_{UCL} is more pronounced during night time when urban surfaces restore the heat that was absorbed during the daytime (Oke, 2011).

Urban heat island is significantly affected by regional climate. Seasonal climate change and even the time of day govern the intensity of the UHI. A study of urban climate in fifty cities located in the USA showed that UHI is heavily affected by local atmospheric condition. Cities exhibited the highest UHI effect when calm and stable conditions prevailed (Mohajerani et al., 2017). The urban heat island (UHI) is particularly enhanced by changes in climate. McCarthy et al. (2010) included an urban land surface model in the HadAM3 Global Climate Model. They showed that regions with high population growth coincide with regions of high urban heat island potential, and climate change may amplify urban heat islands by up to 30% in some locations.

Cities will be exposed to rising temperature due to both climate change and urban heat island effect. According to the Second Assessment Report of the Urban Climate Change Research Network, temperatures in 100 cities around the world are projected to increase by 0.7 to 1.5°C by the 2020s, 1.3 to 3°C by the 2050s and 1.7 to 4.9°C by the 2080s (Rosenzweig et al., 2018). UHI effect will exacerbate this warming in most cities centers as the maximum intensity of UHIucl may reach more than10°C (Kłysik

& Fortuniak, 1999; Mohan et al., 2013; Oke, 1981). During cold winter season, the urban heat has a positive effect on city dwellers as it can reduce building energy demand (Davies et al., 2008; Hirano & Fujita, 2012). However, more negative effects of UHI have been documented. Heat stress is a direct consequence of the UHI that might be devastating during hot summer time. Children, the elderly, and people with existing illnesses may be particularly vulnerable to this stress. Different studies documented a rise in mortality and morbidity due to heat-related stress and linked it to UHI effect in compact cities (Heaviside et al., 2017; Tan et al., 2010; Tomlinson et al., 2011).

Figure 1. Urban heat island illustration: a) Urban heat island isotherms, b) Spatial and temporal features of surface and urban canopy layer urban heat island (solid line represents $UHI_{surface}$ and dashed line represents UHI_{UCL})
Source: United States Environmental Protection Agency

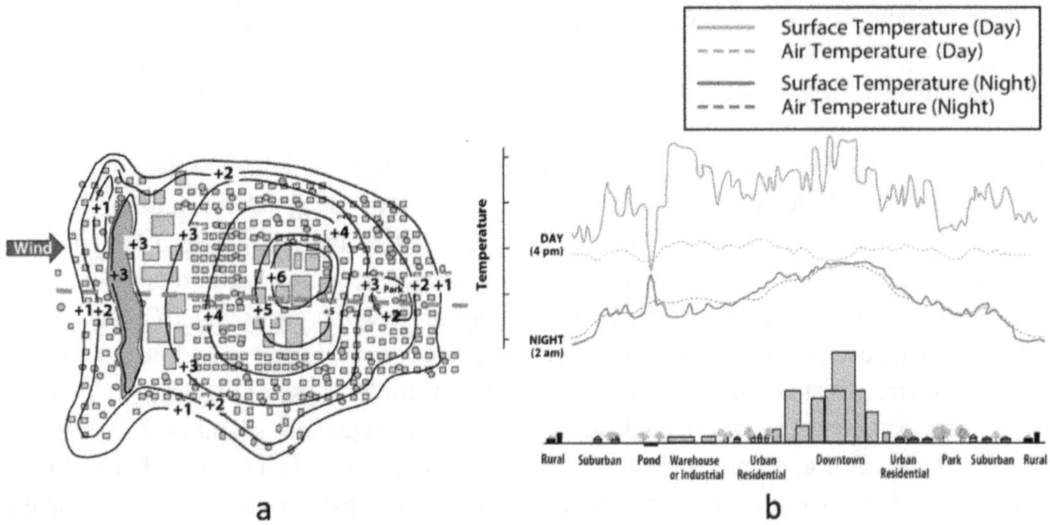

Heat waves also cause an increase in electricity usage as the population uses air conditioning to provide thermal comfort. Akbari et al.(2001) showed that a typical UHIucl may range between $0.5^{\circ}C$ and $3^{\circ}C$ in a warm summer daytime, which may cause an additional 5-10% municipal peak electricity demand. Lachir et al. (2016) showed that in Marrakech, a semiarid city in North Africa, an increase of 1oC in the maximum mean air temperature can cause an increase in energy use of 4.4% that can be explained by the rise in air conditioning needs. A recent review on UHI impacts on building energy consumption stated that UHI could result in a median increase of 19.0% in energy demand for cooling with stronger impacts in urban centers than that in urban periphery (Li et al., 2019).

This increase in energy consumption implies that more fossil fuel is burned, causing higher emissions of green house gases. At the same time, waste heat release from air condition amplifies UHI intensity. Liu et al. (2011) showed that air conditioning of a typical office building cluster contribute to UHI intensity by 0.7°C at mid-day. Hsieh et al., (2007) studied Heat rejection from residential buildings in Taipei and its impact on the outdoor air temperature. He found AC units have raised the outside temperature between $0.5^{0}C$ and 2°C during evenings (7 p.m. to 2 a.m.) with feedback to cooling loads of 10.7% raise. As such, the use of air conditioning to mitigate urban heat island triggers a vicious cycle that results in

UHI amplification, a higher use of air conditioning and energy consumption. This is not considered as a sustainable solution to reduce heat exposure. Therefore other methods and approaches are needed to make future urban areas more sustainable and climate resilient (Lundgren & Kjellstrom, 2013).

URBAN CLIMATE ASSESSMENT: TOOLS AND METHODS

Cities are characterized by their high spatial heterogeneity. All spaces in cities are a unique combination of natural and artificial patches that respond differently to climate forcing. As such, the resulting urban climate and particularly urban heat island are specific to each city and take different shapes within the same city. Thus, local case studies of UHI are required to provide consistent information to integrate urban climate in decision-making for urban development. Different methods have been developed to assess UHI, taking into consideration urban spatial heteroginity (Deilami et al. 2018; Mirzaei & Haghighat, 2010) .These methods can be classified into three categories:

1. Ground-Based Observations

This is the oldest method for monitoring the urban climate; its main benefits are the high accuracy of measurements and the continuity of records. But it presents a limited spatial coverage; generally cities have a single urban weather station which is clearly insufficient to resolve the complex urban environment. However advances in technology and communications improved access to meteorological information, and a new generation of high-resolution Urban Meteorological Networks has emerged (UMNs) (Muller et al., 2013).These have been implemented in several cities around the world and provide useful information for assessing the urban heat island. Azevedo et al. (2016) illustrate the use of this technique to quantify and identify the spatial patterns of the daytime and night-time urban heat island in Birmingham, UK. The study used air temperature measurement from the Birmingham Urban Climate Laboratory urban meteorological network. This near-real time, high-resolution urban meteorological network includes 25 automatic weather stations and 82 low-cost non-standard Wi-Fi air temperature sensors. The study also used remote sensing data to assess surface temperature and analyze surface urban heat island $UHI_{surface}$, meanwhile the UMN data provides the urban heat island in the urban canopy layer UHI_{UCL}. The results are given in figure 2 and show that $UHI_{surface}$ is more important during the daytime (up to 9°C) and is largely more intense than the UHI_{UCL}. The latter is more pronounced during night time with an increase of air temperature in the urban center of up to 1.5°C.

2. Remote Sensing Approach

Thermal Infrared (TIR) imagery represents an alternative to instrumental measurements. Satellite-based TIR sensors provide spatial coverage at low cost. However, they are associated with a lower accuracy than the measurements, and data availability can be limited when the sky is covered by clouds. Satellite TIR sensors measure Top Of Atmosphere (TOA) radiances in the TIR window of the electromagnetic spectrum (wavelength between 8 and 15 μm). Land Surface Temperature (LST) is derived from these data using algorithms based on the inverse of Planck's law, atmospheric corrections and land surface emissivity values (Al-Hamdan et al., 2016; Sánchez-Aparicio et al., 2020). TIR images often include quality control datasets that can help filter good quality data. And thus provide accurate LST.

Figure 2. Urban heat island assessment using urban meteorological networks (for UHI_{UCL}) and MODIS (for $UHI_{surface}$) in Birmingham, UK during June, July, August 2013: a) Land use classes in Birmingham, b) locations of the network stations and sensors, c) Average night-time UHI_{UCL} intensity (18:00–05:59), d) Average daytime UHI_{UCL} intensity (06:00–17:59), e) Average daytime $UHI_{surface}$ intensity (at ~13:30), f) Average night-time $UHI_{surface}$ intensity (at ~01:30)
Source: modified from Azevedo et al., 2016

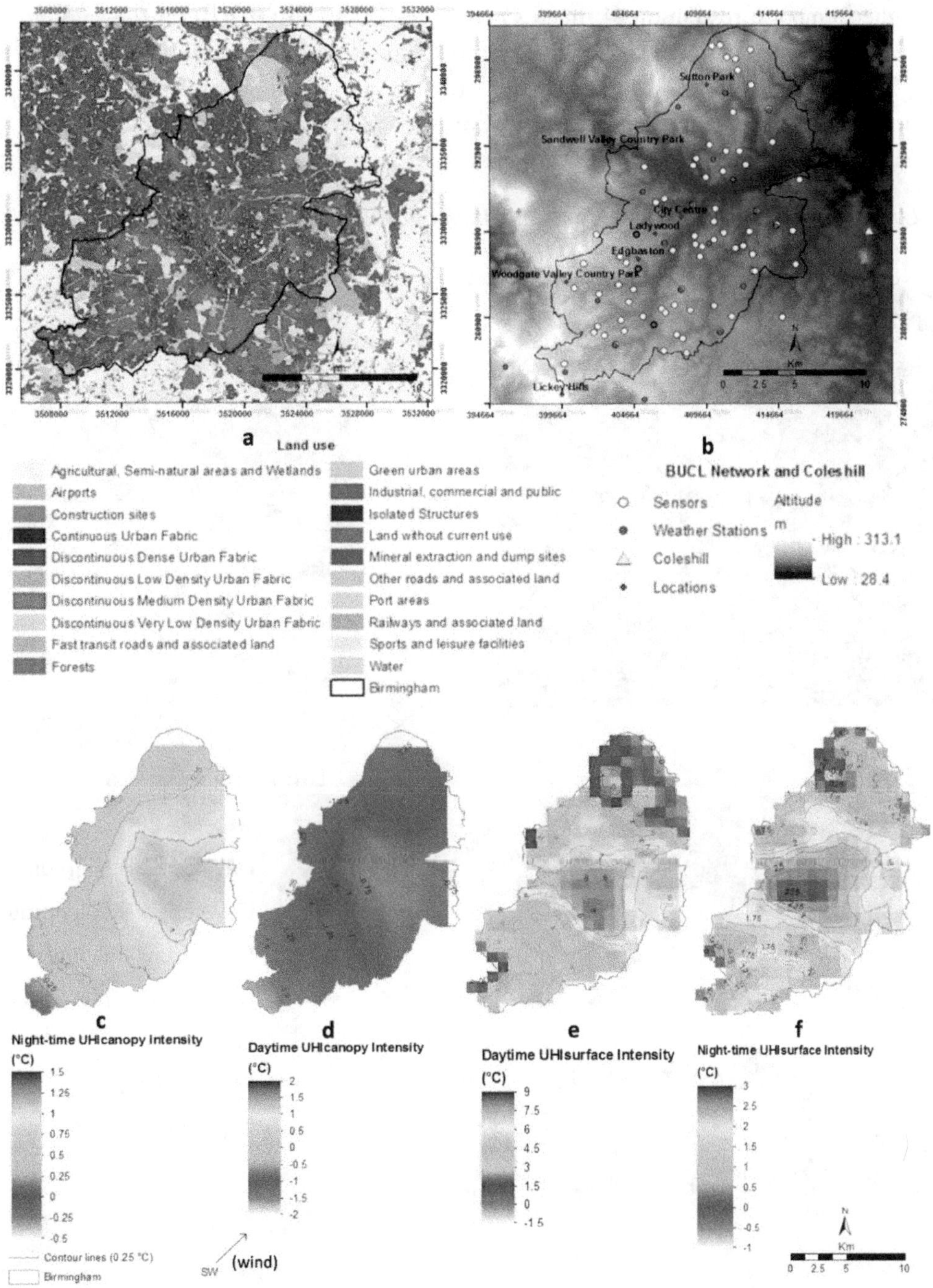

The surface urban heat island (UHI$_{surface}$) was first studied using these techniques in the 1970s (Matson et al., 1978). Since then advances and technical developments in remote sensing enhanced the availability of TIR data for scientific research. Landsat thermal sensors and the MODerate resolution Imaging Spectroradiometer (MODIS) on Terra and Aqua are the most widely used remote sensing instruments for the assessment of UHI$_{surface}$(Zhou et al., 2019). Landsat provides global TIR data at a high spatial resolution (30 to 120 m) and at temporal resolution of 16 days for more than 30 years. MODIS provide data at a higher temporal resolution (4 images per day at 13h30 and 01h30 for the satellite Aqua and 10h30 and 22h30 for Terra) but at a spatial resolution of 1km. Figure 2 (e and f) illustrates the assessment of UHI$_{surface}$ in Birmingham using Modis LST. Figure 3 shows an example of land surface temperature derived from Landsat 5 Tematic Mapper over a 5km2 area in Washington DC.

Figure 3. UHIsurface assessment using landsat 5 thematic mapper over a 5Km² area in Washington DC
Source: modified from Al-Hamdan et al., 2016

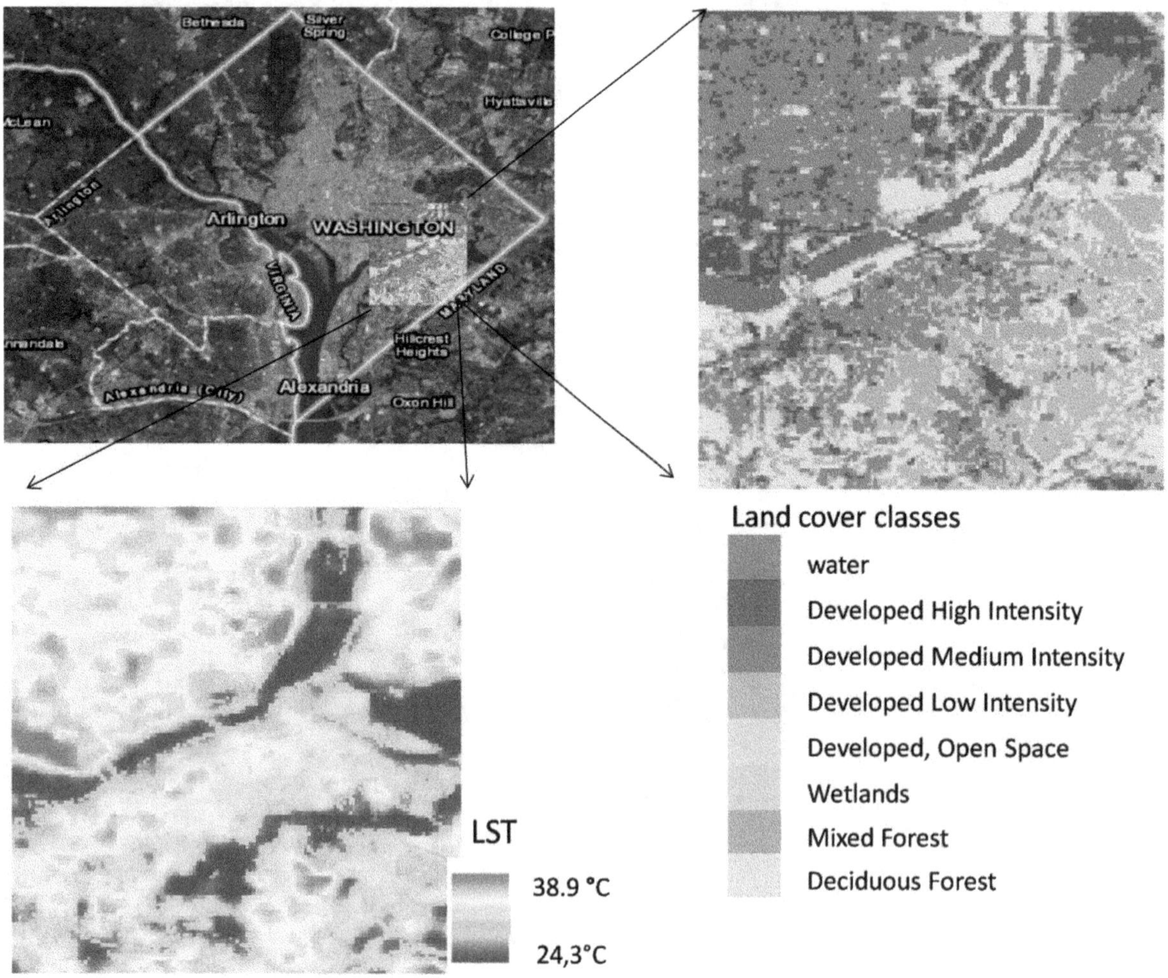

Currently, different satellites with multiple TIR sensors provide global high resolution datasets, at different spatial and temporal resolutions. Most of these satellite datasets are freely available and can be downloaded at the U.S. Geological Survey (USGS) Global Visualization Viewer (http://glovis.usgs.gov), or the *USGS* Earth Explorer data portal (https://earthexplorer.usgs.gov/or). Table 1, shows information on currently operational TIR sensors, providing publicly available global data that can be used to assess UHI.

Table 1. Characteristics of currently operational sensor providing global data to drive LST

Sensor	Satellite	Spatial resolution	Temporal resolution	Starting year	More information
Advanced Very High Resolution Radiometer (AVHRR)	NOAA's Polar Orbiting Environmental Satellites (POES)	1.1 km	12 hours	1978	(Kidwell, 1995)
Enhanced Thematic Mapper Plus (ETM+)	Landsat 7	60 m	16 days	1999*	(USGS, 2019a)
MODerate resolution Imaging Spectroradiometer (MODIS)	Aqua and Terra	1Km	12 hours (for each satellite)	2002 (aqua), 2000 (terra)	(Wan, 2013)
Advanced Spaceborne Thermal Emission and Reflection Radiometer (ASTER)	Terra	90 m	12 hours	2000	(JPL, 2001)
The Spinning Enhanced Visible and InfraRed Imager (SEVIRI)	MSG (Météosat Second Generation)	3 km	15 minutes	2002	(Aminou, 2002)
Thermal Infrared Sensor (TIRS).	Landsat 8	100 m	16 days	2013	(USGS, 2019b)
Sea and Land Surface Temperature Radiometer (SLSTR)	Sentinel 3	1 Km	1 day	2016	(Polehampton et al., 2021)
* Since 2003, ETM+ images have 22% data gaps because of a failure of its **Scan Line Corrector**					

Thermal infrared imagery can also be acquired using airborne platforms. Sensors placed in helicopters. And unmanned Aerial Vehicles (UAVs) can provide a marvelous view of land surface temperature in urban area at a very high resolution. These techniques, however, are generally only available for small areas and at significant cost to the end user.

3. Modeling Approaches

UHI was studied using first energy budget models since the 1970s. These models provide physical explanation of UHI and help to understand its driving parameters such as urban forms, construction materials, regional climate and anthropogenic activities. The major benefit of a computer model simulation of UHI is the possibility to perform comparative analyses based on different scenarios, to forecast future UHI and to evaluate its mitigation strategies. Also a wide range of spatial and temporal resolution can be achieved depending on the model inputs and the available computing capacity. Figure 4 shows an example of urban heat island simulation in Marrakech, a semi-arid city in North Africa. Lachir et al. (2016) combined Landsat and MODIS data in the Simple Biosphere model (SiB2) to simulate surface temperature (figure 4-b) and assess the urban heat island (figure 4-c). Simulations were conducted for the land cover in 2010 (control), a pre-urban land cover scenario (PU) and for future urban development at different rates (25%, 50%, 75% and 100% of the study area). Future projections were done according to a business as usual scenarios (BAU) and a smart growth scenario (SMG) that allows the urban expansion to occur first in bare lands, with all remaining bare lands converted to evergreen orchards (figure 4-d).

Figure 4. Assessment of urban growth impact on land surface temperature in Marrakech during the vegetation growing season (February-Mars-Avril) using the Simple biosphere model
Source: Lachir et al., 2016

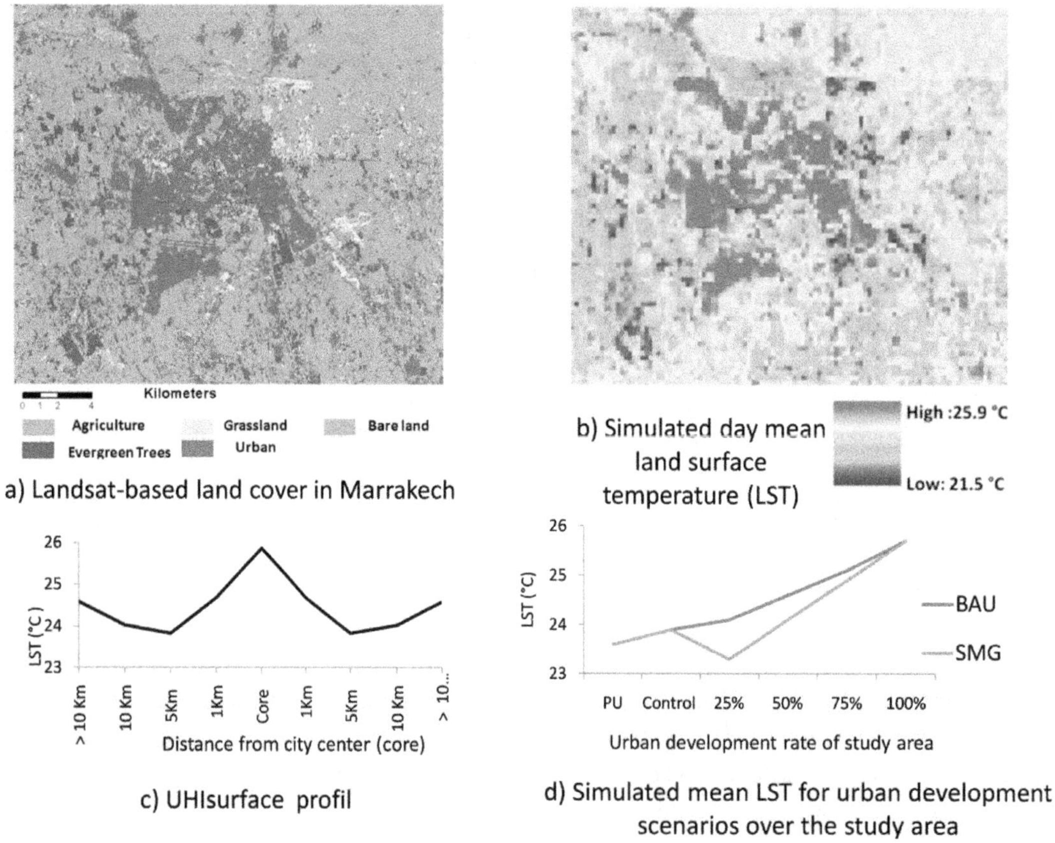

Two different modeling approaches have evolved:

Energy Balance Models (EBMs):

These models have been broadly developed after the work of (Oke, 1982). They investigate the energy conservation of an urban canopy layer using remote sensing and observation data to characterize the land surface and validate the model outputs. This method main disadvantage is the poor presentation of airflow around the buildings. The airflow model in EBMs is greatly simplified or even decoupled from the energy budget equations. The Simple Biosphere model SiB2 is a land surface model that includes an Energy Balance Model, developed by (Sellers et al., 1986) for use within General Circulation Models. It was then modified by Bounoua et al. (2009) to simulate urban climate in cities in North Africa and the

USA (Bounoua et al., 2015; Lachir et al., 2016; Zhang et al., 2015). The Town Energy Balance (TEB) is Another popular Energy Balance Model for urban climate simulations, developed by (Valéry Masson, 2000). The TEB model is open-source and available on the website: http://redmine.cnrm-game-meteo.fr/projects/teb

Computational Fluid Dynamics (CFD):

These techniques, dealing with the governing equations of the flow (Navier–Stokes equations), resolve urban land-atmosphere interactions within detailed urban geometry where detailed analysis of airflow and temperature fields can be achieved at a finer scale. As such, a high resolution urban 3D model and high computing capacity is required to run CFD models. This limits considerably the simulation domain size to few hundred meters compared EBMs that can simulate the energy balance at different scales (neighborhood, city, country, global). CFD are extensively used since the 1990s in different urban climate studies, particularly to assess wind flow, thermal comfort and pollutant dispersion around buildings (Toparlar et al., 2017). ENVI-met (Bruse, 2004) is among popular software based on a microclimate CFD model that can simulate interactions between urban surfaces, vegetation and the atmosphere. Figure 5 illustrates the use of ENVI-met to simulate streetscape temperatures and to assess the mitigation role of urban trees.

Figure 5. ENVI-met: Potential temperature simulations at 2 m height at 04:00 pm in Bolzano, Italy. Scenario 1 accounts for vegetated streetscapes and scenario 2 assumes that vegetation is removed from the streetscapes
Source: modified from Russo & Zerbe, 2016

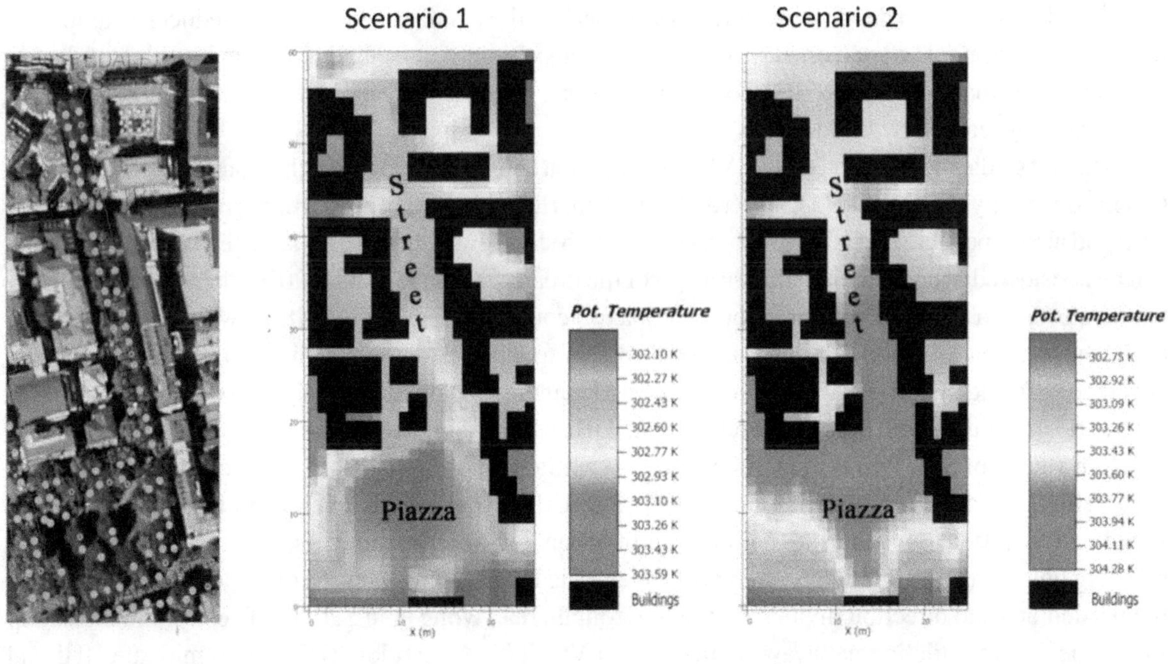

INTEGRATION OF URBAN HEAT ISLAND MITIGATION MEASURES IN URBAN SPATIAL PLANNING

Urban heat island modeling has contributed significantly to develop and assess efficient UHI mitigation measures. These are directly related to urban planning and design as they tackle urban forms and functions. Hart & Sailor (2009) summarized the factors that might be considered during the urban planning process to improve the urban microclimate and reduce the UHI intensity. These include city canyon geometry, building density, building height to width ratio, sky view factor, roads and traffic density, building and surface materials, and the use of water and green space. These factors have been extensively studied, and literature demonstrated the efficiency of a wide range of UHI mitigation strategies. However there is no consensus on a unique and perfect combination of these strategies that can be generalized to all cities because of the irregularity of the UHI effect and its high dependence on each city's characteristics. As such UHI is considered as an issue best tackled with localized research and mitigation efforts (Mohajerani et al., 2017).

In this section we will illustrate popular mitigation measures found in literature categorized into three groups.

1. The Form and Layout of Buildings and Streets

Urban morphology affects considerably radiative and heat exchange. Sky-View Factor (SVF), defined as the fraction of the sky able to be seen from within canyons (Johnson & Watson, 1984), is a key factor that sets both solar access during the day and heat loss by infrared radiation during the night. A good surrogate to sky view factor is the street height to width ratio (aspect ratio) (H/W). Although, complex urban environment is better described using SVF. This is illustrated in figure 6 as it shows examples of SVF calculation. In a deep street canyon (large H/W), the small sky view factor reduces heat loss by radiation (in comparison to an open space with a larger sky view factor) and results in less night radiative cooling. Also small SVF induce less convective cooling as wind velocity is reduced due to the obstruction to air flow caused by buildings.

Different studies demonstrated that SVF is a significant determining factor of the air temperature distribution inside the city (Unger, 2009). Chen et al. (2012) reported an inversely proportional relationship between SVF and air temperature elevation in urban sites. A high sky view decrease heat storage in urban sitting and can be considered as a good mitigation strategy of the urban heat island. Yuan (2010) demonstrated that air temperature, on a clear and still-air night (at 3:00a.m.) could decrease about 0.48 °C with a 0.1 increase in SVF in urban areas of Hong Kong. Also, he linked SVF to more common planning parameters such as site coverage ratio and building height to provide Hong Kong urban planners a straightforward and applicable geometric guideline to increase SVF and mitigate UHI while maintaining land use.

High values of SVF will also facilitate Urban ventilation that remains a key mechanism for cooling urban climate and UHI mitigation (Gu et al., 2020). Urban ventilation depends on local climate (wind speed and direction) as well as a combination of urban geometry parameters, such as the aspect ratio, building height and the building frontal area index. The latter is the total area of building facades facing the particular wind direction divided by the horizontal area (Wong et al., 2010). These parameters help locate the main ventilation pathways, called Urban Ventilation Corridors (UVCs), to mitigate UHI and improve air quality by accelerating the diffusion of air pollution. Many cities around the world are developing plans for urban ventilation corridors to avoid the deterioration of the urban wind environment caused by disorganized urban densification and sprawl.

Figure 6. Sky view factor (SVF) calculation for three intersections in Madrid, Spain
Source: Aram et al., 2020

China's government paid significant attention to the development of UVCs that becomes the focus of planning climate resilient cities. A national technical guide for urban ventilation corridors is developed to serve as an aid for decision-making in the initial stage of town planning and urban design (Ren et al., 2018). Figure 7 shows the establishment of ventilation corridors in Changchun City in China. In this study S. Chang et al. (2018) used computational fluid dynamics (CFD) to assess wind speed and direction within the city and a geographic information system (GIS) to combine wind environment with UHI and air pollution data to ensure that ventilation corridors are placed within the areas of the city that need them most. The proposed UVCs benefit from the continuity of green spaces and water bodies in Changchun and allow fresh air to reach the city center to reduce the UHI effect and to accelerate pollution diffusion.

2. Urban Green Spaces

Green spaces are the most documented UHI mitigation measure; several studies demonstrate that increasing the vegetative cover in urban areas can decrease outdoor temperatures, enhance thermal comfort and reduce energy use along with other benefits such as runoff and pollution reduction and carbon sequestration. This cooling effect is known in literature as the Park Cool Island (PCI) or more recently as the Green Space Cool Island (GCI). This is described by the cooling effect distance (CED), which specify the area where air temperature is influenced by the green space, and the cooling effect intensity (CEI) defined as the temperature difference between the green space interior and its surrounding built-up areas, beyond the range of the green space influence (Chang et al., 2007). Aram et al.(2019) conducted an extensive review of studies on urban green spaces cooling effects. Their results show that large parks (areas of more than 10 ha) have the most significant cooling effect with a CEI of 1–2 °C and a CED over 350 m from the park boundary. The case studies considered in (Aram et al., 2019) and that are providing information about green space sizes and forms are summarized in figure 8.

Figure 7. Construction of urban ventilation corridors in Changchun, China
Source: Chang et al., 2018

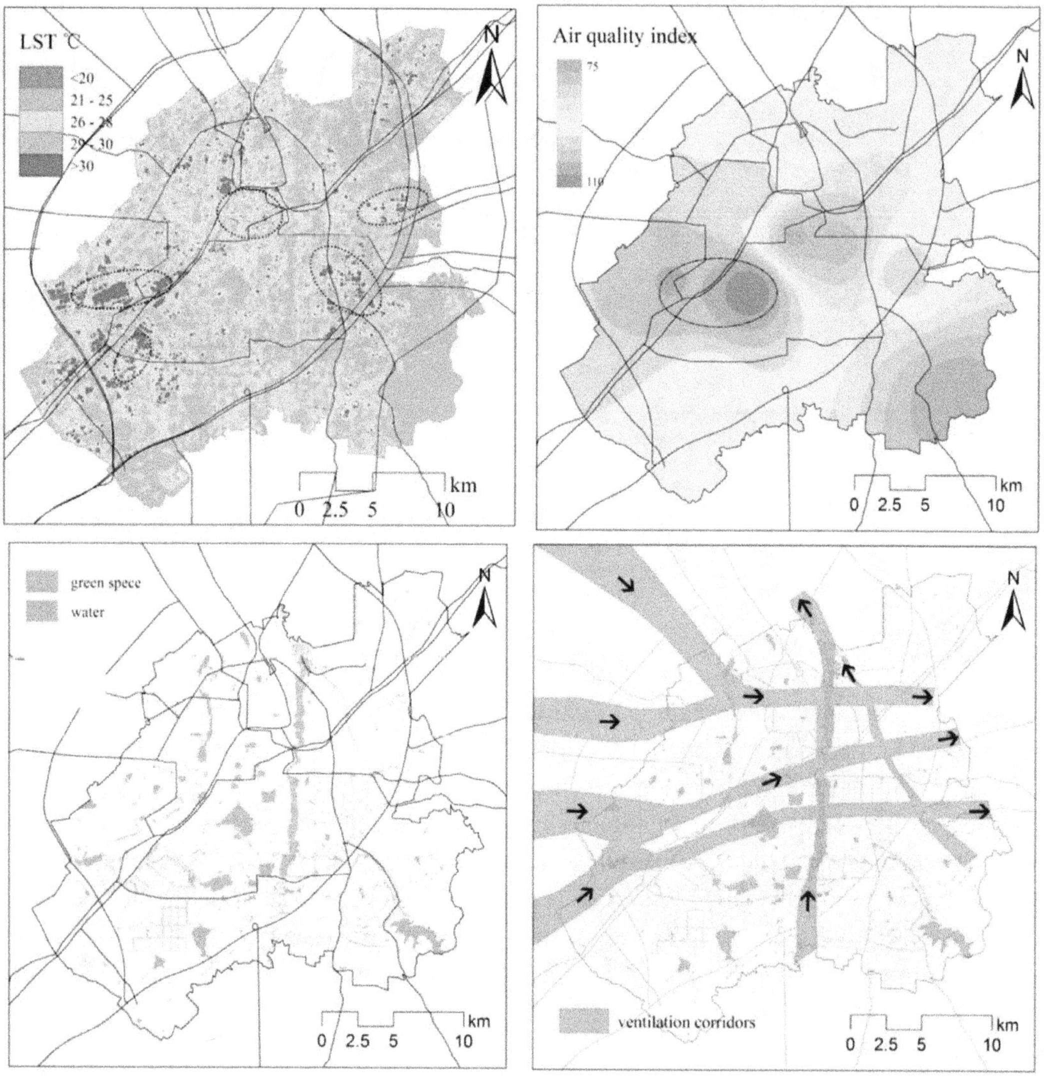

The cooling effect of urban green space is found to be positively correlated with its leaf area index (LAI) (Xiao et al., 2018). This is defined as the total leaf area per unit ground surface area. It characterizes the vegetation canopy and it quantifies its transpiration and degree of shading. Trees with a dense canopy will provide better cooling effects, followed by shrubs then lawns. As such a stronger emphasis should be placed on the selection of plant species during the process of urban green space design. The integration of water bodies should also be considered as it can enhance the PCI (Gunawardena et al., 2017). Larger water bodies, with simple geometry and a dense surrounding tree cover can offer better cooling effects (Du et al., 2016).

Building green surfaces such as green walls or green roofs are proved to enhance thermal comfort and reduce energy use of individual buildings (Costanzo et al., 2016). However their contribution to urban heat island mitigation at a city or a neighborhood scale is less significant. Their cooling effect at the pedestrian-level air temperature was estimated in a residential neighborhood of subtropical Hong Kong to 0.4 -1.7 °C but it was limited to immediate surrounding buildings (Peng & Jim, 2013). Ng et al. (2012) Also demonstrated that green roofs in Hong Kong are ineffective for thermal comfort near the ground and that planting sidewalk trees results in a better cooling effect. Nevertheless building green surfaces may impact indirectly UHI at a larger scale as buildings will require less air-conditioning to achieve indoor thermal comfort and will release less waste heat.

Figure 8. Urban green space cooling effect
Source: Aram et al., 2019

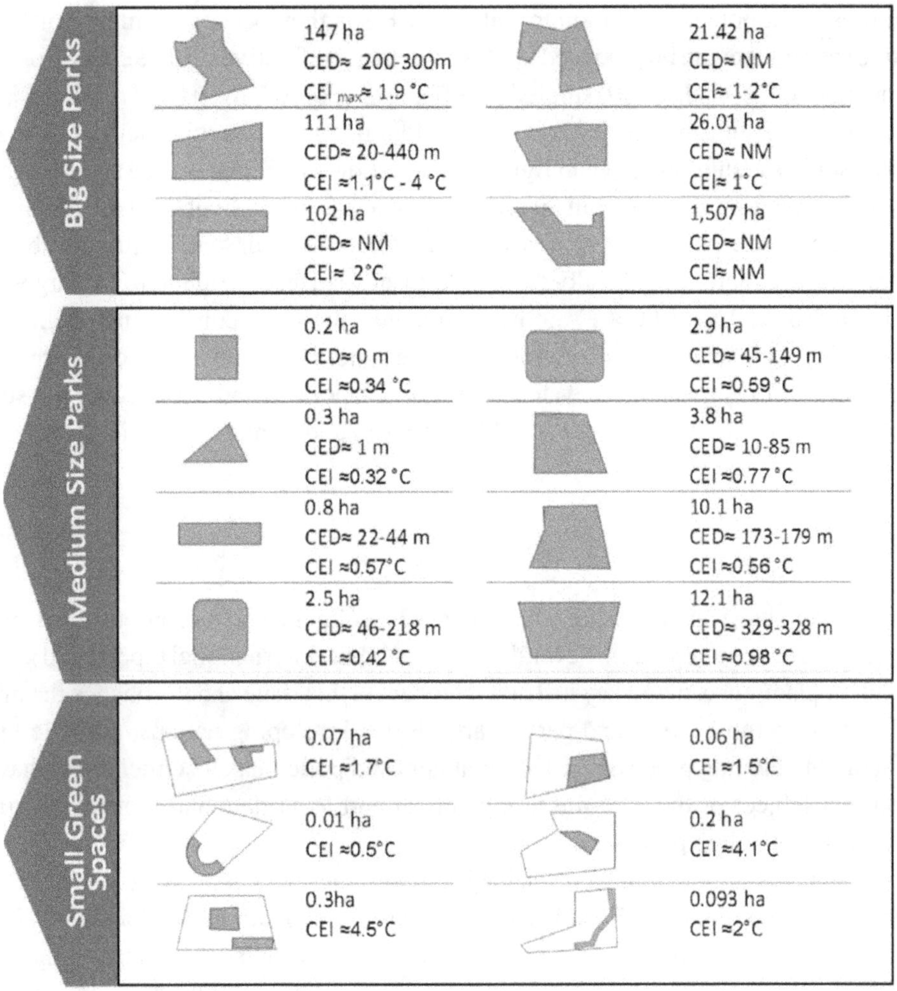

3. Urban Construction Materials

Cities are mostly covered by impervious, heat-absorbent and dark-colored surfaces. e.g. Akbari & Rose (2008) showed that four metropolitan areas in the USA are covered by 19–25% roofs and 29–39% paved surfaces. These surfaces alter significantly the energy budget and increase the UHI intensity. The use of alternative construction materials is widely studied as an UHI mitigation measure. Cool roofs and cool pavements are the most documented solutions. These are characterized by a higher albedo which helps reduce the solar energy absorption, decreases the surface temperature and the amount of sensible heat released to the atmosphere. However this solution may result in excessive glare and important maintenance requirements of cool paints. Cool roofs were shown to be more effective than green roofs to mitigate UHI. A study compared green roofs and cool roofs (with albedo higher than 0.65) in different climatic conditions in Italy. It shows that surface temperature may decrease up to 25 °C with the cool roof and up to 20 °C with the green roof. It also proved strong reductions in the released sensible heat fluxes: 40% to 75% in case of green roofs and more than 75% for cool roofs (Costanzo et al., 2016). Cool roofs also reduce cooling load and the waste heat released by air conditioning.

Pavement presents a potential solution to mitigate UHI as they occupy an important area of cities. Cool (reflective) pavements can be produced at a low cost using reflective surface coating or light-colored aggregates and binder. Sen et al. (2019) studied the effects of cool pavements and roofs on UHI in an area covered at 35% by impervious surfaces (roads and buildings) in the Phoenix Metropolitan Area in Arizona, USA. The study's results are given in figure 9. They show that replacing the aged asphalt pavement with new one led to maximum increase in air temperature at a 2 m height of 0.50 ∘C. Meanwhile the use of more reflective concrete pavements decreased the neighborhood air temperature by about 0.20–0.40 ∘C. an extreme scenario increasing the albedo of all artificial surfaces resulted in the largest decrease in air temperature, by about 0.80–1.00 ∘C. Another pavement solution to enhance urban climate is permeable pavements. These are an efficient solution to reduce runoff and control floods in urban areas. Also they can mitigate urban heat when rain water in the soil evaporates and cools down the surface of the pavements. However this solution is more suitable for warm and humid climates (Santamouris, 2013).

CONCLUSION

Cities around the word have an increasing concern on urban climate and its consequences on life quality and public health and economic wealth. Several research studies, international reports and global agendas confirm that urban planning can be a key instrument to tackle this issue and to enhance the urban climate condition. However, in most cities, and particularly in the developing nations, climatic knowledge is still lacking in urban planning processes. Urban planners and policymakers either do not have sufficient knowledge on this subject or they are missing practical and local data and tools to integrate climate information in urban planning processes.

This chapter attempts to disseminate the scientific knowledge on Urban Heat Island, the most documented urban climate modification. It explains how urban forms, materials and activities influence this phenomenon and it presents examples of urban spatial planning measures to mitigate its impacts. However, provided information are not enough to fully understand how cities can fight the global issue of climate change. Further urban climate aspects need to be considered into urban planning such as floods, drought, climate impacts on urban ecosystems and carbon emission and sequestration. Also, to include climate into the already complicated urban planning processes, urban planners and designers need to be provided with simple applicable rules regarding what types of urban designs best mitigate undesirable local microclimatic conditions. Therefore, it is a necessity today to develop more exhaustive and local technical guides at city level, to study in detail each city's specific urban climate and to transform the complex scientific concepts into straightforward regulations and guidelines for the urban spatial planning and design.

Figure 9. Air temperature differences at 2 m height from the existing pavements scenario, compared to different pavement and roof scenarios in Phoenix, Arizona
Source: Sen et al. 2019

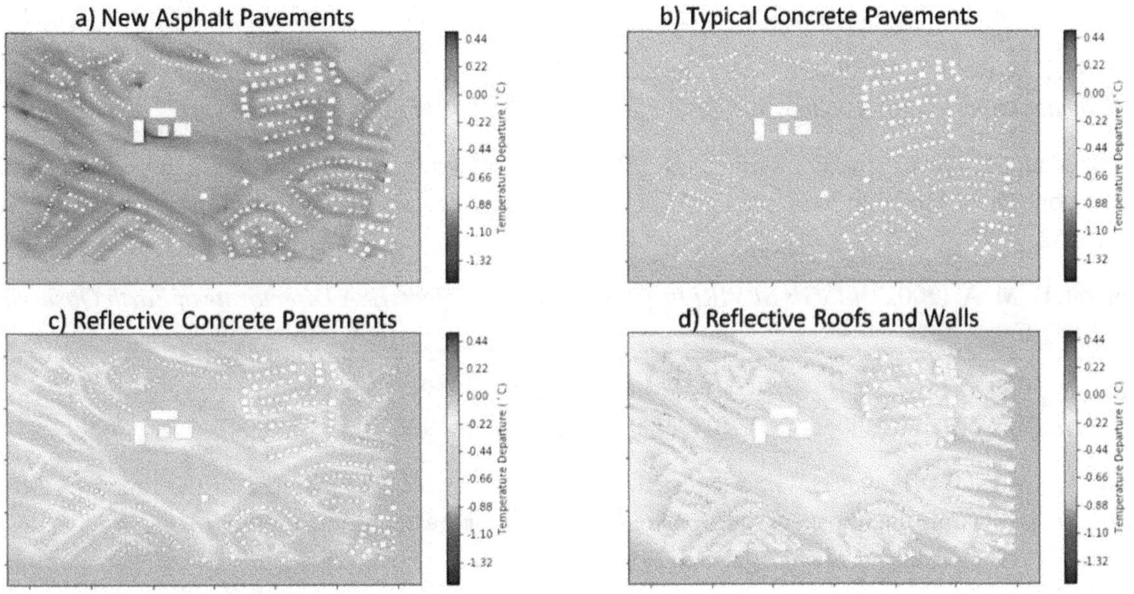

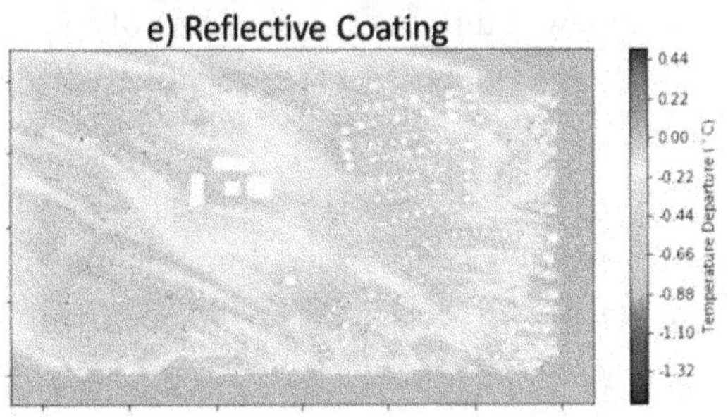

Scenario		Albedo
Existing Pavements		0.22
New Asphalt Pavements	(Pavements modified)	0.05
Typical Concrete Pavements	(Pavements modified)	0.35
Reflective Concrete Pavements	(Pavements modified)	0.50
Reflective Roofs and Walls	(Roofs and walls modified)	0.60
Reflective Coating	(Pavements, roofs, and walls modified)	0.60

Urban climate maps (UCMaps) are a good example to achieve this objective. This concept was first developed in Germany in the 1970s. It analyzes climate conditions in urban areas and provides spatial tools to help planners, developers, and policy makers to understand the needs and potential problems related to climate, and to incorporate this knowledge into urban planning and design (Ren et al., 2011). For this, UCMaps have two main components: the urban climate analysis maps (UCAnMaps) that spatialize and visualize urban climate assessment, and the urban climatic planning recommendation maps (UC-ReMaps) that provide planning instructions. The latter specifies which parts of the city require improvements and what measures should be considered. Great efforts need to be done by country governments and the international scientific community to generalize the development and the use of UCMpas worldwide. This will be a big step to build sustainable and climate resilient cities.

REFERENCES

Akbari, H., Pomerantz, M., & Taha, H. (2001). Cool Surfaces and Shade Trees to Reduce Energy Use and Improve Air Quality in Urban Areas. *Solar Energy*, *70*(3), 295–310. doi:10.1016/S0038-092X(00)00089-X

Akbari, H., & Rose, L. S. (2008). Urban Surfaces and Heat Island Mitigation Potentials. *Human-Living Environment Society English Journal*, *11*(2), 85–101. doi:10.1618/jhes.11.85

Al-Hamdan, M., Quattrochi, D., Bounoua, L., Lachir, A., & Zhang, P. (2016). Using Landsat, MODIS, and a Biophysical Model to Evaluate LST in Urban Centers. *Remote Sensing*, *8*(11), 952. doi:10.3390/rs8110952

Aminou, D. M. A. (2002). *MSG's SEVIRI Instrument, MSG Project, ESA Directorate of Earth Observation*. ESTEC.

Aram, F., Higueras García, E., Solgi, E., & Mansournia, S. (2019). Urban green space cooling effect in cities. *Heliyon*, *5*(4), e01339. doi:10.1016/j.heliyon.2019.e01339 PMID:31008380

Aram, F., Solgi, E., Garcia, E. H., & Mosavi, A. (2020). Urban heat resilience at the time of global warming : Evaluating the impact of the urban parks on outdoor thermal comfort. *Environmental Sciences Europe*, *32*(1), 117. doi:10.118612302-020-00393-8

Arnfield, A. (2003). Two decades of urban climate research : A review of turbulence, exchanges of energy and water, and the urban heat island. *International Journal of Climatology*, *23*(1), 1–26. doi:10.1002/joc.859

Azevedo, J. A., Chapman, L., & Muller, C. L. (2016). Quantifying the Daytime and Night-Time Urban Heat Island in Birmingham, UK: A Comparison of Satellite Derived Land Surface Temperature and High Resolution Air Temperature Observations. *Remote Sensing*, *8*(2), 153. doi:10.3390/rs8020153

Bounoua, L., Safia, A., Masek, J., Peters-Lidard, C., & Imhoff, M. L. (2009). Impact of Urban Growth on Surface Climate : A Case Study in Oran, Algeria. *Journal of Applied Meteorology and Climatology*, *48*(2), 217–231. doi:10.1175/2008JAMC2044.1

Bounoua, L., Zhang, P., Mostovoy, G., Thome, K., Masek, J., Imhoff, M., Shepherd, M., Quattrochi, D., Santanello, J., Silva, J., Wolfe, R., & Toure, A. M. (2015). Impact of urbanization on US surface climate. *Environmental Research Letters*, *10*(8), 084010. doi:10.1088/1748-9326/10/8/084010

Bruse, M. (2004). *ENVI-met 3.0: Updated Model Overview*. Academic Press.

Chang, C.-R., Li, M.-H., & Chang, S.-D. (2007). A preliminary study on the local cool-island intensity of Taipei city parks. *Landscape and Urban Planning*, *80*(4), 386–395. doi:10.1016/j.landurbplan.2006.09.005

Chang, S., Jiang, Q., & Zhao, Y. (2018). Integrating CFD and GIS into the Development of Urban Ventilation Corridors : A Case Study in Changchun City, China. *Sustainability*, *10*(6), 1814. doi:10.3390u10061814

Chen, L., Ng, E., An, X., Ren, C., Lee, M., Wang, U., & He, Z. (2012). Sky view factor analysis of street canyons and its implications for daytime intra-urban air temperature differentials in high-rise, high-density urban areas of Hong Kong : A GIS-based simulation approach. *International Journal of Climatology*, *32*(1), 121–136. doi:10.1002/joc.2243

Costanzo, V., Evola, G., & Marletta, L. (2016). Energy savings in buildings or UHI mitigation? Comparison between green roofs and cool roofs. *Energy and Building*, *114*, 247–255. doi:10.1016/j.enbuild.2015.04.053

Davies, M., Steadman, P., & Oreszczyn, T. (2008). Strategies for the modification of the urban climate and the consequent impact on building energy use. *Energy Policy*, *36*(12), 4548–4551. doi:10.1016/j.enpol.2008.09.013

Deilami, K., Kamruzzaman, M., & Liu, Y. (2018). Urban heat island effect : A systematic review of spatio-temporal factors, data, methods, and mitigation measures. *International Journal of Applied Earth Observation and Geoinformation*, *67*, 30–42. doi:10.1016/j.jag.2017.12.009

Du, H., Song, X., Jiang, H., Kan, Z., Wang, Z., & Cai, Y. (2016). Research on the cooling island effects of water body : A case study of Shanghai, China. *Ecological Indicators*, *67*, 31–38. doi:10.1016/j.ecolind.2016.02.040

Gu, K., Fang, Y., Qian, Z., Sun, Z., & Wang, A. (2020). Spatial planning for urban ventilation corridors by urban climatology. *Ecosystem Health and Sustainability*, *6*(1), 1747946. doi:10.1080/20964129.2020.1747946

Gunawardena, K. R., Wells, M. J., & Kershaw, T. (2017). Utilising green and bluespace to mitigate urban heat island intensity. *The Science of the Total Environment*, *584-585*, 1040–1055. doi:10.1016/j.scitotenv.2017.01.158 PMID:28161043

Hart, M. A., & Sailor, D. J. (2009). Quantifying the influence of land-use and surface characteristics on spatial variability in the urban heat island. *Theoretical and Applied Climatology*, *95*(3-4), 397–406. doi:10.100700704-008-0017-5

Heaviside, C., Macintyre, H., & Vardoulakis, S. (2017). The Urban Heat Island : Implications for Health in a Changing Environment. *Current Environmental Health Reports*, *4*(3), 296–305. doi:10.100740572-017-0150-3 PMID:28695487

Hirano, Y., & Fujita, T. (2012). Evaluation of the impact of the urban heat island on residential and commercial energy consumption in Tokyo. *Energy*, *37*(1), 371–383. doi:10.1016/j.energy.2011.11.018

Howard, L. (1833). *The Climate of London: Deduced from meteorological observations made in the metropolis and at various places around it.* Academic Press.

Hsieh, C.-M., Aramaki, T., & Hanaki, K. (2007). The feedback of heat rejection to air conditioning load during the nighttime in subtropical climate. *Energy and Building*, *39*(11), 1175–1182. doi:10.1016/j.enbuild.2006.06.016

Johnson, G. T., & Watson, I. D. (1984). The Determination of View-Factors in Urban Canyons. *Journal of Applied Meteorology and Climatology*, *23*(2), 329–335. doi:10.1175/1520-0450(1984)023<0329:TDOVFI>2.0.CO;2

JPL. (2001). *ASTER Higher Level User Guide, Version 2.0.* https://asterweb.jpl.nasa.gov/content/03_data/04_Documents/ASTERHigherLevelUserGuideVer2May01.pdf

Kidwell, K. B. (1995). *NOAA Polar Orbiter Data Users Guide : (TIROS-N, NOAA-6, NOAA-7, NOAA-8, NOAA-9, NOAA-10, NOAA-11, NOAA-12, NOAA-13, and NOAA-14)*. National Oceanic and Atmospheric Administration, National Environmental Satellite, Data, and Information Service, National Climatic Data Center, Satellite Data Services Division.

Kłysik, K., & Fortuniak, K. (1999). Temporal and spatial characteristics of the urban heat island of Łódź, Poland. *Atmospheric Environment, 33*(24), 3885–3895. doi:10.1016/S1352-2310(99)00131-4

Lachir, A., Bounoua, L., Zhang, P., Thome, K., & Messouli, M. (2016). Modeling the Urban Impact on Semiarid Surface Climate : A Case Study in Marrakech, Morocco. *Canadian Journal of Remote Sensing, 42*(4), 379–395. doi:10.1080/07038992.2016.1194746

Li, X., Zhou, Y., Yu, S., Jia, G., Li, H., & Li, W. (2019). Urban heat island impacts on building energy consumption : A review of approaches and findings. *Energy, 174*, 407–419. doi:10.1016/j.energy.2019.02.183

Liu, J., Ma, F., & Li, Y. (2011). The Effect of Anthropogenic Heat on Local Heat Island Intensity and the Performance of Air Conditioning Systems. *Advanced Materials Research, 250-253*, 2975–2978. . doi:10.4028/www.scientific.net/AMR.250-253.2975

Long, J., & Rice, J. (2018). From sustainable urbanism to climate urbanism. *Urban Studies (Edinburgh, Scotland), 56*(5), 004209801877084. doi:10.1177/0042098018770846

Lundgren, K., & Kjellstrom, T. (2013). Sustainability Challenges from Climate Change and Air Conditioning Use in Urban Areas. *Sustainability, 5*(7), 3116–3128. doi:10.3390u5073116

Masson, V. (2000). A Physically-Based Scheme For The Urban Energy Budget In Atmospheric Models. *Boundary-Layer Meteorology, 94*(3), 357–397. doi:10.1023/A:1002463829265

Masson, V. (2006). Urban surface modeling and the meso-scale impact of cities. *Theoretical and Applied Climatology, 84*(1-3), 35–45. doi:10.100700704-005-0142-3

Matson, M., Mcclain, E. P., McGinnis, D. F. Jr, & Pritchard, J. A. (1978). Satellite Detection of Urban Heat Islands. *Monthly Weather Review, 106*(12), 1725–1734. doi:10.1175/1520-0493(1978)106<1725:SDOUHI>2.0.CO;2

McCarthy, M. P., Best, M. J., & Betts, R. A. (2010). Climate change in cities due to global warming and urban effects. *Geophysical Research Letters, 37*(9), n/a. Advance online publication. doi:10.1029/2010GL042845

Miller, J. D., & Hutchins, M. (2017). The impacts of urbanisation and climate change on urban flooding and urban water quality : A review of the evidence concerning the United Kingdom. *Journal of Hydrology. Regional Studies, 12*, 345–362. doi:10.1016/j.ejrh.2017.06.006

Mirzaei, P. A., & Haghighat, F. (2010). Approaches to study Urban Heat Island – Abilities and limitations. *Building and Environment, 45*(10), 2192–2201. doi:10.1016/j.buildenv.2010.04.001

Mohajerani, A., Bakaric, J., & Jeffrey-Bailey, T. (2017). The urban heat island effect, its causes, and mitigation, with reference to the thermal properties of asphalt concrete. *Journal of Environmental Management, 197*, 522–538. doi:10.1016/j.jenvman.2017.03.095 PMID:28412623

Mohan, M., Kikegawa, Y., Gurjar, B. R., Bhati, S., & Kolli, N. R. (2013). Assessment of urban heat island effect for different land use–land cover from micrometeorological measurements and remote sensing data for megacity Delhi. *Theoretical and Applied Climatology, 112*(3), 647–658. doi:10.100700704-012-0758-z

Muller, C. L., Chapman, L., Grimmond, C. S. B., Young, D. T., & Cai, X. (2013). Sensors and the city : A review of urban meteorological networks. *International Journal of Climatology, 33*(7), 1585–1600. doi:10.1002/joc.3678

Myrup, L. O. (1969). A Numerical Model of the Urban Heat Island. *Journal of Applied Meteorology and Climatology, 8*(6), 908–918. doi:10.1175/1520-0450(1969)008<0908:ANMOTU>2.0.CO;2

Ng, E., Chen, L., Wang, Y., & Yuan, C. (2012). A study on the cooling effects of greening in a high-density city : An experience from Hong Kong. *Building and Environment, 47*, 256–271. doi:10.1016/j.buildenv.2011.07.014

Oke, T. R. (1981). Canyon geometry and the nocturnal urban heat island : Comparison of scale model and field observations. *Journal of Climatology, 1*(3), 237–254. doi:10.1002/joc.3370010304

Oke, T. R. (1982). The energetic basis of the urban heat island. *Quarterly Journal of the Royal Meteorological Society, 108*(455), 1–24. doi:10.1002/qj.49710845502

Oke, T. R. (2011). Urban heat island. In *The Handbook of Urban Ecology* (pp. 120–131). Routledge.

Peng, L. L. H., & Jim, C. Y. (2013). Green-Roof Effects on Neighborhood Microclimate and Human Thermal Sensation. *Energies, 6*(2), 598–618. doi:10.3390/en6020598

Polehampton, Smith, Ghent, Wooster, Bruniquel, & Dransfeld. (2021). *Copernicus Sentinel-3 SLSTR Land User Handbook*. Academic Press.

Ren, C., Ng, E. Y., & Katzschner, L. (2011). Urban climatic map studies : A review. *International Journal of Climatology, 31*(15), 2213–2233. doi:10.1002/joc.2237

Ren, C., Yang, R., Cheng, C., Xing, P., Fang, X., Zhang, S., Wang, H., Shi, Y., Zhang, X., Kwok, Y. T., & Ng, E. (2018). Creating breathing cities by adopting urban ventilation assessment and wind corridor plan – The implementation in Chinese cities. *Journal of Wind Engineering and Industrial Aerodynamics, 182*, 170–188. doi:10.1016/j.jweia.2018.09.023

Rosenzweig, C., Solecki, W., Romero-Lankao, P., Mehrotra, S., Dhakal, S., & Ali Ibrahim, S. (2018). *Second Assessment Report of the Urban Climate Change Research Network*. https://climate-adapt.eea.europa.eu/metadata/publications/second-assessment-report-of-the-urban-climate-change-research-network/uccrn_2015_secondassessmentreport.pdf

Russo, A., & Zerbe, S. (2016). Quantifying the local-scale ecosystem services provided by urban treed streetscapes in Bolzano, Italy. *AIMS Environmental Science, 3*(1), 58–76. doi:10.3934/environsci.2016.1.58

Sánchez-Aparicio, M., Andrés-Anaya, P., Del Pozo, S., & Lagüela, S. (2020). Retrieving Land Surface Temperature from Satellite Imagery with a Novel Combined Strategy. *Remote Sensing, 12*(2), 277. doi:10.3390/rs12020277

Santamouris, M. (2013). Using cool pavements as a mitigation strategy to fight urban heat island—A review of the actual developments. *Renewable & Sustainable Energy Reviews, 26*, 224–240. doi:10.1016/j.rser.2013.05.047

Sellers, P. J., Mintz, Y., Sud, Y. C., & Dalcher, A. (1986). A Simple Biosphere Model (SIB) for Use within General Circulation Models. *Journal of the Atmospheric Sciences, 43*(6), 505–531. doi:10.1175/1520-0469(1986)043<0505:ASBMFU>2.0.CO;2

Sen, S., Roesler, J., Ruddell, B., & Middel, A. (2019). Cool Pavement Strategies for Urban Heat Island Mitigation in Suburban Phoenix, Arizona. *Sustainability, 11*(16), 4452. doi:10.3390u11164452

Tan, J., Zheng, Y., Tang, X., Guo, C., Li, L., Song, G., Zhen, X., Yuan, D., Kalkstein, A. J., Li, F., & Chen, H. (2010). The urban heat island and its impact on heat waves and human health in Shanghai. *International Journal of Biometeorology, 54*(1), 75–84. doi:10.100700484-009-0256-x PMID:19727842

Tomlinson, C. J., Chapman, L., Thornes, J. E., & Baker, C. J. (2011). Including the urban heat island in spatial heat health risk assessment strategies : A case study for Birmingham, UK. *International Journal of Health Geographics, 10*(1), 42. doi:10.1186/1476-072X-10-42 PMID:21682872

Toparlar, Y., Blocken, B., Maiheu, B., & van Heijst, G. J. F. (2017). A review on the CFD analysis of urban microclimate. *Renewable & Sustainable Energy Reviews, 80*, 1613–1640. doi:10.1016/j.rser.2017.05.248

UN Habitat. (2010). *Sustainable Urban Development Through Expanding Equitable Access to Land, Housing, Basic Services and Infrastructure; Dialogue on the Special Theme for the Twenty-Third Session of the Governing Council; Report by the Executive Director*. Rio de Janeiro: UN Habitat.

Unger, J. (2009). Connection between urban heat island and sky view factor approximated by a software tool on a 3D urban database. *International Journal of Environment and Pollution, 36*(1/2/3), 59. doi:10.1504/IJEP.2009.021817

United Nations. (2017). New urban agenda : H III: Habitat III: Quito 17-20 October 2016. In *United Nations Conference on Housing and Sustainable Urban Development*. United Nations.

United Nations. (2019). *World urbanization prospects 2018 : Highlights*. Author.

United Nations. (2020). *The value of sustainable urbanization*. Author.

US EPA. (2014). *Learn About Heat Islands* [Overviews and Factsheets]. US EPA. https://www.epa.gov/heatislands/learn-about-heat-islands

USGS. (2019a). *Landsat 7 (L7) Data Users Handbook*. Author.

USGS. (2019b). *Landsat 8 (L8) Data Users Handbook*. Author.

Wan, Z. (2013). *MODIS Land Surface Temperature Products*. Academic Press.

Wong, M. S., Nichol, J. E., Ng, E. Y. Y., Guilbert, E., Kwok, K. H., To, P. H., & Wang, J. Z. (2010). Gis Techniques For Mapping Urban Ventilation. *Using Frontal Area Index And Least Cost Path Analysis., 38*, 6.

Xiao, X. D., Dong, L., Yan, H., Yang, N., & Xiong, Y. (2018). The influence of the spatial characteristics of urban green space on the urban heat island effect in Suzhou Industrial Park. *Sustainable Cities and Society*, *40*, 428–439. doi:10.1016/j.scs.2018.04.002

Yuan, C. (2010). Mitigating urban heat island effects of high density cities : A study at Hong Kong. *Architectural Journal*, 120-123.

Zhang, P., Bounoua, L., Thome, K., Wolfe, R., & Imhoff, M. (2015). Modeling Surface Climate in U.S. Cities Using Simple Biosphere Model SiB2. *Canadian Journal of Remote Sensing*, *41*(6), 525–535. doi:10.1080/07038992.2015.1110013

Zhou, D., Xiao, J., Bonafoni, S., Berger, C., Deilami, K., Zhou, Y., Frolking, S., Yao, R., Qiao, Z., & Sobrino, J. A. (2019). Satellite Remote Sensing of Surface Urban Heat Islands: Progress, Challenges, and Perspectives. *Remote Sensing*, *11*(1), 48. doi:10.3390/rs11010048

Chapter 9
Urban Flood Management Under Changing Climate:
A Land Use-Based Study of Kochi Urban Area, Kerala

Muhammed Sulfikkar Ahamed
Resilience Innovation Knowledge Academy, India

Shyni Anilkumar
National Institute of Technology, Calicut, India

ABSTRACT

Climate change and the associated phenomenon have put major cities and their surroundings at multi-dimensional risk patterns because of hazards, with flooding being a major hazard in the Asian Peninsula. With authorities such as National Disaster Management Authority, India reporting multiple urban local bodies to be under flood risk, it is essential to prioritize flood risk management in the urban planning process in India. Kochi, the commercial capital of Kerala, India has been frequently affected by flooding events. Various factors have been attributed to the flood risk of Kochi Corporation, which requires validation. Against this backdrop, the study focuses on comprehending significant factors attributed to the vulnerability of settlements in the study region and promoting a way forward based on lessons learned and good practices across the world. This is achieved by analyzing significant databases and computations using GIS. The research outcome would help define strategies for sustainable land-use-based development, promoting effective flood management in the Kochi urban area.

INTRODUCTION

The aim of the chapter is to study urban flooding and, to assess the role of urbanization, urban management policies in the growing flooding events that question the sustainable development goal of sustainable towns and cities. It is no denial that urbanization has become a common phenomenon with projections suggesting that two-third of the world's population would end up living in cities by 2050 (UNDESA,

DOI: 10.4018/978-1-7998-8331-9.ch009

2018). While cities grow spatially, it transforms the socio-economic aspects of the citizens, and results in disparities, owing to the unscientific urban development strategies. The alarming effect of such planning measures, in the form of urban flooding, health hazards, and so on, disowns the city's call for sustainability. While these disasters re-surge the debate of sustainable cities, scientific comprehension of the specific disaster problem seems to be deficient, as recommended by Sendai Framework for disaster risk reduction 2015-30 (UNDRR, 2015).

The authors have worked on understanding these aspects, to comprehend these urban challenges from the perspective of flood hazards. The authors have undertaken an evidence-based research point of view utilizing primary and secondary data sources to promote strategical interventions, to achieve the goals of frameworks such as Sendai Framework and, New Urban Agenda. It is believed that by the end of the chapter, the readers would be educated on the importance of evidence-based policy-making to address the long-term implications of natural disasters. In general; the chapter addresses the promotion of words into actions for SFDRR, which involves significant relearning along with the likes of disaster risk management. The chapter centers around the principle of resilient cities and the importance of evidence-based decision-making in long-term sustainability, with a case-based study on urban flooding in Kochi city, a prominent growth center in India.

BACKGROUND

The world has been urbanizing at higher rate post industrialization. Urbanization refers to the process of migration of people from rural settlements to the centers of growth. Reports suggest that these centers are responsible for 80% of the world's GDP. These growth patterns were complemented by globalization policies such as the 1991 liberalization, privatization and globalization policy of India that opened doors for more cities and urban agglomeration. United Nations Report of 2018 on global demographics suggests that at present, 55% of the global populations live in the cities, and is expected to rise to 68% by 2050. The present statistics show that these trends have been similar in all countries, irrespective of their income level-based cohorts. Figure 1 depicts the present urban population and the projected urban population by 2050. The report reiterates that the major growth centers by 2050 would be the lower cohorts, namely low middle income and, low-income countries, owing to the rising market trends (UNDESA, 2018).

These growth patterns are characterized by large-scale migration, which results in the demand for land for residential, commercial, and other activities. New markets, infrastructures, and innovations have emerged as a result of these growth patterns, which have benefitted the cities, as well as the rural regions. This is mainly facilitated by the urban-rural linkage that helps in the exchange of knowledge, products, and services. These have also resulted in new centers of production, around cities, which has helped in economic, infrastructural growth. While the cities have promoted large-scale integration, the development pattern triggered frequent disasters driven by poor environmental management, uncontrolled development policies, and large-scale land conversions. Centre for Research on the Epidemiology of Disasters states that during 2008-17, the world registered an average of 345 disasters every year and accumulated losses in millions. This is significant as the impact of disasters could be amplified in the urban centers because of large populations and growth patterns. Take for example, The National Disaster Management Authority reports that urbanization could result in the amplified flood peaks of 1.8 times the normal, resulting in large scale impacts (NDMA, 2010). The table 1 below suggests the various events of disasters that have registered widespread losses. The shifting disaster patterns have in turn resulted in to international dialogues on disaster risk management and realized the need to develop implementation frameworks of the management strategies.

One of the major developments was the identification and promotion of disaster risk management in the place of disaster management. In disaster management, risk refers to the potential disaster losses, in the livelihoods, assets, and services, which could occur to a particular community or a society over some specified future period. Risk signifies the possibility of adverse effects in the future. Disaster risk management (DRM) is the systematic process of utilizing administrative directives, organizations, and operational skills and capacities to implement strategies, policies, and improved coping capacities to lessen the adverse impacts of hazards and associated risk patterns. The management is a complex process that includes mitigation, prevention, preparedness, and capacity building. DRM principles have been widely promoted in the contemporary world owing to the complex socio-economic-cultural-political system prevailing in the world order, as it promotes proactive risk management. This is different from disaster management as it focuses on reactive planning with more focus on response, and recovery.

Figure 1. Present and projected urban population of various cohorts (UNDESA, 2018)
Source: UNDESA, 2018

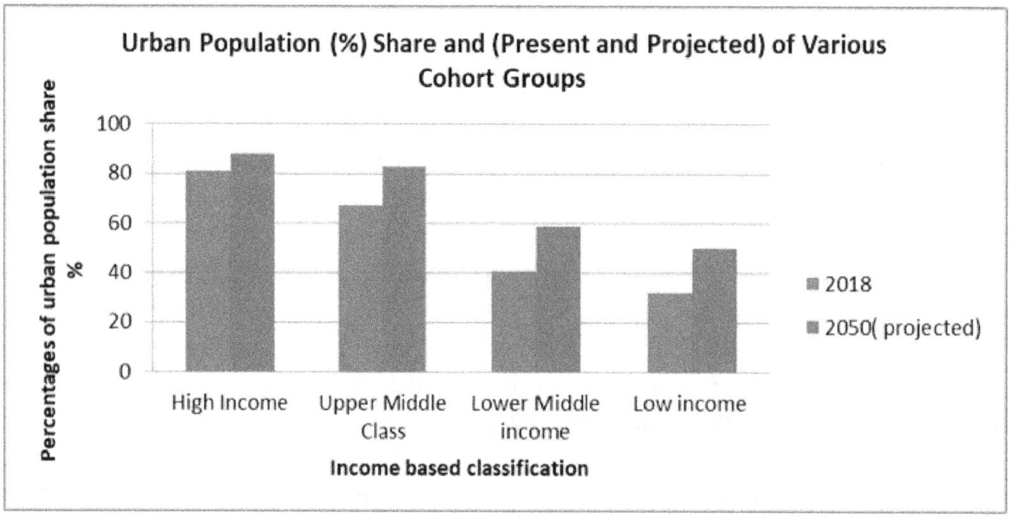

Table 1. Major disasters and impact in the world

Event (Year)	Major Losses
Indian Ocean Earthquake and Tsunami (2004)	2,00,000-3,10,000 deaths
USA Hurricane Katrina (2005)	80.7 billion USD lost
Haiti Earthquake (2010)	3,00,000 deaths
East Africa Drought (2011)	11.5 million affected
Japan Earthquake and Tsunami (2011)	210 billion USD lost

Source:(Rinkesh, 2021)

One of the most important frameworks in disaster risk management is the Sendai Framework for Disaster Risk Reduction 2015-30 (UNDRR, 2015). The framework was a successor to the Hyogo Framework of 2005 and focuses on reducing disaster risk. The Sendai Framework thrust on the prevention of new and reduce existing disaster risk through the implementation of integrated and inclusive economic, structural, legal, social, health, cultural, educational, environmental, technological, political, and institutional measures that prevent and reduce hazard exposure and vulnerability to disaster, increase preparedness for response and recovery, and thus strengthen resilience. The framework has listed four priority areas, namely

1. Understanding disaster risk
2. Strengthening disaster risk governance for disaster risk management:
3. Investing in disaster risk reduction for resilience
4. Enhancing disaster preparedness for effective response, and to "Build Back Better" in recovery, rehabilitation, and reconstruction

The Sendai Framework promotes risk-informed planning of urban areas and, suggests evidence-based policy making on various dimensions of risk such as vulnerability, capacity, and so on, for proactive disaster management. These suggestions have been redefined at global and national and local levels, for effective delivery .Some of the major pointers in the Sendai Framework are as follows:

Table 2. Major pointers of Sendai framework of 2015

Priority Areas	Major Pointers
Priority Action 1: Understanding disaster risk	• Promote data collection, analysis and dissemination for various users • Promoting science based methodologies and learning at international and regional level management • To develop, update and disseminate location based real-time data using GIS for enhanced risk mapping • Strengthen long-term investment for innovations in multi hazard risk management
Priority Action 2: Strengthening disaster risk governance for disaster risk management	• To mainstream and integrate sectoral, national and local disaster risk reduction strategies and plans across timelines • To carry out periodic assessment, and promote discourses • To promote transboundary cooperation for disaster risk management
Priority Action 3: Investing in disaster risk reduction for resilience	• To promote structural and non-structural disaster resilience measures at national and local levels, such as disaster risk transfer mechanism and so on • To promote coherence with various frameworks and sustainable goals
Priority Action 4: Enhancing disaster preparedness for effective response, and to "Build Back Better" in recovery, rehabilitation, and reconstruction	• To prepare and promote preparedness plans, people-centered multi hazard, multi sectoral early warning systems • To promote resilience to new and existing infrastructures for their safety • To promote standard codes and practices to support disaster preparedness

Source:(UNDRR, 2015)

While the framework suggests various principles and practices for risk management, such as identification and assessment of floodplains, safe human settlements and so on, to be advocated for disaster risk management in the case of flooding, there is a need to inspect these for further interventions. The study focuses on addressing the 'Priority action 1 -understanding disaster risk' of the Sendai Framework in the case of urban flooding, which is being analyzed further in the chapter.

URBAN FLOOD RISK MANAGEMENT - A SCRUTINY

Flooding refers to the covering or inundation of normal, dry land with a large amount of water, and thereby affecting the livability of the settlements. Flooding is a major threat to all forms of settlements from rural to urban, and can result in cascading hazards such as health hazards and is a result of both natural and anthropogenic factors. Centre for Research on the Epidemiology of Disasters(CRED, 2018) suggests that flooding affects 50% of the total vulnerable communities and, has the second-most mortality rate, and occurrence rate in the world. The report suggests that Asia is highly affected by the impact of flooding, owing to the landmass and the large population. The Asia Pacific Disaster Report of 2019(UN-ESCAP, 2019) suggests that Asian countries get impacted majorly from floods and other disasters with 41 percent of the total absolute annual average losses in China and India.

There are various forms of flooding, such as

- Coastal flooding: Flooding in coastal areas due to storm upsurges or high tides
- Pluvial/ Urban flooding/ Flash flooding:
- Fluvial/ Riverine flooding: Flooding in riverine zones, due to the exceedance in water volume above river capacity, mostly due to heavy rainfall.

Historically, flooding in urban centers are attributed to excessive rainwater runoff, poor drainage and infiltration, and improper land use management and urban flooding is the most challenging owing to the significant role of cities in the country's functioning. The Sendai Framework of 2015 suggests various measures such as mainstreaming of disaster risk management of floodplains and delineation of safe human settlements (UNDRR, 2015). In theoretical researches, flood forecasting of cities and towns is designed based on hydro-meteorological models that rely on historic data, and thereby promote the return period of floods. These models work well for fluvial floods, but the present challenges such as improper land use management, climate change have affected the overall return period of floods and have led to phenomena such as urban flash flood which are unusual and affects the city functioning, alongside challenging SDGs such as SDG-11 for 'sustainable cities and towns'. The chapter focuses on urban flooding, which affects cities.

Urban Flooding- Causes, Challenges and Possible Risks

Urban flooding is a major challenge as the world moves towards a pro-urbanized scenario, with projections suggesting that global urban communities would cross 60% by 2050. FEMA defines urban flooding as the inundation of properties in a densely populated built environment because of heavy showers on impervious surfaces that affects the carrying capacity of the drainage systems. The urban projections also reflect the growth pattern in Asian countries which forms the part of upper and lower middle income countries, and poses a significant challenge, as these cities are most vulnerable to urban flooding. The National Disaster Management Report of India, 2019 around 5161 urban local bodies in India are prone to flooding. This is important because Census of India 2011 states that around 377 million plus population (31%) lives in cities. The number of Indian cities has also grown to 7935 in 2011 from 1827 in 1901.

Various factors, both natural and anthropogenic, play a pivotal role in the urban flood hazard, and it challenges a wide variety of fields in urban livelihood. The following section details the possible causes and challenges of urban flood risk management. There is a need to study the major causes, challenges and possible impacts of the urban flood risk on stakeholders for effective flood risk management.

Causes of Urban Flooding

There are many reasons for urban flooding, as these regions are growth centers. A literature review was conducted and significant factors were outlined for the urban flooding.

Urbanization and land use conversions play a major role in urban flooding (Jha et al., 2012). The extent of urbanization in regulating flood risk is complex. Urbanization affects the overall rainwater movement and storages, as it results in the increase of built-up areas and reduction of green covers, which used to be a part of natural sponges for water retention and renewal. This increases the hydrological response time and results in increased flood risk. NDMA, India reports that the urbanization results in the increase of flood peaks to almost eight times and flood volume to six times than normal rates which could be detrimental to the settlements. Rapid urbanization also affects the hydrological models, due to the large flows generated due to increased built-up areas and can result in the occurrence of small floods up to 10 times the normal. The study by Hollins(Hollis, 1975) states that rapid urbanization, accompanied by 30% built area, could reduce the 100 years flood simulation interval by half. The situation is further complexed by larger apartments replacing smaller homes and settlements.

Drainage and waste mismanagement is another major factor contributing to the urban flooding. Drainage basins play a major role in regulating the flood risk by directing the water out of neighborhoods. The increasing urbanization pattern, complex with aging drainage systems results in poor storm water management. This is further complexed by poor waste management that results in the blockages of drainage systems in the region, aggravating the flood risk. A recent report on urban flooding in Detroit city, USA stated that around 70% of the total drainage systems are poorly maintained, aggravating the flood risk. India faces the same issue of improper drainage management(University of Maryland College Center for Disaster Resilience et al., 2018). Post Disasters Needs Assessment of Kerala, 2018 stated that poor drainage management was one of the major reasons for flooding in the state of Kerala in 2018 that resulted in widespread damages of around 27000 crore rupees. Similar learning could be drawn in the case of the Mumbai floods of 2005, which had an aged drainage system. These factors will, in turn, affect the capacity, and increase the water detention time, which will affect the cities and their functioning. (United Nations Developmental Program, 2018)

Floodplain mismanagement is a significant factor in urban flooding. Floodplains are low-lying, designated regions of flooding and the rate of floodplain conversions speak about improper development patterns and possible vulnerabilities associated with this associated development model(Park & Lee, 2019). Most of the cities have undergone large-scale conversions, and most of them have come at the expense of floodplains, and associated waterways such as lakes and canals. One of the most fascinating findings of the Chitale Committee post-2005 Mumbai floods was the occupation of floodplains, which aggravated the flood disaster in the city. Similar learning was also shared in the case of Dhaka City post-1998 flooding. (NDMA, 2010)

Besides these factors, climate change-induced rainfall patterns influence the flood hazards in the city. Increasing precipitations, as a result of longer positive Indian Ocean Dipole, is a major challenge in the East African- South Asian zones, which include India, Pakistan, and China, which are highly vulnerable to floods. The regions have seen some of the most devastating experiences such as the Kerala Floods of 2018, India, where excess rainfall was a major contributor. The Kerala State Action Plan For Climate Change(Government of Kerala & Department of Environment and Climate Change, 2014) states that the cities in central Kerala including Kochi, the financial capital are projected to deviate by +20 mm-30mm from its present average rainfall, which will aggravate the present conditions. Similar learning could also be found in the case of the 2005 Mumbai floods, where the city received excess rainfall. All these factors generally contribute to the flood vulnerability of a region and can aggravate the flood risk.

Impacts due to Urban Flooding

Urban flooding affects the cities and results in large-scale losses. There are direct and indirect losses associated with urban flooding in cities. Flooding damages the basic infrastructure facilities and affected the functioning of lifeline/ critical infrastructure systems during crises. These include important road linkages, hospitals, schools, and other shelter facilities, and so on. These impacts could also extend to building infrastructures, where the flooding would result in damage of structures and, could affect the utility facilities, and the basement parking areas, which could result in serious damages such as electrocution, loss of humans. The most affected structures are the ones in the existing floodplains. There are various forms of structural damages due to water such as structural deterioration in the form of cracks due to water penetration, structural damage due to water pressure and chemically contaminated water, and so on. (Jha et al., 2012)

Indirect impacts of floods are decreased capacity of lifeline support due to the damages of infrastructures, disruptions in the supply chain, markets, customer pool, and disruption of international and national services, in the case of flooded airports. There is a major socio-economic impact of flooding, as it results in reduced capacities of urban poor communities, as they would be the most affected due to geographical positioning. The urban poor would face challenges in the rebuilding process as it would create economic constraints for the community, which could result in the poor progression towards sustainable goals in the due process. Flooding could also result in traumas and heart attacks amongst the affected communities, and is more prominent in flash floods as per reports. Cascading impacts due to flooding are a significant impact. Flooding could affect chemical facilities in urban centers and result in chemical pollution, and leakages which could result in NATECH or Natural Hazards Triggering Technological Disasters. Another cascading effect of flooding is the health hazards in the form of communicable diseases, such as Rat-bite fever which could affect the community and could result in systemic failures such as health and governance.

Kerala and Floods- An Assessment

Kerala is the south western state of India, with a population of 3.33 crore populations, and covering a total of 38863 square kilometres in area. The population density of Kerala stands at 859 people per square kilometer, one of the highest in the state. The region has over 44 monsoon-fed rivers running through it, some of them are River Periyar which runs 244 kilometers and cover multiple districts such as Ernakulam, Idukki and so on, and River Bharathapuzha. The state receives an annual average rainfall of 3000 mm every year during six month long monsoon period (June-Nov), which is controlled by the south west and north east monsoons.

Figure 2. Map of Kerala, India
Source: Created by Authors, 2020

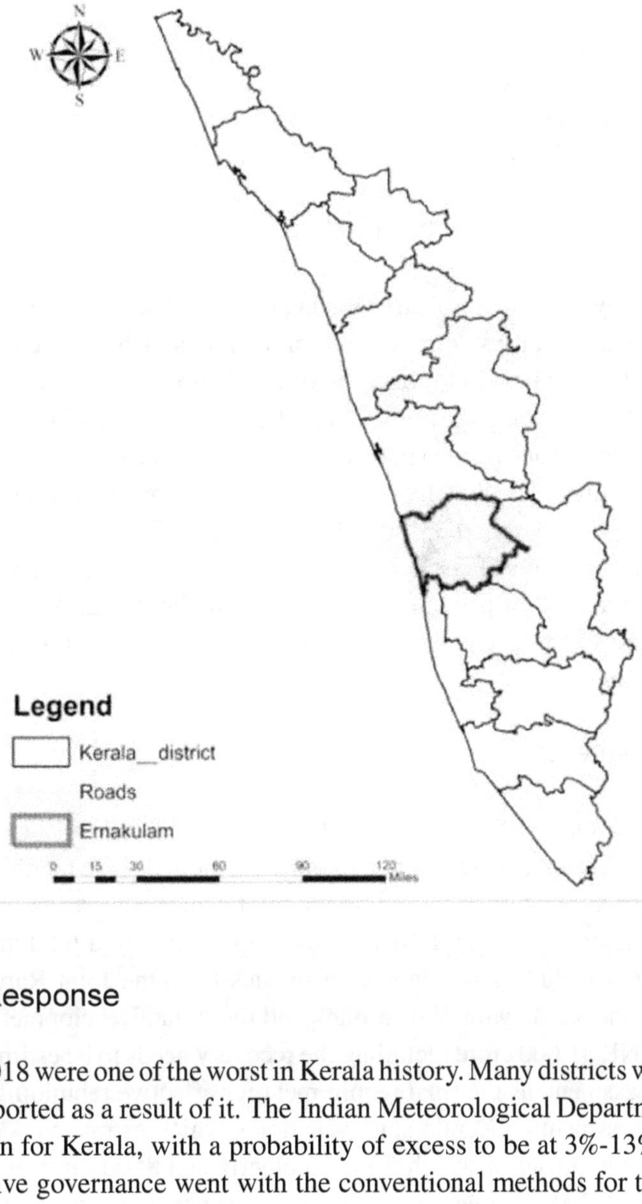

2018 Floods and Response

The Kerala floods of 2018 were one of the worst in Kerala history. Many districts were affected, and large-scale damages were reported as a result of it. The Indian Meteorological Department in 2018 forecasted a normal rainfall season for Kerala, with a probability of excess to be at 3%-13%. Based on the report, the Kerala administrative governance went with the conventional methods for the upcoming monsoon such as drainage cleaning, and basic precautions(Memorandum, 2018). On contrary to the forecast, the state during July-August received a rainfall of 2346.6 mm, which is an excess of 42% which used to average at 1649.5 mm of rainfall, which affected major districts and resulted in cascading disasters such as landslides and other disasters (Joseph et al., 2020). Various systems related to transportation and demographics were affected during the flood scenario. In Kerala, the public transportation system had collapsed in many parts of the state, such as the city of Kochi, which is the commercial capital of the state. Cochin International Airport in Kochi had to suspend all its flight services as well as Kochi Metro. Bus services to Kochi were affected after parts of national highways connecting the city inundated and interstate train services were hugely hit.

The state had an immediate response to the disaster scenario by deploying various measures and rescue teams such as

- Kerala Fire and Rescue Services
- National Disaster Response Force (NDRF)
- Army, Naval, and Coast guards
- Emergency/ 24 hrs. Helplines
- Camps and Shelters
- Relief Collection Points (Humanity Road, 2018)

Public participation played a pivotal role in reducing the casualties and, promoting effective response measures. Along with government systems, various communities such as the coastal fishermen community stepped forward and were pivotal in the rescue of 65000 populations. Critical infrastructures such as the Kochi Metro started operating as a relief life, allowing free rides for relief workers (Humanity Road, 2018). Relief materials from various parts of the country, and nearby districts helped in response and reduced possible casualties. The Indian Navy set up a relief camp at an aircraft hangar, INS Garuda, Kochi, and airlifts were carried out for the flood-affected people. The Navy set up a shelter in a primary school in Aluva for survivor support Private Hospital infrastructures played a pivotal role in extending the services to cope with a surge of patients from neighboring hospitals. Along with, telecommunication services were made free of cost to the flood-affected sites of Kerala for rescue. These helped in improving the flood response.

Post Disaster Assessments

Post-disaster Needs Assessment (PDNA) was conducted by a consortium comprising of World Bank, Asian Development Bank, Kerala government and other partner firms (United Nations Developmental Program, 2018). The total estimations concerning the total damages were computed to be around INR 10,557 crore and total losses to be at INR 16,163 crore, amounting to a total disaster effect of around INR 26,720 crore without including the damage estimates from the Joint Rapid Damage and Needs Assessment (JRDNA) conducted by the World Bank and the Asian Development Bank (ADB). PDNA estimated a net total of INR 31,000 crore including the recovery needs to be estimated by the Joint Rapid Damage and Needs Assessment, as the final requirement for effective rebuilding and recovery.

A human impact assessment of Kerala state was done, which estimated 433 people died and 1.4 million people in relief camps out of which 20% are elderly and 8600 disabled. The study found that 10 percent of death from tribal communities which is of serious concern. It was reported that around 6.7 million were left with no access to pipe water facilities after the disaster and 400000 with no toilet access, which could result in cascading hazards such as health. Along various documents of survivors were also lost during the disaster, which created stresses around the community.

Conclusion

The floods of 2018 were a major learning curve for the Kerala community, and a lot of factors played into the disaster management. One of the good practices was the community led emergency response management and cooperation. Community was the epicenter of rescue missions, and coordinated with various organizations to deliver services to the affected communities in the flood affected zones. The community coordinated with the help of digital media, and physical supports such as coastal fishermen communities and helped in reducing the casualties. Another factor that helped in the rebuilding was the worldwide exposure, which helped in aids and services to the community. The government stepped in to deliver efficient services to the affected communities in the state to address the physical and mental challenges concerning the affected. Government played a major role in reducing the disaster impacts and supporting the affected in rebuilding, along with issuing memorandum for new land management policies.

The major challenges outlined for the disaster risk management are:

- Poor land, drainage management plans
- Poor liaison between community and administrative services
- Community capacity building measures such as community risk mapping, resilience training, mock drills, etc. were absent for the Kerala scenario (Joseph et al., 2020)

The floods of 2018 showcased the poor progression towards addressing the priority areas of Sendai Framework, such as Priority-1 Understanding the disaster risk. This is important as the flood return periods are decreasing, and new challenges such as urban flooding because of urbanization is on the rise. Pivoted on these issues, the authors have studied Kochi city of Kerala, as a research study to address the priority-1 of Sendai Principle

KOCHI MUNICIPAL CORPORATION AND URBAN FLOODING- A CASE BASED ANALYSIS

Kochi Municipal Corporation is located in Ernakulam, the central district of Kerala India. The municipal corporation is a part of Kochi Urban Agglomeration, which is the largest agglomeration in the state. The region is the most densified zone of the state with a population density of 6667 persons per sq. km against a state average of 819 people per square kilometer and has an overall population of 633330 within its 94.88 square kilometers of area (Census, 2011). The city is divided into 74 wards, which is governed by the Cochin Municipal Authority. The regional and, administrative maps of Kochi are shown in the figure 3 and 4.

Geographically, the city is a low lying region, sloping east to west, with an average mean sea elevation of 7.5 m above MSL (Sowmya et al., 2015). The region consists of a complex set of rivers, backwaters, canals, and sea which plays a vital role in the drainage of the region. The western part of the region is bounded by the Arabian Sea, and the central area contains the lake Vembanad, which is an essential drainage system. The region has tributaries from River Periyar, which joins the Vembanad, and then to the Arabian Sea. The Vembanad Lake forms a part of National Waterway-3, which connects the districts of Kollam in the south to Kottapuram in the Northern Kerala. The region has a tropical climate with the monsoon season during June-November.

The region holds an important role in the economics of the state as well as the country and is considered the economic capital of the state. The region contributes heavily to the state GDP, and the cumulative contribution from the district stands at 15% of the total GDP .Kochi is home to various industrial belts and has the largest shipbuilding facility in India, the Cochin Shipyard. Besides, the region has the High Court of Kerala, which serves Kerala and Lakshadweep Islands. The Port of Cochin is located in the city, which is the busiest port in the state. Besides, the region has good physical and, social infrastructures such as the Kochi Metro, which makes it one of the most livable cities in the region(Town and Country Planning GOK, 2009).

Urban Flooding in Kochi

Kochi Municipal Corporation is challenged by continuous urban flooding intensifying each year coupled with climate changes. The corporation has reported multiple events of flooding in the last few years, affecting the normal functioning of the city. The floods have become a subject of public concern, and judicial and legislative sections have addressed their concerns regarding the incidents, exacerbated by poor management. The corporation has undertaken certain measures such as Operation Breakthrough 2019, to clean the drainage system, and canals in the corporation post-2019 events. The floods of 2020 prove that these measures have fewer impacts on the system, as it is complexed by large-scale land conversions and improper regulatory measures, both by the public and policy-makers(M K Sunil Kumar, 2021). Earlier studies in 2003 by Cochin University of Science and Technology (Ridge, 2003) had implied that improper city management, which involved building over water bodies along with canal mismanagement, would lead to issues regarding flooding in the long term, which is a present day reality. This is complexed by rising precipitation, which affects the overall district. The authors advocated a 4-year study on the rainfall pattern to address the studies suggesting the possible variance in the future and found out that there is a significant growth in the monthly average rainfall during June-Nov in the years 2018 and 2019 (See figure-3). This could be implied to the long-standing IOD or Indian Ocean Diploe, which affects the East African and South Asian Zones, and thereby result in heavy showers. These shreds of evidence underpin the temporal simulation study that suggested that the North Central Districts including Kochi Municipal Region will show a variance of +20-30 mm (Government of Kerala & Department of Environment and Climate Change, 2014). Recurring floods of Kochi have put thrust on research studies that address the core issues of land use, floodplain mismanagement, that empowers polices and strategies that address the long-term management of these studies.

Figure 3. Four year assessment on the rainfall pattern (2016-19) of Kochi municipal corporation
Source: NASA, 2020

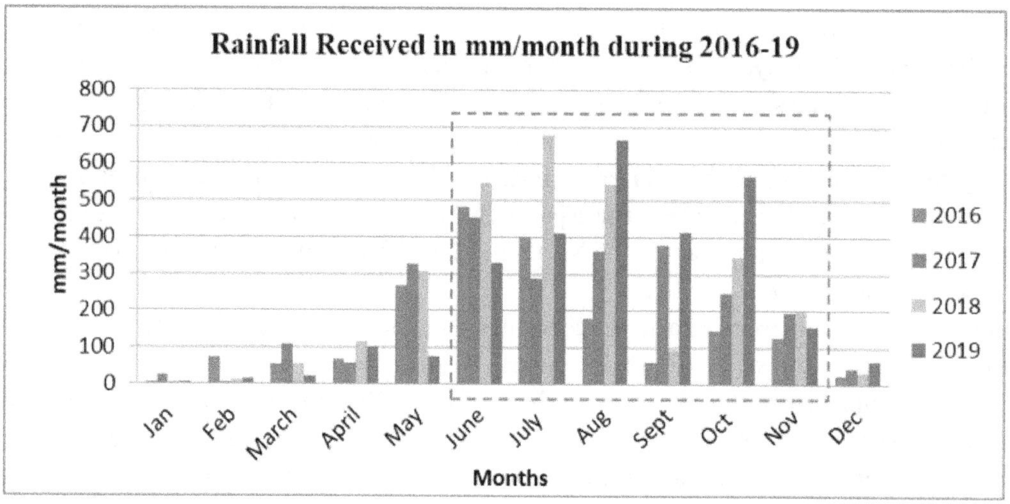

Livable cities are a necessity for the community as cities become more globalized and markets are more concentrated, wherein impacts on one economy will affect the arrangement of other economies and people. In this scenario, a city of global importance such as Kochi and its issue of flooding are inevitable. The city has been challenged by the flood scenario mainly due to man-made and natural factors and it is high time that a methodology is developed to understand the overall regions of vulnerability and to plan accordingly. To study further and address the Sendai Framework, the authors carried out a detailed study on urban flooding in Kochi.

Figure 4.The regional setting of Kochi municipal corporation
Source: Created by Authors, 2020

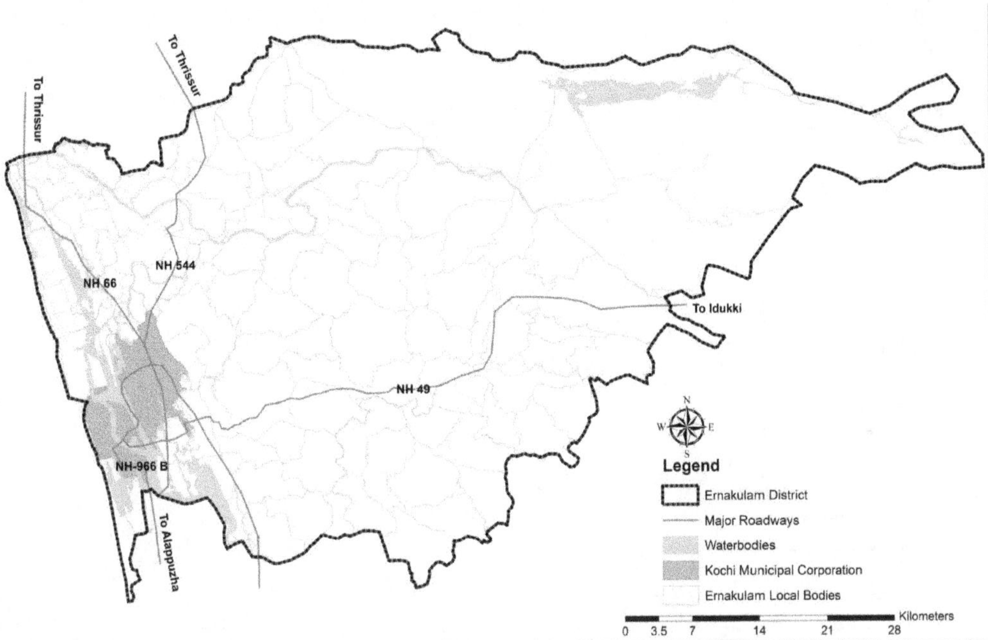

Aim, Objectives and Methodology

The aim of the study is to understand key issues and challenges within Kochi urban settlement aggravating the flood risks. The study undertakes two major objectives

- To understand and merge significant factors attributed to the vulnerability of the urban settlement under study
- To promote a way forward based on lessons learned and good practices across the world

The study follows a mixed-method approach, wherein general urban development factors which alter the vulnerability of the region are studied for strategies. The research methodology involves the identification of issues of urban flood risk management wherein generic studies are implemented to understand the urban development related as well as vulnerability related issues and challenges. The resultant data is used to understand and tabulate various forms of vulnerability issues for Kochi Corporation. This is further applied to the site for understanding the site-specific vulnerabilities and issues. The outcome would be strategical solutions that could be done to mitigate the crisis of vulnerability and to ensure effective flood risk management.

Figure 5. Map showing the Cochin corporation wards
Source: Created by Authors, 2020

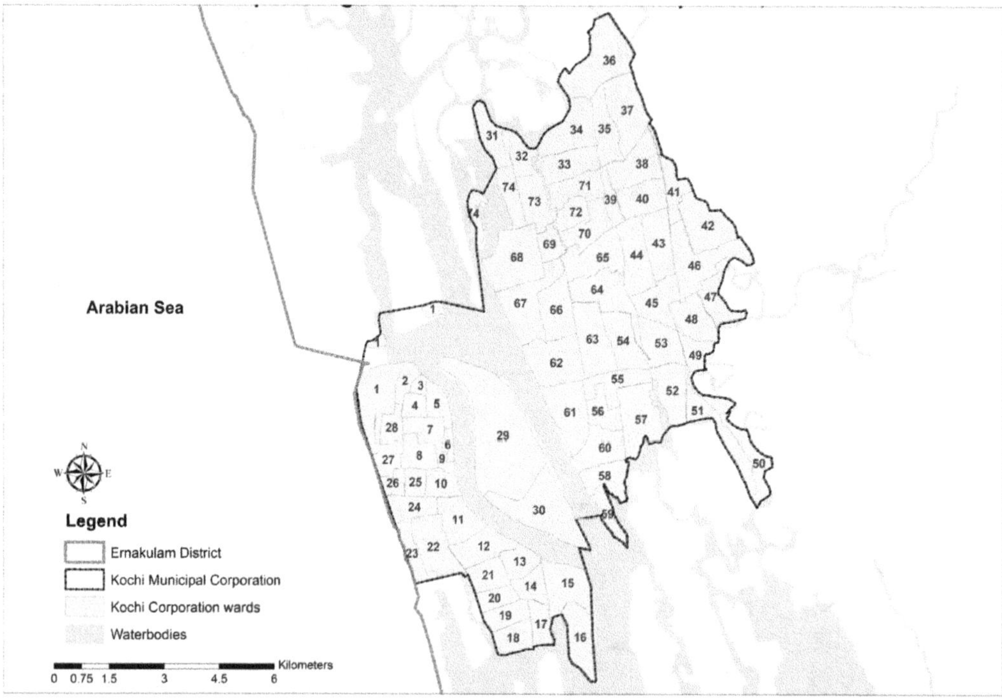

Key Vulnerability Parameters of Kochi Urban Flooding

Based on the detailed literature review as discussed, the study analyzed four major reasons for urban flooding in Kochi Municipal Corporation. The four major factors that were analyzed are as follows:

- Land use conversions and rising built-up patterns
- Drainage and waste management
- Floodplain mismanagement
- Rainfall/ Precipitation patterns.

The table 3 discusses the consolidated parameters and the references.

Figure 6. Methodology for the study
Source: Created by Authors, 2020

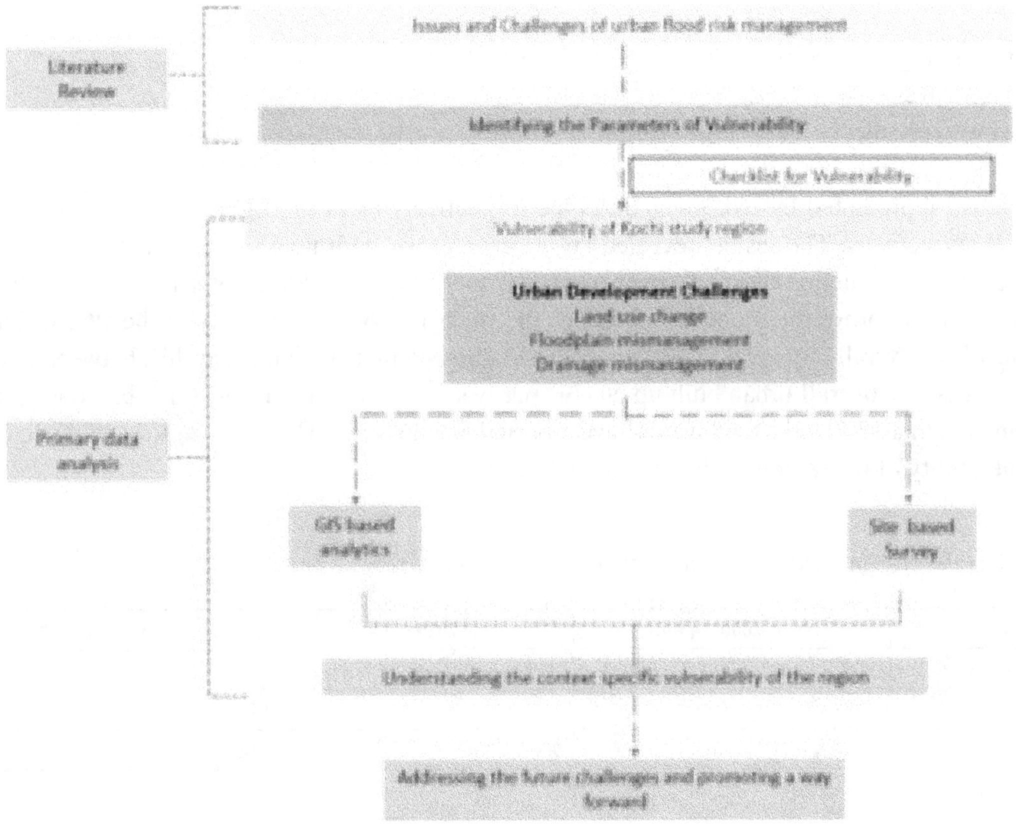

Table 3. Consolidated parameters for urban flood vulnerability

Parameters	References
Land use conversion	(Oak Ridge National Laboratory, 2003) (Sowmya et al., 2015) (NDMA, 2010), (Park & Lee, 2019) (Elboshy et al., 2018)(United Nations Developmental Program, 2018)(Jha et al., 2012)(Bollin et al., 2003)
Flood plain mismanagement	
Drainage and waste management	
Rainfall Patterns	

Source: (Created by Authors, 2020)

ANALYSIS AND RESULTS

Based on the literature study, the researchers attempted to address the influence of urban development patterns in the urban flooding of Kochi Corporation. The study focused on satellite imagery analysis to gather significant details about the extent of urban development challenges. The study initially addressed the concern of the **growth of built-up regions** because of urban pressure as it influences the overall runoff, and can result in urban flooding. Kochi Corporation is highly challenged by the urban demand, because of its strategic and, economic importance, which has resulted in a shift in the overall land share. To study the overall land share shift, Land Sat images (8 and 7) retrieved from USGS (November, 2019 and October, 2000 respectively) were used for land use classification in GIS. Three major classifications were done:

- Built up region,
- Green cover
- and water bodies

These are represented by red, green and blue respectively as in the figure 7 and 8 below of 2000 and 2019 respectively. The results revealed an increase in the built up region, from 55 to 70.7 square kilometers and, a reduction of green cover from 24.5 to 9.8 square kilometers, which adds to the flood risk, as it could improve the runoffs, and reduce the water percolation capacity of the city, which adds concerns of the overall city level management. The table 4 depicts the land use shift between 2000 and 2019. Besides, the overall urban built up change per ward was computed using GIS based analytics. It was found out that 87% of the total wards have reported conversions to build up regions, which increases the vulnerability of the region. This is explained in figure 9

Table 4. Table showing the conversion rate of various land uses

Type	Year -2000 (sq.km)	Year-2019 (sq.km)	Conversion Rate
Built up area	55.5	70.7	27%
Water bodies	14.88	14.3	-4%
Green cover	25.5	9.8	-61%

Source:(Created by Authors, 2020)

Figure 7. Classified land use map of Kochi 2000
Source: Created by Authors, 2020

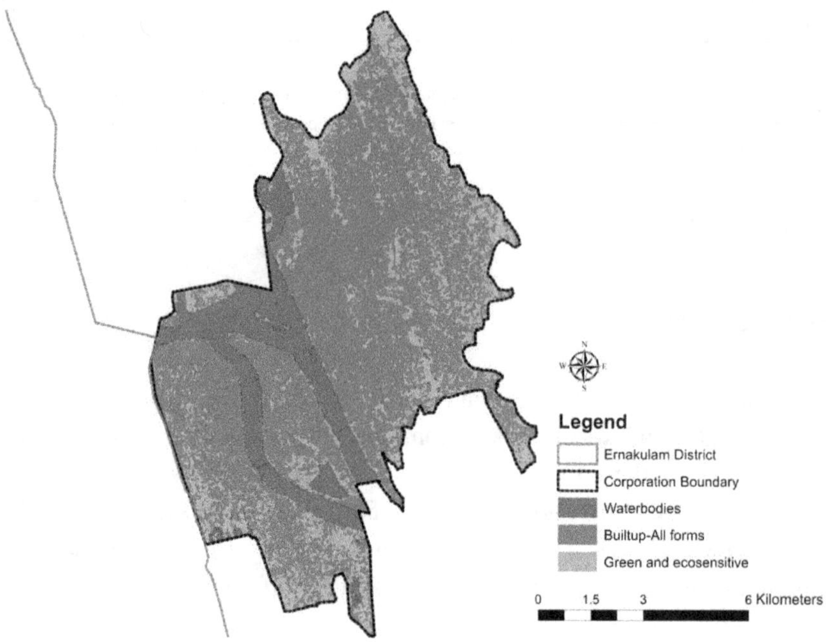

Figure 8. Classified land use map of Kochi 2019
Source: Created by Authors, 2020

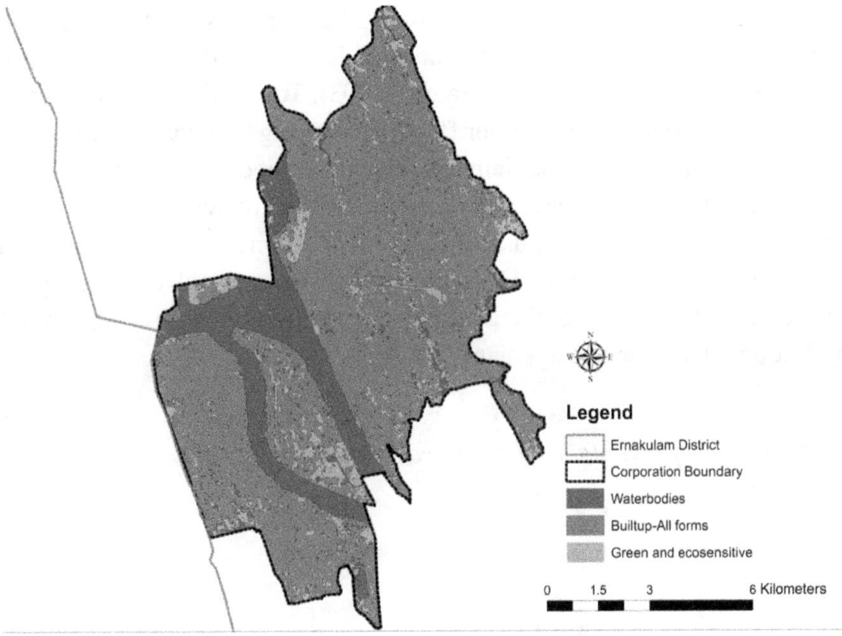

Figure 9. Final rate of change of built up over 2000-19
Source: Created by Authors, 2020

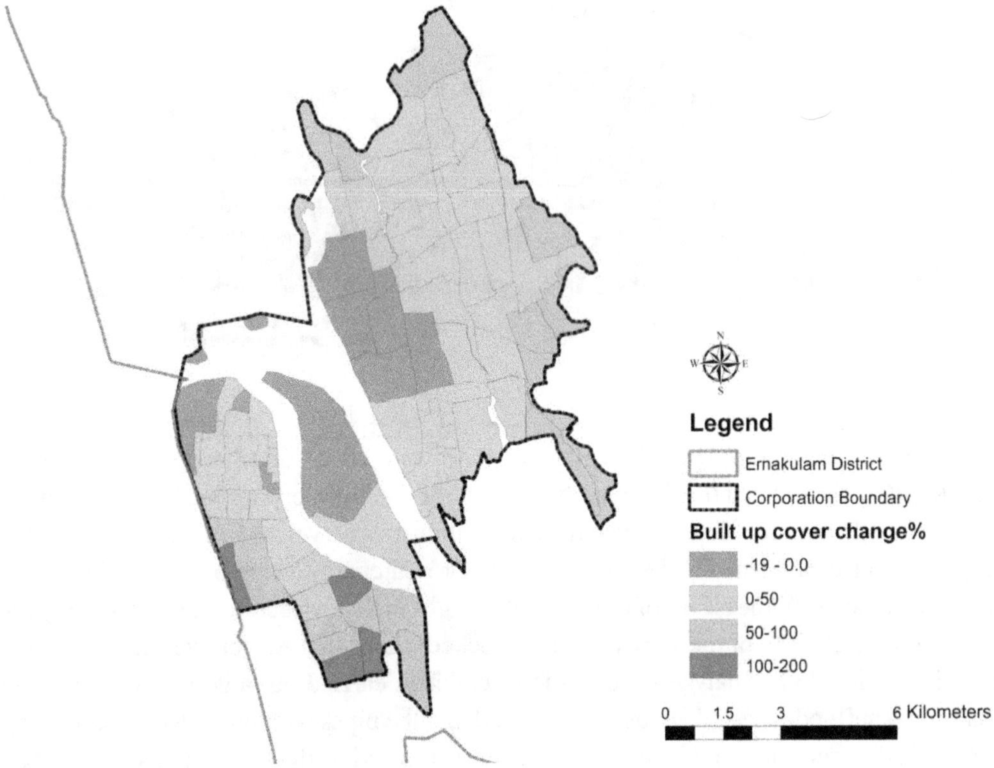

Flood Plain Mismanagement: Is a major concern in the Kochi Corporation. As per the state disaster management authority, it is found that close to 25 square kilometers of the region are susceptible to flooding, owing to the presence of canals and Vembanad Lake (represented in figure 10). The authors identified the land use conversions in the flood plains of Kochi Corporation based on various reports and temporal images(Oak Ridge National Laboratory, 2003). To scrutinize the hypothesis, the conversions in built-up regions are again computed for floodplains using GIS analytics, as flood plain-related constructions could result in vulnerable settlements which are directly under the flood risk. GIS-based analytics indicated that the overall conversion in the floodplains follows a similar trend to the built-up conversion which indicates unsustainable use of eco-sensitive zones for construction, aggravating the flood risk. This is explained in the figure 11 below. Furthermore, on advocating a primary survey of the region, it was found out that most of the constructions do not follow the buffer requirement (12m) as prescribed by National Disaster Management Authority.

Figure 10. Map showing the flood plains of Kochi corporation
Source: Created by Authors

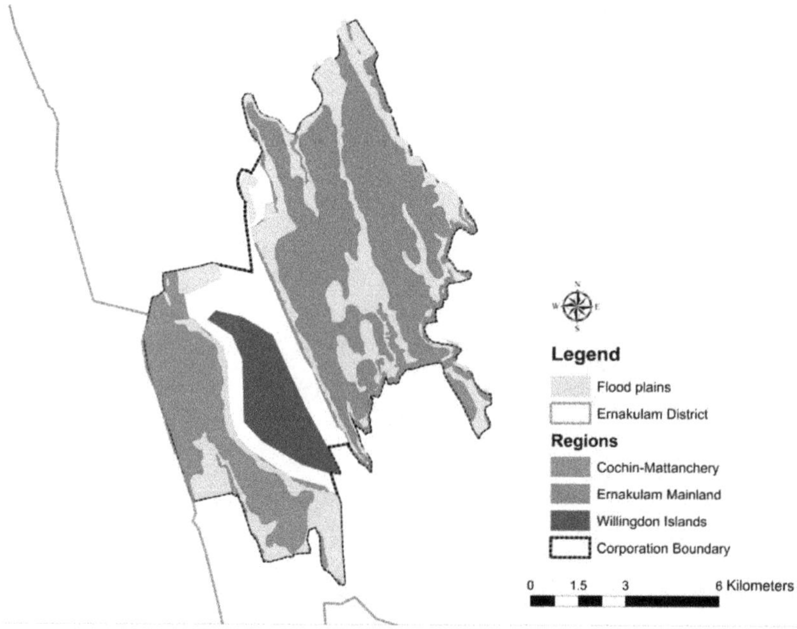

Drainage Mismanagement: Drainages play a major role in urban flood management. Various studies concerning Kochi indicated that the drainage systems in the Cochin Corporation have been mismanaged over the years, mostly because of negligence from the authorities as well as from the public in due course of time (Special Correspondent The Hindu, 2020). The researchers carried out a stakeholder survey of ward counselors to identify the issues of drainage blockages in the respective wards, whereby the results revealed that more than 50% of the wards contain blocked drains that exacerbate the flood risk. This is represented in figure 13. To analyze further, the researchers carried out a household survey based on random sampling at flood-affected wards that revealed that the pipes with untreated wastewater from the kitchen are directly connected to the city drainage system. Besides, the household survey revealed that

30% of the surveyed households do not utilize the corporation waste collection system, which could be associated to the drainage issues. Moreover, 90% of the roads observed in the surveyed wards had their drains dumped with wastes or partially blocked as indicated in the figure 14 below. These factors will influence the overall risk patterns of the study region.

Figure 11. Final rate of change of built up over the floodplains/flood zones of Kochi corporation 2000-19
Source: Created by Authors, 2020

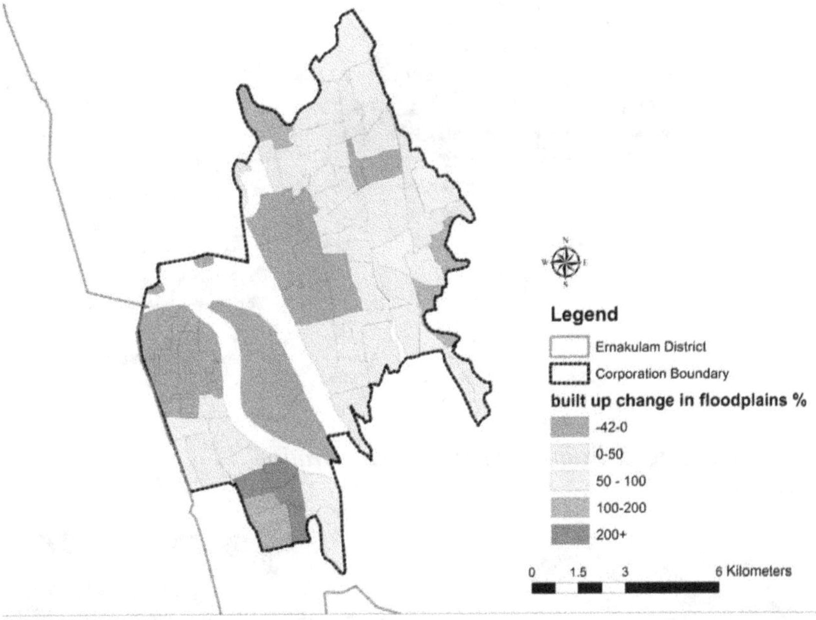

Figure 12. Figure showing vulnerable constructions along the water bodies
Source: Created by Authors, 2020

Figure 13. Map showing wards with existing drainage blockages
Source: Created by Authors, 2020

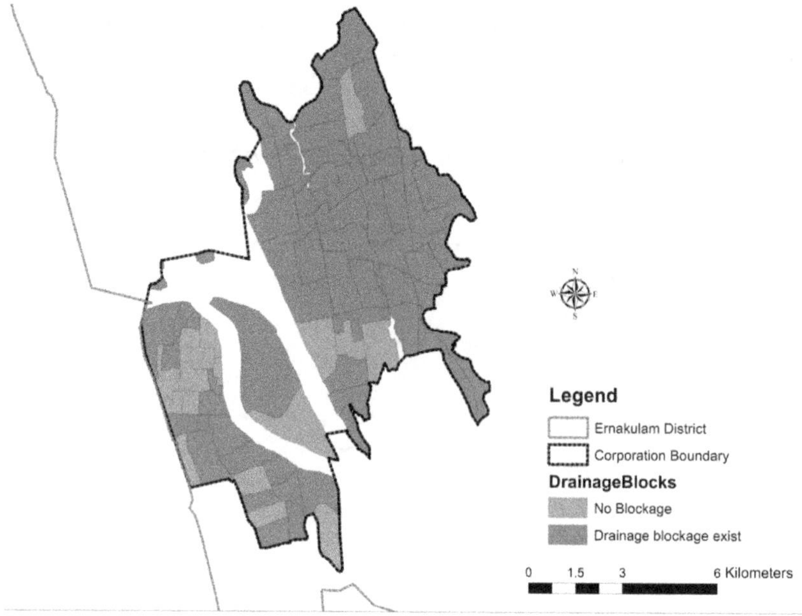

Figure 14. Blocked and damaged drains of the Kochi corporation
Source: Created by Authors, 2020

Furthermore, the monthly precipitation during monsoon (June- November) of the concurrent years from 2016-19 has been on the rise as discussed above in figure 3, which implies that the overall management needs faster policies and solutions as the overall precipitation models have been modified with climate change. This is important as the city has reported recurring incidents of urban flooding during 2018 and 2019. These complexities call for comprehensive solutions to address the challenges of Kochi Municipal Corporation.

SOLUTIONS AND RECOMMENDATIONS

The research realized that the Cochin urban region has undergone a significant change in its land use, resulting to large-scale floodplain mismanagement and consequent increase in surface runoff. The study further identifies poor drainage scenario, which is characterized by aged, poor design, and management, as factors aggravating flood risks. The study recommends the following steps to address the flood risk in the study region.

- The researcher suggest utilizing effective measures such as land-use planning that discourage built-up in eco-sensitive zones, complex modeling for run-off management and promoting ward level flood management plans, which can be facilitated by ward counselors and other representatives. The land-use plans should also incorporate specific drainage and waste management/monitoring scheme, for better resilience and capacity.
- To enhance risk-informed-planning, the researchers suggest promoting practices such as enhanced flood resilient structural measures for existing infrastructures, and measures such as community training programs at the household level in the flood risk zones.
- The authors further identify the need to advocate planning measures to address the impacts at regional, local and household level and aggregate these measures in the form of long term, medium and short term interventions

CONCLUSION

Urban flooding is a major challenge that can disrupt the functioning of societies, and urban growth centers, and can eventually challenge the sustainability, and Sendai Framework goals. The study suggests the possibilities of science and technology in the policy making, and promotes evidence-based planning measures, that can be scaled across various sectors. Based on the study, it is imperative that a complex mechanism of responsibility sharing be derived for urban flood risk management that is supported by public, scientific communities and the policymakers. The researcher suggest utilizing platforms, whereby the policy makers and researchers share knowledge and good practices to address the risk patterns, in the context of climate change. To address the flood risk in an effective way, the policy makers should promote decentralized planning measures, and should facilitate measures for incorporating science, technology and, innovation into the governance. Furthermore, more studies on flood risk transfer mechanisms such as insurances should be advocated for promoting long-term sustainability.

REFERENCES

Bollin, C., Cárdenas, C., Hahn, H., & Vatsa, S. K. (2003). *Disaster Risk Management by Communities and Local Governments*. Secretary, Regional Policy Dialogue, Inter-American Development Bank. https://www.researchgate.net/publication/254421276_Disaster_Risk_Management_by_Communities_and_Local_Governments

Census. (2011). *Census of India 2011 Kerala Series-33 Part Xii-B District Census Handbook Ernakulam Village And Town Wise Primary Census Abstract (Pca) Directorate Of Census Operations Kerala*. https://www.censusindia.gov.in/2011census/dchb/3208_PART_B_ERNAKULAM.pdf

CRED. (2018). Natural Disasters 2018. In *Emergency Events Database, Center for Research on the Epidemiology of Disasters*. https://www.emdat.be/

Elboshy, B., Kanae, S., Gamaleldin, M., Ayad, H., Osaragi, T., & Elbarki, W. (2018). A framework for pluvial flood risk assessment in Alexandria considering the coping capacity. *Environment Systems & Decisions, 39*(1), 77–94. doi:10.100710669-018-9684-7

Government of Kerala, & Department of Environment and Climate Change. (2014). *Kerala State Action Plan on Climate Change*. http://www.moef.gov.in/sites/default/files/KERALA STATE ACTION PLAN ON CLIMATE CHANGE.pdf

Hollis, G. E. (1975). The effect of urbanization on floods of different recurrence interval. *Water Resources Research, 11*(3), 431–435. doi:10.1029/WR011i003p00431

Humanity Road. (2018). *2018 India Kerala Flooding* (Vol. 501, Issue c). https://reliefweb.int/sites/reliefweb.int/files/resources/Humanity Road Sitrep No 1%2C Kerala India Floods%2C 16Aug2018%281%29.pdf

Jha, A. K., Bloch, R., & Lamond, J. (2012). *Cities and Flooding A Guide to Integrated Urban Flood Risk Management for the 21st Century*. The World Bank. https://openknowledge.worldbank.org/handle/10986/2241

Joseph, J. K., Anand, D., Prajeesh, P., Zacharias, A., Varghese, A. G., Pradeepkumar, A. P., & Baiju, K. R. (2020). Community resilience mechanism in an unexpected extreme weather event: An analysis of the Kerala floods of 2018, India. *International Journal of Disaster Risk Reduction, 49*, 101741. doi:10.1016/j.ijdrr.2020.101741

Sunil Kumar, M. K. (2021). *Operation Breakthrough fails to end waterlogging in Kochi _ Kochi News - Times of India*. https://timesofindia.indiatimes.com/city/kochi/op-breakthrough-fails-to-end-waterlogging-in-city/articleshow/82985016.cms

Memorandum, A. (2018). *State Relief Commissioner, Disaster Management (Additional Chief Secretary) Government of Kerala. August.*

NASA. (2020). *POWER Data Access Viewer*. https://power.larc.nasa.gov/data-access-viewer/

NDMA. (2010). *National Disaster Management Guidelines: Management of Urban Flooding*. A publication of the National Disaster Management Authority, Government of India.

Oak Ridge National Laboratory. (2003). *Possible Vulnerabilities of Cochin, India, to Climate Change Impacts and Response Strategies to Increase Resilience*. Author.

Park, K., & Lee, M. H. (2019). The development and application of the urban flood risk assessment model for reflecting upon urban planning elements. *Water (Switzerland), 11*(5), 1–17. doi:10.3390/w11050920

Rinkesh. (2021). World's 10 Worst Natural Disasters - conserve-energy-future. *Conserve-Energy-Future*. https://www.conserve-energy-future.com/10-worst-natural-disasters.php

Sowmya, K., John, C. M., & Shrivasthava, N. K. (2015). Urban flood vulnerability zoning of Cochin City, southwest coast of India, using remote sensing and GIS. *Natural Hazards*, *75*(2), 1271–1286. doi:10.100711069-014-1372-4

Special Correspondent the Hindu. (2020, August). *Kochi Corporation_ Water doesn't drain from Kochi streets _ Kochi News - Times of India*. https://www.thehindu.com/news/cities/Kochi/kochi-suffers-as-drains-overflow/article32299553.ece#

Town and Country Planning GOK. (2009). *Kochi City Development Plan*. https://cochinmunicipalcorporation.kerala.gov.in/web/guest/development-plan

UN-ESCAP. (2019). *The Disaster Riskscape Across Asia-Pacific: Pathways for Resilience, Inclusion and Empowerment*. United Nations. https://www.unescap.org/publications/asia-pacific-disaster-report-2019

UNDESA. (2018). *World Urbanization Prospects: The 2018 Revision (ST/ESA/SER.A/420)*. (Vol. 12). United Nations. https://population.un.org/wup/Publications/Files/WUP2018-Report.pdf

UNDRR. (2015). *Sendai Framework for Disaster Risk Reduction 2015-2030* (Issue 1). https://www.preventionweb.net/files/43291_sendaiframeworkfordrren.pdf

UNDRR. (2020). *Hazard Definition & Classification Review: Technical Report*. United Nations Office for Disaster Risk Reduction (UNDRR). https://www.undrr.org/publication/hazard-definition-and-classification-review

UNISDR. (2009). Terminology on Disaster Risk Reduction. In *United Nations Office for Disaster Risk Reduction*. UNDRR. doi:10.1021/cen-v064n005.p003

United Nations Developmental Program. (2018). *Kerala: Post Disaster Needs Assessment-Floods and Landslides August 2018*. Author.

University of Maryland College Center for Disaster Resilience. (2018). *The Growing Threat of Urban Flooding: A National Challenge*. https://cdr.umd.edu/urban-flooding-report

ADDITIONAL READING

CRED. (2018). Natural Disasters 2018. In Emergency Events Database, Center for Research on the Epidemiology of Disasters. https://www.emdat.be/

Jha, A. K., Bloch, R., & Lamond, J. (2012). Cities and Flooding: A Guide to Integrated Urban Flood Risk Management for the 21st Century. The World Bank. https://openknowledge.worldbank.org/handle/10986/2241

UNDESA. (2018). *World Urbanization Prospects: The 2018 Revision (ST/ESA/SER.A/420)*. (Vol. 12). United Nations. https://population.un.org/wup/Publications/Files/WUP2018-Report.pdf

UNDRR. (2015). Sendai Framework for Disaster Risk Reduction 2015-2030 (Issue 1). https://www.preventionweb.net/files/43291_sendaiframeworkfordrren.pdf

UNDRR. (2020). Hazard Definition & Classification Review: Technical Report. In United Nations Office for Disaster Risk Reduction (UNDRR). https://www.undrr.org/publication/hazard-definition-and-classification-review

UN-ESCAP. (2019). The Disaster Riskscape Across Asia-Pacific: Pathways for Resilience, Inclusion and Empowerment. United Nations. https://www.unescap.org/publications/asia-pacific-disaster-report-2019

UNISDR. (2009). Terminology on Disaster Risk Reduction. In United Nations Office for Disaster Risk Reduction (UNDRR). doi:10.1021/cen-v064n005.p003

United Nations Department of Economic and Social Affairs Population Division. (2018). World Urbanization Prospects: The 2018 Revision (ST/ESA/SER.A/420) (Vol. 12). United Nations. https://population.un.org/wup/Publications/Files/WUP2018-Report.pdf

KEY TERMS AND DEFINITIONS

Climate Change: The United Nations Framework Convention on Climate Change (UNFCCC) defines climate change as a change of climatic conditions, attributed directly or indirectly to the human activities, that affects the global atmospheric constituents, and results in natural variations over comparable time periods. Climate change increases uncertainty, which increases risk patterns.

Disaster: A serious disruption of the functioning of a community or a society which involves widespread human, material, economic or environmental losses and impacts, which exceeds the ability of the affected community or society to cope using its own resources is known as disaster (UNISDR, 2009).

Disaster Risk Management: Disaster risk management is the systematic process of using administrative directives, organizations, and operational skills and capacities to implement strategies, policies, and improved coping capacities in order to lessen the adverse impacts of hazards and the possibility of disaster. The management is a complex process which includes mitigation, prevention, preparedness and capacity building (UNISDR, 2009).

Hazard: Hazards refer to a process or activity that may cause damage to human life, socio-economic disruptions, or environmental damages. There are three criteria for hazard classification, which was established by the working group of Hazard Report, UNDRR (2020) has published three criteria for hazards: 1) the hazard has the potential to impact a community, 2) proactive and reactive measures exist for the hazard, and 3) the hazard has measurable spatial and temporal components (UNDRR, 2020).

Resilience: Resilience refers to the ability of a system, community or society exposed to hazards to resist, absorb, accommodate to and recover from the effects of a hazard in a timely and efficient manner, including through the preservation and restoration of its essential basic structures and functions (UNISDR, 2009).

Risk: Risk is defined as the potential disaster losses, in lives, health status livelihoods, assets and services, which could occur to a particular community or a society over some specified future time period. Disaster risk signifies the possibility of adverse effects in the future (UNISDR, 2009).

Vulnerability: Vulnerability refers to the characteristics and circumstances of a community, system or asset that make it susceptible to the damaging effects of a hazard (UNISDR, 2009). Vulnerability involves a combination of factors that determine the degree to which a person or community, livelihood, assets, ecosystem are put at risk by a discrete and identifiable event in the nature.

Chapter 10
Water Resource Management:
Geodata Analysis Through Google Earth Engine for Water Resources Management and Spatial Planning

Satya Prakash
https://orcid.org/0000-0001-6633-7480
Sharda University, India

Pinakana Sai Deepak
https://orcid.org/0000-0002-6351-950X
Sharda University, India

ABSTRACT

Water is an essential component for the survival of mankind and for balancing the ecosystem and livelihood. The world is experiencing a scarcity of water, both in terms of quality and quantity. Although there are several in-situ measurement techniques, they seem insufficient for large areas involving several parameters. Analysis of satellite images for estimating the quality and quantity of natural water has become an accepted tool for better spatial planning. With the increase in variety, volume, and velocity of satellite image, a tool for faster and accurate processing of the data is needed. Google Earth Engine (GEE) is one such cloud-based geo-big data platform. This chapter reviews the work of several researchers worldwide who have used and demonstrated the capability of satellite images with other geo-big data such as elevation, landcover, etc. for water resource management on the GEE platform. It can be concluded from the review work that GEE can help in estimating the water quality parameters with reasonable accuracy, comparable to the in-situ measurement, albeit quickly.

DOI: 10.4018/978-1-7998-8331-9.ch010

INTRODUCTION

Water, either in liquid or solid form, is an essential element for the survival of mankind. It is essential for balancing the ecosystem and livelihood. Due to factors like rapid population increase and industrialization, mankind has been facing the problem of water scarcity (Okello et al., 2015). The resource is either being polluted and/or depleted due to excessive use for domestic, industrial and agricultural purposes (Edokpayi et al., 2017). It would be a herculean task to handle the problem of water scarcity in the future because the population is going to increase and thus the increase of water demand. Persistent and precise monitoring of all types of water resources for their proper management is crucial. It has become essential to monitor hydro-morphology continuously to overcome the problem of water scarcity. This can only be done if the water quantity and quality of water are monitored regularly. Not only for drinking and daily use, but the water quality monitoring also helps in identifying water for irrigation, watering of livestock, water sports etc. (Bartram & Balance, 1998). Although this method is effective, it is not suitable for large geographical extent, volume monitoring as well estimating several parameters simultaneously. In this regard, satellite images can reinforce the abilities of water resources researchers and decision-makers to monitor water bodies for their quality and quantity, more effectively (Yigit Avdan et al., 2019). The climate and human actions affect the location and variability of both inland and coastal surface water (Chen et al., 2020). Although global datasets of the location of the surface water and its seasonal variations have been produced from different available resources and statistical extrapolation of regional data, measuring the long-term changes at high resolution remains a challenge.

BACKGROUND

Water Body Monitoring using Remote Sensing

From balloonist taking pictures of the ground using a photo camera as early as 1840 to the launch of WorldView-4 in 2016 with a spatial resolution of 31cm, remote sensing has come a long way. Spatial resolutions have started increasing due to the development in sensor technologies resulting in larger file sizes of the image for the same unit area on the ground straining the existing data processing resources. This multiplication in the volume of information has necessitated the need for automated cloud-based processing of the images. An increase in the number of platforms, number and type of sensors and frequently launching of new satellites has led to an increase in the 3Vs of big-data viz, volume (high volume of data), variety (from different sources/ sensors/ formats) and velocity (speed of emerging data stream) (Chi et al., 2016). In this regard, Google Earth Engine (GEE), developed towards the end of 2010, provides a cloud computing platform addressing the challenges of geo-big data analysis.

Advances in high-resolution remote sensing techniques and smart ICT have contributed to the proliferation of geo-big data in the field of water resources (Sun & Scanlon, 2019). Data about water resources ticks all the boxes of 3Vs for big data necessitating the use of GEE for processing the data for quantifying the qualitative parameters of water bodies. It has been reported that water quality parameters such as coloured dissolved organic matter, Chlorophyll-*a* and pollutants, Secchi disk depth, turbidity, total suspended solids, temperature, total phosphorous, surface salinity, dissolved oxygen, biochemical oxygen demand and chemical oxygen demand can be directly or indirectly estimated from the satellite image analysis (Gholizadeh et al., 2016). The analysis and interpretation of satellite images also help in estimating the trends and relative reservoir volume thus mapping the water supply vulnerability.

Although satellite images provide reasonable results for estimating the hydrological parameters and is a good alternative to the conventional methods, the method of satellite remote sensing have some drawbacks such as large datasets, downloading takes time and processing needs high-end resources. These are some of the reasons, why satellite data, although available for free (not all but majorly, which can be used for water resources monitoring), have not been used extensively in hydrological studies. These drawbacks necessitated the need for a system of faster processing without the need of downloading the images. For such a scenario, Google Earth Engine came as a saviour.

GOOGLE EARTH ENGINE

Google Earth Engine (GEE), a cloud computing platform rolled out by Google in 2010, provides a platform by giving access to the free satellite image datasets from several agencies, and algorithms of digital image processing, vector processing and visualization, machine learning (Google, 2021a). The benefits of GEE in water resource applications include efficient computational methods to speed image processing and model development or application tasks, API and developer environment for model prototyping, and ease of sharing datasets through JSON and workspaces (Hansen, 2015).

GEE contains a large catalogue of datasets. Some of the datasets available in GEE are complete Landsat series (1972 – present), MODIS images (2000 – present), Sentinel satellites (2014 – present), Advanced Land Observing Satellite (ALOS) etc. (Google, 2021b). The complete list of satellite data available in GEE is exhaustively listed in Appendix under Table-IV by Amani et al., 2020. These datasets are freely accessible and can be processed online to get the desired results with the help of Earth Engine (EE) Code Editor. The EE Code Editor can be processed either by JavaScript programming language or Python. GEE users not only can use the data of the datasets but can also upload their data by using the Asset feature. The factors of GEE which makes it stand out from the other remote sensing processing applications are, its ability to process the data which is much faster than other applications as there is no need of downloading the images which indeed increases the processing speed and most of the data available in Google Earth Engine has already been preprocessed and corrected (Amani et al., 2020). GEE also allows the processing of Sentinel-1 GRD (Ground Range Detected) images, but has its limitations in processing SLC (Single Look Complex) Synthetic Aperture Radar (SAR) data and generating interferograms. The applications of GEE are wide open. Figure 1 lists some of the applications, where researchers have used it, of GEE in different fields, including hydrology.

The volume and variability, with season and usage, of water, needs to be studied at the regional and global level for proper planning, development and management of water resources. This chapter aims at comprehensively exploring the usage of satellite images for hydrology related studies for spatial planning. There are mainly two sections of the chapter, first one details the methods for estimating the quality parameters of water using satellite images and GEE, whereas the second deals with the determination of the volume of water in the reservoir and its variability over time. The chapter also lists the applications of satellite remote sensing and the use of GEE for determining different parameters of water quality and how researchers have exploited the algorithms of GEE for determining the volume of water in the reservoirs. It has been shown that the results for both water quality and quantity can be determined using remote sensing techniques and especially using GEE with considerable accuracy.

Figure 1. GEE applications
Source: Amani et al., 2020

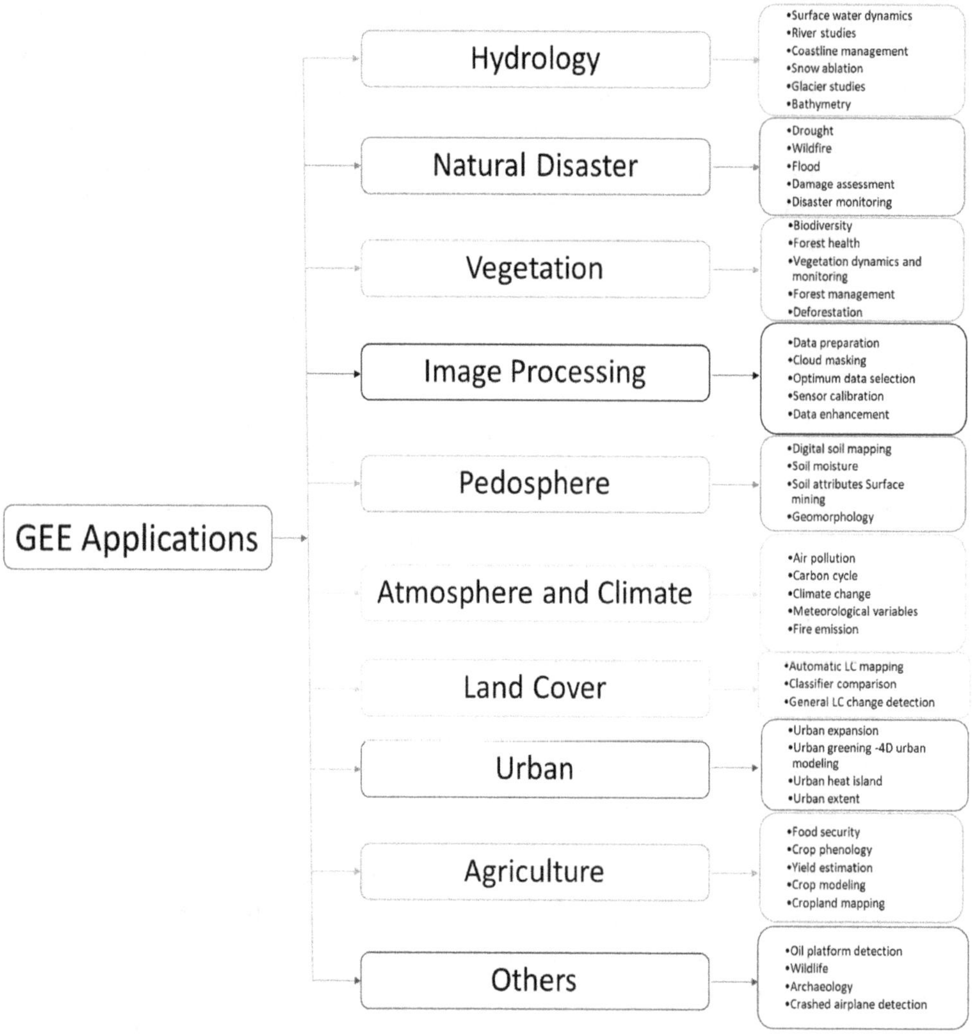

Water Quality

Water quality is the biological, physical and chemical characteristics of the water and is the primary factor that decides its use for drinking, food supply, recreation or any other usage (Bartram & Balance, 1998). It can be measured in terms of Ammonia content, Biological Oxygen Demand (BOD), Chemical Oxygen Demand (COD), chloride, conductivity, turbidity, chlorophyll-*a*, harmful algae, pollution-sediment, submerged habitat, temperature, metals, pH value, dissolved oxygen, nutrients (primarily phosphorus and nitrogen), and many other contaminants (Babin et al., 2015). The assessment of water quality helps in identifying pollutants in specific, and the sources of pollution (Giri et al., 2020). The monitoring of water quality programs started long back in 1976 (Gwynne, 1982). The monitored variables can be segregated into three categories i.e., basic, optional, and of global significance. The basic group consists of those variables which are vital for water quality assessment and are measured at all stations. The optional group consists of only those parameters which are important in some areas and whose selection for measurement depends on particular circumstances, whereas the global significance consists of variables of those toxic substances which are of long-term global significance (Gwynne, 1982). The monitoring of these variables is usually done by in-situ data collection. The method of in-situ data collection provides high accuracy. Some of the methods of in-situ data collection require more and it only offers the results at a definite place from where the sample is taken. This makes the use of this method of measuring the quality of large water bodies inaccurate and may result in inappropriate interpretation and generalizations (Gholizadeh et al., 2016). To reduce the workload and the bias that arises out of this, there has been wide use of sensors for the measurement of water quality. Starting from the use of pH meter around the 1930s to the use of molecularly imprinted sensors and microwave spectroscopy, many sensors were put into work to examine the quality of water (Yaroshenko et al., 2020). These methods of detection are cost-effective. However, some sensors require continuous availability of power and need frequent maintenance for uninterrupted application. Moreover, these sensors need to be replaced in case of malfunctioning and/or after their life period and need to be upgraded to newer devices. Other limitation includes damage of sensors due to natural disasters or by aquatic animals (Xu et al., 2020). These disadvantages necessities the use of technology that can monitor the water quality remotely, without human intervention along with being reasonably accurate, comparable with the in-situ methods. The need was also for a less complicated method overcoming the disadvantages of other processes. Satellite remote sensing methods of indirect estimation of water quality parameters paved a way for measuring water quality, overcoming the majority of the disadvantages of the conventional methods.

There are two primary techniques to model the water quality using remote sensing. The first one is empirical modelling which is statistical whereas the second one i.e., semi-analytical is based on the physics of light interactions with the water surface (Babin et al., 2015). These developments led to innovative, regional monitoring of inland surface water, including algal pigments and suspended sediment (Vogt & Vogt, 2016). Many of the satellite images which are useful for determining water quality, are available for free and is available to users worldwide, within a reasonable amount of time after the acquisition (Jiang et al., 2017). Commercial and free image porceissing software can be used to process the downloaded satellite images. However, GEE has come up with several algorithms, made available to the users for free, along with the satellite data from LANDSAT and other sensors, which can be used to determine the water quality. The algorithms of GEE can further be customized by the users to ascertain additional water quality parameters which may be added to the library of GEE for others to use.

Water has certain reflectance characteristics, the scattering and absorbance which are directly or indirectly related to the water quality parameters like turbidity, suspended matter, temperature, pH, and dissolved coloured organic matter. With the knowledge about the optical characteristics of the parameters, the quantitative values can be measured and recorded based on the light reflectance which is measured by the remote sensing instruments (UN-SPIDER, 2020) as shown in Figure 2. There are several satellites and sensors used to assess the water quality of a specific place. Some of these sensors/wavelengths and the parameters have been listed by Gholizadeh et al., 2016.

There are nine indicators available to assess the water quality out of which four can be spectrally captured. They are chlorophyll-*a*, coloured dissolved organic matter content, turbidity and total suspended solids (Graf et al., 2019). Various studies have used different algorithms in GEE to measure the indicators which can be spectrally captured.

Figure 2. Water quality vis-a-vis wavelengths and satellite sensors
Source: Greb and Daniela, 2018

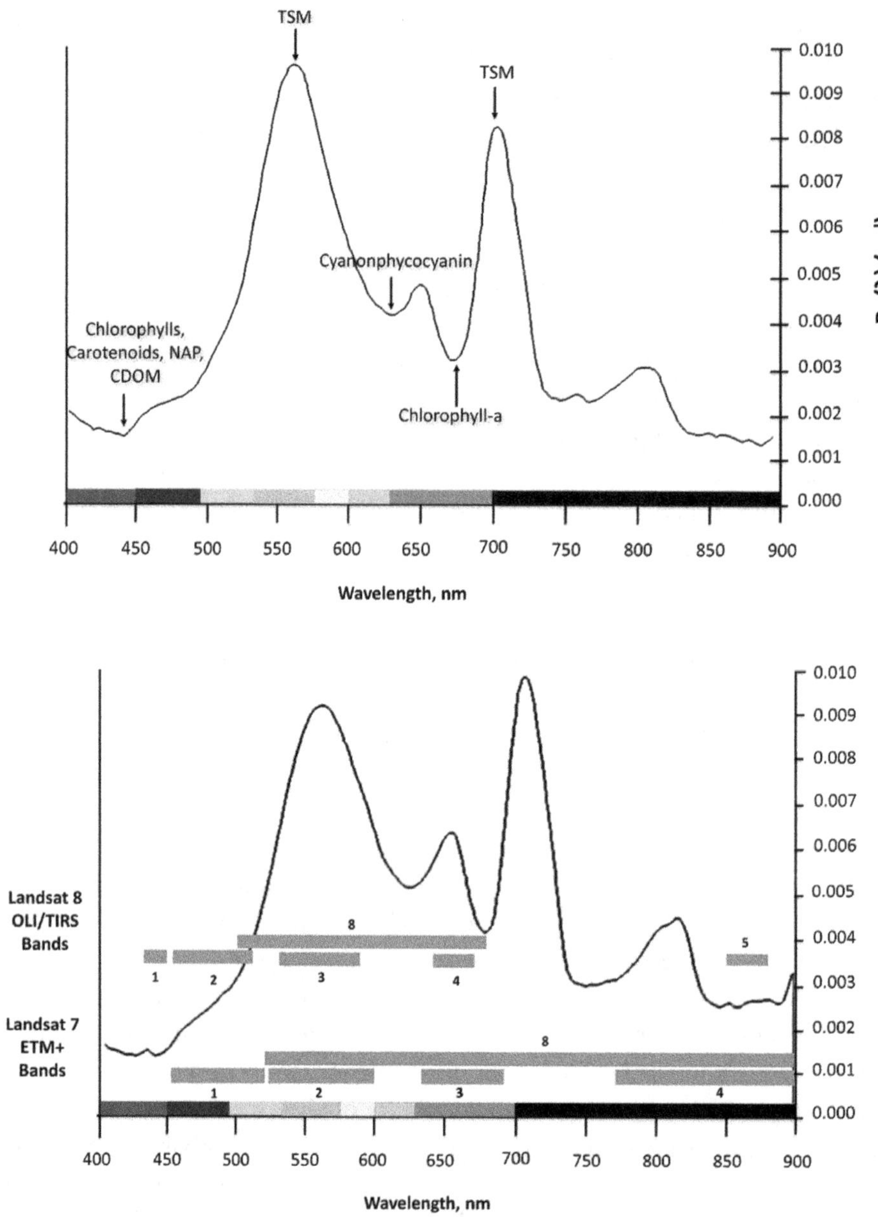

Mohamed, 2015 used the 300 meters spatial resolution Medium Resolution Imaging Spectrometer (MERIS) images of July-Oct 2009, to map the water quality of Lake Manzalah located in Egypt with the help of basic ENVISAT toolbox (A) ATSR and MERIS (BEAM) software version 4.2. The inbuilt algorithms of BEAM software were used to map the water quality of the lake. They measured turbidity (TUR) and Chlorophyll-*a* (CHL) and compared the results with the in-situ data and concluded that formulating regression models based on Top of Atmosphere (TOA) reflectance is a valid approach to generate Earth Observation (EO) based water quality information in an operational setting. Ouma et al.,

2020 compared the efficiency of Sentinel-2A/MSI (S2A) and Landsat-8 (L8)/OLI (Operational Land Imager) satellite sensors in estimating Chl-*a*, turbidity and Total Suspended Solids (TSS) in Chebara Dam, Kenya. EMRM – empirical multivariate regression model was used to retrieve the water quality parameters. The study results showed that in visible and NIR bands, the accuracy for Chl-*a* and TSS was better for Landsat-8 whereas for estimating turbidity, both the satellites provided similar results. Anderson et al., 2020 evaluated the use of hyperspectral data from Hyperion to estimate Chl-*a* levels and compare with the in-situ data for Lake Atitlan in Guatemala. The authors concluded that the use of OCx – ocean colour algorithm for the calculation of chlorophyll-*a* as the most suitable one. Poddar et al., 2019 have used two bio-optical Ocean Color algorithms OC-2 (2-bands) and OC-3 (3-bands) to access the performance of Chl-*a* for waters of the northern coastal Bay of Bengal using Landsat-8 OLI and Sentinel-2 MSI. They concluded that the Sentinel-2 MSI over-estimated Chl-*a* most of the time when compared with OLI data and OC-2 performed better than OC-3.

Chlorophyll - a

Amongst all water quality parameters, Chlorophyll-*a* (Chl-*a*) is one of the most studied parameters using remote sensing techniques (Chawla et al., 2020). Chl-*a* is the dominant pigment in the green algae among others like Cyanophyta, Chrysophyta etc. (Gorde & Jadhav, 2013). Measurement of chlorophyll in water depends upon the empirical relationship between reflectance in narrow bands and chlorophyll. The in-situ measured data results in a spectrum with increasing reflectance, with an increase in chlorophyll levels (Devi et al., 2015). The presence of different Chl-*a* concentrations in water leads to spectral reflectance curve with absorption in the blue (~0.4 μm) and red (~0.7 μm) bands, and reflectance in green (~0.5 μm) and near-infrared (~0.8 μm) bands (Chawla et al., 2020). Kimambo et al., 2019 measured the Chl-*a* levels with the use of remote sensing and the in-situ data (2017-18) of the Mindu dam, Tanzania. The data of Landsat-7 & 8 (1999-2017) have been used to measure the chlorophyll-*a*, where the data was collected using GEE datasets. The algorithm used in this study to measure Chl-*a* is OC-2 for both Landsat-7 & 8. The results of the in-situ data measurements showed that there was an increase in the levels of Chl-*a* and the Landsat-7 data also shows the same increase. The in-situ data ranged from 9.43 to 34.52 μg/L which falls within the range of OC-2 for Landsat-8, whereas the Landsat-7 data mostly fell under < 10 μg/L.

Malahlela et al., 2018 used Landsat-8 OLI for estimating Chl-*a* concentrations in Vaal Dam in South Africa using the NIR-Red bands for the period 2014-16 employing stepwise logistic regression (SLR). The authors also used MODIS data for validating Chl-*a* concentrations and they found an overall accuracy of 80% for 2014 images, 65% for 2015 Landsat images. Hansen et al., 2015 made an estimation model of Chl-*a* for Lake Michigan, USA. A general algorithm, as suggested by Fuller et al., 2007, was used with the help of GEE in the study. Time series analysis was done to calculate the amount of increase of Chl-*a* in the area of interest. Kalale, 2019 conducted an estimation and mapping of turbidity in the Lower Charles River, USA with the use of Landsat-8 OLI satellite imagery in GEE. A traditional approach of fitting simple equation using regression techniques was used in this study. The best retrieval algorithm was found by empirically deriving a model from all combinations. The best performing model was found to be a combination of blue and green reflectance with R^2 and RMSE values as 0.841 Formazin Nephelometric Units (FNUs) and 0.838 FNUs respectively. This model was used to visualize the spatial distribution of turbidity in the study region using Google Earth Engine. A model validation dataset was used to further assess the performance of the model which resulted in R^2 and RMSE values of 0.832 FNUs and 0.430 FNUs respectively. Several other empirical algorithms were developed to determine the Chl-*a*. Some of them are NDCI – Normalized difference chlorophyll index algorithm (Mishra & Mishra, 2012), MCI – Maximum chlorophyll index algorithm (Binding et al., 2013), SABI – Surface algal bloom index algorithm (Alawadi, 2010), FLH – Fluorescence line-height algorithm (Zhao et al., 2010) and NGRDI – Normalized green-red difference index algorithm (Feng et al., 2014). Several datasets of chlorophyll are available in GEE which can be used for studying the biology of coastal zones (Google, 2021d). Huovinen et al., 2019 estimated different water quality parameters including Chl-*a* for Lake Panguipulli, Chile using Landsat-5 TM, 7 ETM+, 8 OLI and Sentinel-2A/B MSI and validated with the in-situ data. The authors used the ocean Chlorophyll two-band (OC2v2) algorithm to calculate Chl-*a* which is based on the blue/green ratio of Sentinel data. This algorithm overestimated Chl-*a* levels slightly in the summer spring. They concluded that the estimation of Chl-*a* was somewhat less sensitive to detect seasonal variation due to a slight overestimation of the low chlorophyll levels of summer and spring. In near-shore areas, especially in the vicinity of river run-off, the accuracy of satellite-derived chlorophyll prediction is likely influenced by turbidity.

These studies show that Chl-*a* can be estimated, to a greater extent with reasonable accuracy, using the satellite images in GEE. However, as of now the values are over-estimated/under-estimated (depending upon the satellite sensor used) and several other factors affects the estimation of Chl-*a* which need to be studied and the values adjusted accordingly. Using the existing algorithms in GEE, reasonable accuracy may be obtained for the generalized study of any region. For local studies, more experimentation with the in-situ data is needed.

Turbidity/ TSS

Another important parameter to study the water quality is turbidity. Turbidity is the light scattering property of water which is caused by the presence of suspended and dissolved matter such as clay, silt, finely divided organic matter, plankton and other microscopic organisms, organic acids and dyes (Nas et al., 2010). Turbidity can be related to TSS and Secchi disk depth (zSD), measuring either the attenuation of a beam of light passing through water or, the scattering by particulate matter usually as detected within 90^0 of the light beams with units of nephelometric turbidity units (NTU). The relationship between TSS and turbidity is variable in space which varies with the change in particle size distribution, shape, composition and reflective index (Matthews, 2021). Notable relationships have been reported between turbidity and suspended sediments and radiance or reflectance from spectral Landsat bands. It has been reported that the wavelengths ranging from 700 to 800 nm (NIR) were most productive to determine turbidity in surface waters (Chebud et al., 2012). Turbidity presence in water bodies increases reflectance from red and NIR regions of the electromagnetic spectrum (Chawla et al., 2020). Graf et al., (2019) have generated the turbidity map of Lake Volta, Ghana. They have used the Landsat-8 observation in GEE by applying a regression formula developed by Liu and Wang, 2019. They have selected the traditional method development methods i.e., MLR algorithm (Multiple Linear Regression) and GEP algorithm (Gene-Expression Programming) to develop the model. In the developed model when the results of both the algorithms were compared with the actual measured turbidity data, the GEP model gave more accurate and reasonable results than the MLR algorithm. Huovinen et al., 2019 used the single-band, switching algorithm named T_Dogliotti (Dogliotti et al., 2015) using red band at reflectance (red) < 0.05, NIR band for reflectance (red) > 0.07 and linear weighting in the transition zone (0.05–0.07) and compared with the in-situ data of from monitoring database. It was found that the seasonal long-term means of turbidity from satellite imagery did not show significant differences from in situ data in any of the seasons. Balasubramaniana et al., 2020 presented a Statistical, Inherent Optical Property (IOP) -based, and muLti-conditional Inversion proceDure algorithm (SOLID) which can be used for the estimation of satellite-derived TSS for water in rivers, lakes, estuaries and coastal waters. The algorithm was implemented and compared with other algorithms over the Chesapeake Bay, San-Francisco-Bay-Delta Estuary, Lake Okeechobee, and Lake Taihu in the US with the use of Sentinel-2A/B images. The authors concluded that the SOLID model outperformed the state-of-the-art models in assessing the TSS levels when compared with >3500 in-situ databases and can be used for global mapping of TSS. Gu et al., 2020 proposed a turbidity measurement model which is based on a random forest ensemble (RFE). GEE datasets were used to obtain hyperspectral data of 13 bands. An algorithm named RFE-GEE was proposed for river turbidity measurement. The results of the algorithm were compared with nine other recently developed learners and the performance of RFE-GEE was found to be better than all others. The study also concluded that models based on neural networks might be inappropriate for turbidity measurement in large water bodies. When empirical algorithms, semi-empirical algorithms and semi-analytical models are compared to each other, semi-analytical models give more appropriate results while measuring turbidity in ocean waters (Topp et al., 2020).

Colored Dissolved Organic Matter (CDOM)

Coloured/Chromophoric Dissolved Organic Matter (CDOM) is a naturally occurring water-soluble, biogenic, heterogeneous organic substance, that absorbs light in both ultraviolet and visible range. It is in yellow to brown (Usali & Ismail, 2010) and commonly called a yellow substance or Gelbstoff which is the fraction of dissolved organic matter (DOM) pool in the sea. It was found that CDOM is responsible for an increase in the absorption of light with a decrease in the wavelength ranging between 350nm-700nm (Devi et al., 2015). The other water quality parameters like Chl-*a*, TSS, turbidity (Chawla et al., 2020) and especially DOC – dissolved organic carbon (Zhu et al., 2014) were found to be in strong correlation with CDOM. Juhls et al., 2019 compared the in-situ and MERIS extracted the CDOM of large parts of the western and central Laptev Sea shelf, coastal regions around the Lena River Delta and channels of the Lena River in Russia. The authors used the OCRS- Ocean Colour Remote Sensing algorithms, which provides the absorption coefficient of coloured dissolved organic matter, which in turn can be used to estimate DOC levels at a high temporal and spatial resolution over large regions. Five open-source Object-based image Clustering and Retrieval System (OCRS) algorithms were tested and compared out of which the OLCI - (Sentinel-3 Ocean and Land Colour Instrument) Neural Network Swarm (ONNS) algorithm showed the best performance and ONNS derived DOC concentrations showed moderate performance. They conclude that for near coastal areas, an adequate retrieval of DOC by satellite can be used with the help of OCRS ONNS algorithm. Zhu et al., 2014 evaluated over 15 CDOM estimation algorithms and compared them with the in-situ samples taken from Kawkawlin and Saginaw Rivers and Lake Huron, MI, USA. Out of the 15 algorithms, the semi-analytical inversion algorithms named QAA-v4, QAA-v5 and QAA – CDOM consistently gave better results than others across all low, medium and high CDOM level waters. The QAA-CDOM algorithm was the best of the three because it reduces the interference from high concentrations of sediments and Chl-*a* in freshwater environments. According to the authors, the most effective wavelengths/band locations for estimating CDOM could vary depending on the levels of spectral interference from high concentrations of particulate matter in the water column. The results also suggested that including wavelengths >600 nm in the algorithms improves CDOM estimation accuracy significantly, particularly for complex freshwater environments.

Griffin et al., 2018 estimated the daily CDOM using discharge-constituent regression-based models from 2000 to 2013. The Landsat Ecosystem Disturbance Adaptive Processing System (LEDAPS, Version 2) data from Google Earth Engine (GEE) was used for this study which calculates surface reflectance for Landsat 5 TM and 7 ETM+. The regression model named LOADEST – Load Estimator was used and the R^2 values range from 0.45 to 0.89 for the six rivers. The study concludes that the use of this model would increase the frequency of observations despite climate change. Olmanson et al., 2016 compared Landsat 8 and Landsat 7 for measuring CDOM in lakes and they concluded that Landsat 7 helps in finding the long-term changes in CDOM and hence it is better than Landsat 8.

Machine Learning Algorithms for Water Quality Determination

Hafeez et al., 2019 compared several Machine Learning (ML) algorithms i.e., Artificial Neural Network (ANN), Random Forest (RF), Cubist regression (CB), Support Vector Regression (SVR) for estimating the water quality parameters such as Turbidity, Chlorophyll-*a*, Suspended Solids (SS) for coastal waters. Landsat-5, 7 and 8 reflectance data were compared with the in-situ reflectance data to validate the performance of the ML algorithms. The water quality estimation was done for water bodies in Hong Kong using satellite images from 1999 to 2015. Out of the four machine learning algorithms, when compared, ANN exhibited the best performance. Its results included $R \approx 0.9$ and $RMSE \approx 0.2–1.4$ for Chl-*a*, $R \approx 0.9$ and RMSE 0.7–2.6 for SS, and $R \approx 0.85$ and $RMSE \approx 0.9–3.1$ for turbidity. They further conclude that the green band and red bands are more sensitive for predicting Chl-*a*, and the red band and its combination with blue and green bands are sensitive for predicting SS and turbidity. Kim et al., 2014 compared the machine learning algorithms viz., RF, CB and SVR for estimation of Chl-*a* and Suspended Particulate Matter (SPM) concentrations of coastal water quality at selected regions along the west coast of South Korea. The data set of GOCI - Geostationary Ocean Color Images were used to map the estimated concentrations and compared with the 63 in-situ samples. It was found that the SVR gave best results of $R^2 \sim 0.9$ and Cross-Validation (CV) RMSE ~ 1.6–1.7 mg/m^3 (39%) for Chl-*a* concentration, and $R^2 \sim 0.98$ and CV RMSE of 8.8–11.4 g/m^3 (48–63%) for SPM. The study results also stated that the hourly GOCI results showed the concentrations of Chl-*a* and SPM, are influenced by the tidal phases. Ahmed et al., 2019 used machine learning algorithms i.e., regression algorithms for assessing water quality index and classification algorithms for classifying the water bodies. The algorithms include Multiple Linear Regression, Polynomial Regression, Gradient Boosting Algorithm, Ridge Regression, Lasso Regression, Elastic Net Regression, Bagging Classifier, Logistic Regression, Gaussian Naïve Bayes. Jalal et al., 2019 analyzed the performance of the SVM model to the decision tree and concluded that SVM is a better option to monitor water quality. The majority of the studies concluded that the SVM algorithm provides more appropriate results when compared to other machine learning algorithms.

All these ML algorithms as well as the satellite data, except for GOCI, used in the study are available in the GEE platform and can be utilized for estimating water quality parameter. The algorithms for supervised and unsupervised classifications can also be accessed through GEE. The ML algorithms available in GEE for supervised classifications are SVM, RF and CART - the Classification And Regression Tree (Google, 2021e). The water quality assessment can be done by using these algorithms in GEE.

Based upon these studies, it can be concluded that the use of big data such as remote sensing technologies along with the satellite data and the data processing algorithm available in the GEE cloud platform can be used to estimate the quality parameters with fairly accurate results. However, some more research needs to be undertaken to refine the ML models and big data algorithms for better spatial water resource planning.

WATER QUANTITY

Another important parameter of interest regarding the study of water is the spatial distribution of water across the globe. The water quality parameter decides that the available water shall be used for which purpose and thus the water quantity and quality assessment are interrelated. The studies of water quality may need details on amounts of dissolved substances that are added or discharged often from aquatic systems as a result of water inflow or outflow (Boyd, 2015). Measurement of water quantity not only helps in validating the water quality assessment results but also helps in knowing the amount of water available at any point in time and its variability over the period. These data help in flood monitoring, surface water monitoring, drought management, watershed management (Bagchi and Busa, 2011), groundwater monitoring (Singh & Raju, 2019), among many others. The surface and subsurface water resources can be categorized into streamflow, terrestrial water storage and reservoirs (Chawla et al., 2020). Streamflow is the process of the hydrological cycle by which water from the land surface is transported to the oceans (Mishra & Coulibaly, 2009). Terrestrial water storage – TWS is the sum of soil moisture, groundwater, snow and ice, water in biomass, and surface water in lakes, reservoirs, wetlands and river channels (Girotto & Rodell, 2019). Terrestrial water storage plays a major role in the climate system, primarily through the exchange of water and energy at the land surface (Kim et al., 2009). It modulates the hydrological cycle and is the key determinant of water availability and drought indicator (Pokhrel et al., 2021). Measurement of volume in reservoirs provides information that can be critical to climate change and water resource management (Ehsani et al., 2017).

Remote sensing helps in monitoring the large area in less time and with fewer resources. The only said drawback of this technology in measuring the streamflow is that these results which are obtained need to be validated and calibrated with the help of in-situ data. TWS measurement is crucial for sustainable water resources management and even for food security. But the lack of sufficient in-situ data and complexity of modelling the processes governing TWS are the major limiting factors for accurate estimation (Abileah et al., 2011). Some components can be measured directly but the estimation of these components on a large and scale is very difficult and is lacking (Prakash et al., 2014). Remote sensing helps to estimate the TWS components in large scales over large areas in less time and with fairly good accuracy. The data from satellites like GRACE - Gravity Recovery and Climate Experiment helped to produce a map of TWC with unprecedented accuracy (Abileah et al., 2011). Landsat and the MODIS which are optical sensors and SAR sensors like RADARSAT, JERS-1, and ERS are widely used in estimating water volume in reservoirs (Tang et al., 2010).

Streamflow Estimation

Datasets obtained through remote sensing can provide continuous global coverage of discharge estimates, which offers an opportunity to improve the global streamflow network. Although satellite sensors cannot measure streamflow discharge directly, other variables like water level height, velocity, river cross-section area, river width, and surface water slope can be measured by sensors using which the streamflow estimation can be done (Chawla et al., 2020). Sproles et al., 2018 used GEE to compute the SCF – Snow Cover Frequency values, which were used as an input in SnowCloudHydro to forecast streamflow for snow-dominated watersheds towards snow mapping and hydrologic modelling. The authors concluded that GEE helped in lowering the computational, budgetary and logistical barriers in implementing remotely sensed large datasets of snow and support the implementation towards better management of water resources. Sazib et al., 2020 studied the relationship between precipitation, soil moisture, and streamflow for 601 watersheds across the USA using GEE. The data-packs for this study i.e., NASA-USDA global soil moisture product was made available through GEE. The authors aggregated soil moisture and precipitation data sets to monthly composites and then spatially averaged over the watershed boundaries using GEE. They further applied the SVR model and ARIMA - AutoRegressive Integrated Moving Average for streamflow forecasting. The results indicated that the SVR models with antecedent precipitation and streamflow as predictors performed better forecasting, regardless of the watershed class than that with only streamflow input is a better option than the ARIMA model. They conclude that the application of satellite-based soil moisture and precipitation can be used very effectively for improving streamflow modelling and forecasting. Yang et al., 2020 presented an algorithm named RivWidthCloud used in the GEE to automatically extract river and widths from remotely sensed images. Landsat images were used to conduct the study. The results obtained using this algorithm were validated through in-situ data and RMS error = 99.2 m, mean absolute error = 43.1 m, mean bias = -20 m was obtained. This indicates that the algorithm can be used on a large scale to measure the river width. The river width, along with slope, aspect and curvature may help in streamflow estimation. Parameters essential for streamflow estimates such as slope, aspect, and curvature can be obtained from the terrain data. Safanelli et al., 2020 presented an algorithm TAGEE (Terrain Analysis in the GEE) for calculating the terrain attributes like slope, aspect and curvature for different resolutions of elevation data such as SRTM, AW3D30, GTOPO30, GMTED2010 and many others. TAGEE package has an advantage over functions available in GIS packages as having additional outputs such as curvatures and shape index. They can be useful for environmental mapping and modelling studies for streamflow estimates with results comparable with conventional GIS systems.

Terrestrial Water Storage

Wu et al., 2019a developed a fully automated and scalable algorithm, through GEE, to delineate small depressional wetlands at watershed scales to calculate the water storage capacity by integrating fine spatial resolution LiDAR data and multi-temporal (2009–2017) aerial images. NAIP - National Agriculture Imagery Program images and Global surface water products which are available in GEE were used for the processing and no external training datasets were used. The algorithm uses an unsupervised classification approach and water clusters which are extracted by utilizing prior water masks derived from existing global-scale surface water products based on the long-term Landsat archive. The source code for applying the methodology can be found here (Wu et al., 2019b). Zou et al., 2017 analyzed the spatial-temporal variability of open surface water bodies and its relationship with climate and water exploitation for the 1984-2015 period using 16,000 Landsat-5 & 7 images, available on the GEE platform, for Oklahoma, USA. The indices mNDWI, NDVI and EVI were used to classify open surface water bodies and calculate the variation in the area for water storage. Wang et al., 2018 proposed another method utilizing GEE to estimate annual surface water extent as conventional remote sensing methods consume a high amount of manpower, storage and computing resources. The study was done for the Yangtze River, the longest river in China. A total of 2343 Landsat images were analysed on the GEE platform for the years 1990, 2000,

2010, 2017 using the RFC algorithm and accuracy between 86% to 93% was achieved when compared with Joint Research Centre (JRC) Yearly Water Classification History (v1.0) (Pekel et al., 2016). Another study (Xia et al., 2019) using the indices (Zou et al., 2017) monitored the changes in water surface area of Huai river basin, China using GEE by processing 16,760 scenes of Landsat TM/ETM+/OLI for the period 1 January 1989 to 31 December 2017 and quantified the spatiotemporal variability of the surface water area changes over the years. The results were validated (Pekel et al., 2016) and the algorithm was validated (Liao et al., 2014 & Jun et al., 2014) which resulted in 87% accuracy.

Monitoring of Reservoirs

For better water resource management, continuous observation of the reservoirs in terms of storage, water conservation, flood control, water treatment and aquatic environment is needed (Votruba and Broza, 1989). Gao et al., 2012 created time histories of reservoir storage for selected 34 reservoirs globally using five satellite altimeters for 1992-2010 and MODIS data for 2000-2010. The authors compared the satellite-based results with gauge observations at the five largest reservoirs of the USA and reported an NRMSE between 3-15% and R between 0.92-0.99. Although they did not use GEE for the study, they demonstrated the use of satellite-based data for reservoir storage capacity. Moreover, the K-means clustering algorithm used in the study is available on the GEE platform along with the MODIS and satellite altimeter data which can be used for estimating the reservoir storage capacity with considerable accuracy. Bhagwat et al., 2019 processed the freely available Global Surface Water (GSW) Monthly Water History (MWH) product with corresponding water surface median elevation values from three different digital elevation models DEM (TanDEM-X, SRTM and ALOS) for Lake Mead, Lake Powell and reservoirs in California between 1984 and 2015. The GSW data is based on the processing of 3 million images from Landsat 4, 5, 7 and 8 in GEE. The authors estimated the volumetric fluctuations in the reservoirs and compared the results with volumetric variation estimations for Lake Mead and Lake Powell using in-situ storage data. They found that the model could very well be used in other regions with considerable accuracy. Nguy-Robertson et al., 2018 defined the correlation between satellite-based reservoir elevation and aerial extents. Water levels from satellite-based altimeter data (TOPEX/Poseidon, Jason series) were correlated with MODIS and Landsat TM/ETM+/OLI images for 18 sites spread globally, between 2000 to 2015, using cloud-based image processing environment GEE. The results were correlated to in-situ data for six sites with ground-based measurements. They concluded that with the calibration of surface area vs water level relationships, surface area from satellite images can be used to monitor changes in the water level of the reservoirs.

Xavier et al., 2020 investigated the space-time dynamics of Serra da Mesa Reservoir, Brazil processing Landsat 5 TM and Landsat 8 OLI data in GEE for the period 1998 to 2019. The authors used the SWIR 1, Red and Near Infrared (NIR) spectral bands from the sensors TM (RGB 543) and OLI (RGB 654) with corresponding processing reflectance levels to the top of the atmosphere. The satellite-derived estimates of reservoir water volume were compared with the data from Reservoir Monitoring System, managed by National Water Agency. The authors found that the accuracy was over 90%. They concluded that the main feature of GEE i.e., high-performance cloud computing enables quick and accurate assessments of reservoirs in different environments and can be used on large scale for the management of natural resources.

ADVANTAGES AND DISADVANTAGES OF GEE IN WATER RESOURCE MANAGEMENT

From the above studies, it is relatable to how important is the use of satellite images for the assessment of water quality and quantity of water in the reservoirs. Since the conventional method of satellite data processing takes a lot of time and computing as well as human resources, the use of cloud-based big data processing platforms such as Google Earth Engine has started playing an important role in water resource management and spatial planning. Some of the advantages of GEE over conventional satellite data processing can be listed as,

- **Data Availability**: GEE provided a robust platform and processing environment, providing free accessibility to a series of satellite image data which is important in any Geospatial analysis. The data not only includes the data from sensors such as Landsat, Sentinel but also from MODIS, ASTER, among many others. The GEE contains more than 200 common datasets, more than 5 million images and vector-based datasets. The elevation data from GTOPO30, SRTM, USGS GMTED2010, ETOPO1, HYCOM etc., are also available which can be integrated with the processed data to generate realistic models. The complete list of data available has been listed by Amani et al., 2020.
- **Processing Speed**: The most important of all, processing speed can be increased up to great levels with the use of GEE. As the data is readily available for use, there is no need of downloading the data which reduces the time required for any big data analysis. All operations are automatically processed in bulk and parallel computing are performed. Several years of data can be processed quickly and without the need for local high-end computing resources. Compared with traditional image processing tools the GEE can process images quickly and in batches. In addition, its processing power is constructed based on Google Cloud, not subject to time and place.
- **Machine Learning Algorithms**: GEE platform is designed to allow petabyte-scale data processing, analysis, and visualization. It also provides an integrated environment including cloud computing and classical algorithms. Algorithms such as Random Forest, Support Vector Regression, K-Means among many others are available to use on GEE. They have been used in different case studies to evaluate the water quality and quantity and a reasonable accuracy was achieved. It also contains functions for image processing, vector processing, geometrical analysis and visualization algorithms.
- **Code Editor**: The GEE code editor can be processed through either JAVA or Python. The data, which is not available, can be uploaded, and the results of the processed data can be downloaded with the use of the code editor. Users can share data and codes via URLs based on the GEE's convenient mechanism. The platform serves as a uniform access interface to a variety of data types, free from image bands, projection, resolution, or bit-depth, making multi-sensor analysis feasible (Zhao et al., 2010).

However, there are a few limitations to using the GEE to monitor land cover. First, if further analysis is required, it is time-consuming to download large-scale classification results, since the pixel-based map contains several small patches that increase the data volume. Second, there is a lack of entire control of the GEE platform, and the raw images are online, not in-house. Programs conducted by government agencies may be apprehensive about the accessibility in further research. In addition, the algorithm provided by GEE is not fully explained in some cases (Zhao et al., 2010).

The main advantages and limitations of processing through cloud-based geo-big data GEE, API, data and functions are listed by Amani et al., 2020.

FUTURE RESEARCH DIRECTIONS

As the study of water quality and the estimation of availability of the quantity of water for better water resource management at the regional and global level has become imperative with an increase in the water demand, the use of GEE shall increase with time. With more and more users coming onto the platform to use it, several user-defined algorithm and methodology shall be developed which would result in better management of water resources for better spatial planning.

CONCLUSION

Although the use of satellite images for spatial planning of water resources can not be declined as it gives results with reasonable accuracy when compared with the in-situ data. However, there was a limitation of its use due to an increase in the volume, variability and speed at which the data is being captured. Google has come up with a cloud computing geo-big data platform, GEE, for the online processing of satellite images which uses cloud-based data as well as the cloud-based algorithm for faster processing

REFERENCES

Abileah, R., Vignudelli, S., & Scozzari, A. (2011). *A completely remote sensing approach to monitoring reservoirs water volume*. Paper published in Fifteenth International Water Technology Conference IWTC 15 2011, Alexandria, Egypt.

Ahmed, U., Mumtaz, R., Anwar, H., Shah, A. A., Irfan, R., & García-Nieto, J. (2019). Efficient Water Quality Prediction Using Supervised Machine Learning. *Water (Basel)*, *11*(11), 2210. doi:10.3390/w11112210

Alawadi, F. (2010). Detection of surface algal blooms using the newly developed algorithm surface algal bloom Index (SABI). *Remote Sensing of the Ocean, Sea Ice, and Large Water Regions*, *2010*, 782506. Advance online publication. doi:10.1117/12.862096

Amani, M., Ghorbanian, A., Ahmadi, S. A., Kakooei, M., Moghimi, A., Mirmazloumi, S. M., Moghaddam, S. H. A., Mahdavi, S., Ghahremanloo, M., Parsian, S., Wu, Q., & Brisco, B. (2020). Google Earth Engine Cloud Computing Platform for Remote Sensing Big Data Applications: A Comprehensive Review. *IEEE Journal of Selected Topics in Applied Earth Observations and Remote Sensing*, *13*, 5326–5350. doi:10.1109/JSTARS.2020.3021052

Babin, M., Greb, S. R., Dekker, A. D., Arrigo, K. R., Forget, M. H., & the International Ocean-Colour Coordinating Group. (2015). *Earth Observations in Support of Global Water Quality Monitoring*. International Ocean Colour Coordinating Group. https://repository.oceanbestpractices.org/handle/11329/535

Bagchi, D., & Bussa, R. (2011). Application of Remote Sensing in Water Quality and Water Resources Management – An Overview. *Bhujal News Quaterly*. https://hindi.indiawaterportal.org/content/application-remote-sensing-water-quality-and-water-resources-management-overview/content-type-page/53244

Balasubramanian, S. V., Pahlevan, N., Smith, B., Binding, C., Schalles, J., Loisel, H., Gurlin, D., Greb, S., Alikas, K., Randla, M., Bunkei, M., Moses, W., Nguyễn, H., Lehmann, M. K., O'Donnell, D., Ondrusek, M., Han, T. H., Fichot, C. G., Moore, T., & Boss, E. (2020). Robust algorithm for estimating total suspended solids (TSS) in inland and nearshore coastal waters. *Remote Sensing of Environment*, *246*, 111768. doi:10.1016/j.rse.2020.111768

Bartram, J., & Ballance, R. (1998). *Water Quality Monitoring: A Practical Guide to the Design and Implementation of Freshwater Quality Studies and Monitoring Programmes* (1st ed.). CRC Press. https://www.who.int/water_sanitation_health/resourcesquality/wqmchap2.pdf

Bhagwat, T., Klein, I., Huth, J., & Leinenkugel, P. (2019). Volumetric Analysis of Reservoirs in Drought-Prone Areas Using Remote Sensing Products. *Remote Sensing*, *11*(17), 1974. doi:10.3390/rs11171974

Binding, C., Greenberg, T., & Bukata, R. (2013). The MERIS Maximum Chlorophyll Index; its merits and limitations for inland water algal bloom monitoring. *Journal of Great Lakes Research*, *39*, 100–107. doi:10.1016/j.jglr.2013.04.005

Boyd, C. E. (2015). *Water Quality: An Introduction* (2nd ed.). Springer. https://www.springer.com/us/book/9783319330624

Chawla, I., Karthikeyan, L., & Mishra, A. K. (2020). A review of remote sensing applications for water security: Quantity, quality, and extremes. *Journal of Hydrology (Amsterdam)*, *585*, 124826. doi:10.1016/j.jhydrol.2020.124826

Chebud, Y., Naja, G. M., Rivero, R. G., & Melesse, A. M. (2012). Water Quality Monitoring Using Remote Sensing and an Artificial Neural Network. *Water, Air, and Soil Pollution*, *223*(8), 4875–4887. doi:10.100711270-012-1243-0

Chen, L., Zhang, G., Xu, Y. J., Chen, S., Wu, Y., Gao, Z., & Yu, H. (2020). Human Activities and Climate Variability Affecting Inland Water Surface Area in a High Latitude River Basin. *Water (Basel)*, *12*(2), 382. doi:10.3390/w12020382

Chi, M., Plaza, A., Benediktsson, J. A., Sun, Z., Shen, J., & Zhu, Y. (2016). Big Data for Remote Sensing: Challenges and Opportunities. *Proceedings of the IEEE*, *104*(11), 2207–2219. doi:10.1109/JPROC.2016.2598228

Devi, G. K., Ganasri, B., & Dwarakish, G. (2015). Applications of Remote Sensing in Satellite Oceanography: A Review. *Aquatic Procedia*, *4*, 579–584. doi:10.1016/j.aqpro.2015.02.075

Di Tullio, M., Nocchi, F., Camplani, A., Emanuelli, N., Nascetti, A., & Crespi, M. (2018). Copernicus big data and google earth engine for glacier surface velocity field monitoring: Feasibility demonstration on san rafael and san quintin glaciers. *The International Archives of the Photogrammetry, Remote Sensing and Spatial Information Sciences*, 42(3), 289–294. doi:10.5194/isprs-archives-XLII-3-289-2018

Dogliotti, A., Ruddick, K., Nechad, B., Doxaran, D., & Knaeps, E. (2015). A single algorithm to retrieve turbidity from remotely-sensed data in all coastal and estuarine waters. *Remote Sensing of Environment*, *156*, 157–168. doi:10.1016/j.rse.2014.09.020

Edokpayi, J. N., Odiyo, J. O., & Durowoju, O. S. (2017). *Impact of Wastewater on Surface Water Quality in Developing Countries: A Case Study of South Africa*. Water Quality. doi:10.5772/66561

Ehsani, N., Vörösmarty, C. J., Fekete, B. M., & Stakhiv, E. Z. (2017). Reservoir operations under climate change: Storage capacity options to mitigate risk. *Journal of Hydrology (Amsterdam)*, *555*, 435–446. doi:10.1016/j.jhydrol.2017.09.008

Feng, L., Hu, C., Han, X., Chen, X., & Qi, L. (2014). Long-Term Distribution Patterns of Chlorophyll-a Concentration in China's Largest Freshwater Lake: MERIS Full-Resolution Observations with a Practical Approach. *Remote Sensing*, *7*(1), 275–299. doi:10.3390/rs70100275

Flores-Anderson, A. I., Griffin, R., Dix, M., Romero-Oliva, C. S., Ochaeta, G., Skinner-Alvarado, J., & Barreno, F. (2020). Hyperspectral satellite remote sensing of water quality in lake atitlán, Guatemala. *Frontiers in Environmental Science*, *8*, 7. Advance online publication. doi:10.3389/fenvs.2020.00007

Fuller, L. M., & Minnerick, R. J. (2007). *Predicting water quality by relating secchi-disk transparency and chlorophyll a measurements to landsat satellite imagery for Michigan inland lakes, 2001-2006.* https://pubs.usgs.gov/fs/2007/3022/pdf/FS2007-3022.pdf

Gao, H., Birkett, C., & Lettenmaier, D. P. (2012). Global monitoring of large reservoir storage from satellite remote sensing. *Water Resources Research*, *48*(9), 2012WR012063. Advance online publication. doi:10.1029/2012WR012063

Gholizadeh, M., Melesse, A., & Reddi, L. (2016). A Comprehensive Review on Water Quality Parameters Estimation Using Remote Sensing Techniques. *Sensors (Basel)*, *16*(8), 1298. doi:10.339016081298 PMID:27537896

Giri, A., Bharti, V. K., Kalia, S., Arora, A., Balaje, S. S., & Chaurasia, O. P. (2020). A review on water quality and dairy cattle health: A special emphasis on high-altitude region. *Applied Water Science*, *10*(3), 79. Advance online publication. doi:10.100713201-020-1160-0

Girotto, M., & Rodell, M. (2019). Terrestrial water storage. *Extreme Hydroclimatic Events and Multivariate Hazards in a Changing Environment*, 41–64. doi:10.1016/B978-0-12-814899-0.00002-X

Google. (2021a). *API; Google Earth Engine; Google Developers*. Google. https://developers.google.com/earth-engine/api_docs

Google. (2021b). *Datasets; Google Earth Engine; Google Developers*. Google. https://developers.google.com/earth-engine/datasets

Google. (2021c). *Sentinel 1; Google Earth Engine; Google Developers*. Google. https://developers.google.com/earth-engine/guides/sentinel1

Google. (2021d). *Chlorophyll datasets; Google Earth Engine; Google Developers*. Google. https://developers.google.com/earth-engine/datasets/tags/chlorophyll?hl=en

Google. (2021e). *API; Google Earth Engine; Google Developers*. Google. https://developers.google.com/earth-engine/guides/classification

Gorde, S. P., & Jadhav, M. V. (2013). Assessment of Water Quality Parameters: A Review. *International Journal of Engineering Research and Applications*, *3*, 2029–2035. https://www.ijera.com/papers/Vol3_issue6/LV3620292035.pdf

Graf, L., Asante, S. A., & Kovacs-Győri, A. (2019). *Investigating Conflicts between Gold Mining and Water Protection by Remote Sensing Imagery*. Regional Academy for the United Nations., doi:10.13140/RG.2.2.13344.74244

Greb, S., & Daniela, G. (2018). *Remote sensing of water quality in Wisconsin*. Paper presented at Wisconsin Lakes Convention.

Griffin, C., McClelland, J., Frey, K., Fiske, G., & Holmes, R. (2018). Quantifying CDOM and DOC in major Arctic rivers during ice-free conditions using Landsat TM and ETM+ data. *Remote Sensing of Environment*, *209*, 395–409. doi:10.1016/j.rse.2018.02.060

Gu, K., Zhang, Y., & Qiao, J. (2020). Random Forest Ensemble for River Turbidity Measurement From Space Remote Sensing Data. *IEEE Transactions on Instrumentation and Measurement*, *69*(11), 9028–9036. doi:10.1109/TIM.2020.2998615

Gwynne, M. D. (1982). The Global Environment Monitoring System (GEMS) of UNEP. *Environmental Conservation*, *9*(1), 35–41. doi:10.1017/S0376892900019469

Hafeez, S., Wong, M., Ho, H., Nazeer, M., Nichol, J., Abbas, S., Tang, D., Lee, K., & Pun, L. (2019). Comparison of Machine Learning Algorithms for Retrieval of Water Quality Indicators in Case-II Waters: A Case Study of Hong Kong. *Remote Sensing*, *11*(6), 617. doi:10.3390/rs11060617

Hansen, C.H. (2015). *Google Earth Engine as a Platform for Making Remote Sensing of Water Resources a Reality for Monitoring Inland Waters*. doi:10.13140/RG.2.1.3688.1047

Huovinen, P., Ramírez, J., Caputo, L., & Gómez, I. (2019). Mapping of spatial and temporal variation of water characteristics through satellite remote sensing in Lake Panguipulli, Chile. *The Science of the Total Environment*, *679*, 196–208. doi:10.1016/j.scitotenv.2019.04.367 PMID:31082593

Jalal, D., & Ezzedine, T. (2019). Performance analysis of machine learning algorithms for water quality monitoring system. *2019 International Conference on Internet of Things, Embedded Systems and Communications (IINTEC)*. 10.1109/IINTEC48298.2019.9112096

Jiang, L., Schneider, R., Andersen, O., & Bauer-Gottwein, P. (2017). CryoSat-2 Altimetry Applications over Rivers and Lakes. *Water (Basel)*, *9*(3), 211. doi:10.3390/w9030211

Juhls, B., Overduin, P. P., Hölemann, J., Hieronymi, M., Matsuoka, A., Heim, B., & Fischer, J. (2019). Dissolved organic matter at the fluvial–marine transition in the Laptev Sea using in situ data and ocean colour remote sensing. *Biogeosciences*, *16*(13), 2693–2713. doi:10.5194/bg-16-2693-2019

Jun, C., Ban, Y., & Li, S. (2014). Open access to Earth land-cover map. *Nature*, *514*(7523), 434. doi:10.1038/514434c PMID:25341776

Kalele, A. S. (2019). *Estimation and mapping of turbidity in the lower charles river using Landsat 8 oli satellite imagery*. http://hdl.handle.net/2047/D20323946

Kim, H., Yeh, P. J. F., Oki, T., & Kanae, S. (2009). Role of rivers in the seasonal variations of terrestrial water storage over global basins. *Geophysical Research Letters*, *36*(17), L17402. Advance online publication. doi:10.1029/2009GL039006

Kim, Y. H., Im, J., Ha, H. K., Choi, J. K., & Ha, S. (2014). Machine learning approaches to coastal water quality monitoring using GOCI satellite data. *GIScience & Remote Sensing*, *51*(2), 158–174. doi:10.1080/15481603.2014.900983

Kimambo, O. N., Chikoore, H., Gumbo, J. R., & Msagati, T. A. (2019). Retrospective analysis of Chlorophyll-a and its correlation with climate and hydrological variations in Mindu Dam, Morogoro, Tanzania. *Heliyon*, *5*(11), e02834. doi:10.1016/j.heliyon.2019.e02834 PMID:31763484

Liao, A., Chen, L., Chen, J., He, C., Cao, X., Chen, J., Peng, S., Sun, F., & Gong, P. (2014). High-resolution remote sensing mapping of global land water. *Science China. Earth Sciences*, *57*(10), 2305–2316. doi:10.100711430-014-4918-0

Liu & Wang. (2019). Modelling Reservoir Turbidity Using Landsat 8 Satellite Imagery by Gene Expression Programming. *Water*, *11*(7), 1479. doi:10.3390/w11071479

Malahlela, O. E., Oliphant, T., Tsoeleng, L. T., & Mhangara, P. (2018). Mapping chlorophyll-a concentrations in a cyanobacteria- and algae-impacted Vaal Dam using Landsat 8 OLI data. *South African Journal of Science*, *114*(9/10). Advance online publication. doi:10.17159ajs.2018/4841

Matthews, M. W. (2011). A current review of empirical procedures of remote sensing in inland and near-coastal transitional waters. *International Journal of Remote Sensing*, *32*(21), 6855–6899. doi:10.1080/01431161.2010.512947

Mishra, A. K., & Coulibaly, P. (2009). Developments in hydrometric network design: A review. *Reviews of Geophysics*, *47*(2), RG2001. Advance online publication. doi:10.1029/2007RG000243

Mishra, S., & Mishra, D. R. (2012). Normalized difference chlorophyll index: A novel model for remote estimation of chlorophyll-a concentration in turbid productive waters. *Remote Sensing of Environment*, *117*, 394–406. doi:10.1016/j.rse.2011.10.016

Mohamed, M. (2015). Satellite data and real time stations to improve water quality of Lake Manzalah. *Water Science*, *29*(1), 68–76. doi:10.1016/j.wsj.2015.03.002

Nas, B., Ekercin, S., Karabörk, H., Berktay, A., & Mulla, D. J. (2010). An Application of Landsat-5TM Image Data for Water Quality Mapping in Lake Beysehir, Turkey. *Water, Air, and Soil Pollution*, *212*(1–4), 183–197. doi:10.100711270-010-0331-2

Nguy-Robertson, A., May, J., Dartevelle, S., Birkett, C., Lucero, E., Russo, T., Griffin, S., Miller, J., Tetrault, R., & Zentner, M. (2018). Inferring elevation variation of lakes and reservoirs from areal extents: Calibrating with altimeter and in situ data. *Remote Sensing Applications: Society and Environment, 9*, 116–125. doi:10.1016/j.rsase.2018.01.001

Okello, C., Tomasello, B., Greggio, N., Wambiji, N., & Antonellini, M. (2015). Impact of Population Growth and Climate Change on the Freshwater Resources of Lamu Island, Kenya. *Water (Basel), 7*(12), 1264–1290. doi:10.3390/w7031264

Olmanson, L. G., Brezonik, P. L., Finlay, J. C., & Bauer, M. E. (2016). Comparison of Landsat 8 and Landsat 7 for regional measurements of CDOM and water clarity in lakes. *Remote Sensing of Environment, 185*, 119–128. doi:10.1016/j.rse.2016.01.007

Ouma, Y. O., Noor, K., & Herbert, K. (2020). Modelling Reservoir Chlorophyll-a, TSS, and Turbidity Using Sentinel-2A MSI and Landsat-8 OLI Satellite Sensors with Empirical Multivariate Regression. *Journal of Sensors, 2020*, 1–21. doi:10.1155/2020/8858408

Pekel, J. F., Cottam, A., Gorelick, N., & Belward, A. S. (2016). High-resolution mapping of global surface water and its long-term changes. *Nature, 540*(7633), 418–422. doi:10.1038/nature20584 PMID:27926733

Poddar, S., Chacko, N., & Swain, D. (2019). Estimation of chlorophyll-a in northern coastal bay of Bengal using Landsat-8 OLI and sentinel-2 MSI sensors. *Frontiers in Marine Science, 6*, 598. Advance online publication. doi:10.3389/fmars.2019.00598

Pokhrel, Y., Felfelani, F., Satoh, Y., Boulange, J., Burek, P., Gädeke, A., Gerten, D., Gosling, S. N., Grillakis, M., Gudmundsson, L., Hanasaki, N., Kim, H., Koutroulis, A., Liu, J., Papadimitriou, L., Schewe, J., Müller Schmied, H., Stacke, T., Telteu, C. E., ... Wada, Y. (2021). Global terrestrial water storage and drought severity under climate change. *Nature Climate Change, 11*(3), 226–233. doi:10.103841558-020-00972-w

Prakash, S., Gairola, R., Papa, F., & Mitra, A. (2014). An assessment of terrestrial water storage, rainfall and river discharge over Northern India from satellite data. *Current Science, 107*(9), 1582–1586. http://www.jstor.org/stable/24107219

Safanelli, J., Poppiel, R., Ruiz, L., Bonfatti, B., Mello, F., Rizzo, R., & Demattê, J. (2020). Terrain Analysis in Google Earth Engine: A Method Adapted for High-Performance Global-Scale Analysis. *ISPRS International Journal of Geo-Information, 9*(6), 400. doi:10.3390/ijgi9060400

Sazib, N., Bolten, J., & Mladenova, I. (2020). Exploring Spatiotemporal Relations between Soil Moisture, Precipitation, and Streamflow for a Large Set of Watersheds Using Google Earth Engine. *Water (Basel), 12*(5), 1371. doi:10.3390/w12051371

Singh, A., & Raju, A. (2019). Application of Grace Satellite Data for Assessment of Groundwater Resources in Central Ganga Alluvial Plain, Northern India. *Environmental Concerns and Sustainable Development*, 153–162. doi:10.1007/978-981-13-5889-0_7

Sproles, E. A., Crumley, R. L., Nolin, A. W., Mar, E., & Moreno, J. I. L. (2018). SnowCloudHydro—A New Framework for Forecasting Streamflow in Snowy, Data-Scarce Regions. *Remote Sensing, 10*(8), 1276. doi:10.3390/rs10081276

Sun, A. Y., & Scanlon, B. R. (2019). How can Big Data and machine learning benefit environment and water management: A survey of methods, applications, and future directions. *Environmental Research Letters*, *14*(7), 073001. doi:10.1088/1748-9326/ab1b7d

Tang, Q., Gao, H., Yeh, P., Oki, T., Su, F., & Lettenmaier, D. P. (2010). Dynamics of Terrestrial Water Storage Change from Satellite and Surface Observations and Modeling. *Journal of Hydrometeorology*, *11*(1), 156–170. doi:10.1175/2009JHM1152.1

Tang, Z., de Almeida, C., & Fishwick, P. A. (1991). Time series forecasting using neural networks vs. Box- Jenkins methodology. *Simulation*, *57*(5), 303–310. doi:10.1177/003754979105700508

Topp, S. N., Pavelsky, T. M., Jensen, D., Simard, M., & Ross, M. R. V. (2020). Research Trends in the Use of Remote Sensing for Inland Water Quality Science: Moving Towards Multidisciplinary Applications. *Water (Basel)*, *12*(1), 169. doi:10.3390/w12010169

UN-SPIDER. (2020). *Data application of the month: Water quality monitoring*. UN-SPIDER Knowledge Portal. https://www.un-spider.org/links-and-resources/data-sources/daotm-water-quality

Usali, N., & Ismail, M. H. (2010). Use of Remote Sensing and GIS in Monitoring Water Quality. *Journal of Sustainable Development*, *3*(3). Advance online publication. doi:10.5539/jsd.v3n3p228

Vogt, M. C., & Vogt, M. E. (2016). Research Article: Near-Remote Sensing of Water Turbidity Using Small Unmanned Aircraft Systems. *Environmental Practice*, *18*(1), 18–31. doi:10.1017/S1466046615000459

Wang, C., Jia, M., Chen, N., & Wang, W. (2018). Long-Term Surface Water Dynamics Analysis Based on Landsat Imagery and the Google Earth Engine Platform: A Case Study in the Middle Yangtze River Basin. *Remote Sensing*, *10*(10), 1635. doi:10.3390/rs10101635

Wu, Q., Lane, C. R., Li, X., Zhao, K., Zhou, Y., Clinton, N., DeVries, B., Golden, H. E., & Lang, M. W. (2019). Integrating LiDAR data and multi-temporal aerial imagery to map wetland inundation dynamics using Google Earth Engine. *Remote Sensing of Environment*, *228*, 1–13. doi:10.1016/j.rse.2019.04.015 PMID:33776151

Wu, Q. R., Lane, C., Li, X., Zhao, K., Zhou, Y., Clinton, N., DeVries, B. E., Golden, H., & Lang, M. W. (2019). Mapping wetland hydrological dynamics using Google Earth Engine (Version 1). figshare. https://github.com/giswqs/Wetland-Hydro-GEE doi:10.6084/m9.figshare.8864921.v1

Xavier, G. D. O., Almeida, T. D., Oliveira, C. M. M. D., Oliveira, P. D. S. D., Costa, V. H. B., & Granado, L. M. A. (2020). Estimate and evaluation of reservoir metrics in Serra da Mesa dam (GO) using the Google Earth Engine platform. Ambiente e Agua - An Interdisciplinary. *The Journal of Applied Science*, *15*(5), 1. doi:10.4136/ambi-agua.2584

Xia, H., Zhao, J., Qin, Y., Yang, J., Cui, Y., Song, H., Ma, L., Jin, N., & Meng, Q. (2019). Changes in Water Surface Area during 1989–2017 in the Huai River Basin using Landsat Data and Google Earth Engine. *Remote Sensing*, *11*(15), 1824. doi:10.3390/rs11151824

Xu, G., Shi, Y., Sun, X., & Shen, W. (2019). Internet of Things in Marine Environment Monitoring: A Review. *Sensors (Basel)*, *19*(7), 1711. doi:10.339019071711 PMID:30974791

Yang, X., Pavelsky, T. M., Allen, G. H., & Donchyts, G. (2020). RivWidthCloud: An Automated Google Earth Engine Algorithm for River Width Extraction From Remotely Sensed Imagery. *IEEE Geoscience and Remote Sensing Letters*, *17*(2), 217–221. doi:10.1109/LGRS.2019.2920225

Yaroshenko, I., Kirsanov, D., Marjanovic, M., Lieberzeit, P. A., Korostynska, O., Mason, A., Frau, I., & Legin, A. (2020). Real-Time Water Quality Monitoring with Chemical Sensors. *Sensors (Basel)*, *20*(12), 3432. doi:10.339020123432 PMID:32560552

Yigit Avdan, Z., Kaplan, G., Goncu, S., & Avdan, U. (2019). Monitoring the Water Quality of Small Water Bodies Using High-Resolution Remote Sensing Data. *ISPRS International Journal of Geo-Information*, *8*(12), 553. doi:10.3390/ijgi8120553

Zhao, D., Xing, X., Liu, Y., Yang, J., & Wang, L. (2010). The relation of chlorophyll-a concentration with the reflectance peak near 700 nm in algae-dominated waters and sensitivity of fluorescence algorithms for detecting algal bloom. *International Journal of Remote Sensing*, *31*(1), 39–48. doi:10.1080/01431160902882512

Zhu, W., Yu, Q., Tian, Y. Q., Becker, B. L., Zheng, T., & Carrick, H. J. (2014). An assessment of remote sensing algorithms for colored dissolved organic matter in complex freshwater environments. *Remote Sensing of Environment*, *140*, 766–778. doi:10.1016/j.rse.2013.10.015

Zou, Z., Dong, J., Menarguez, M. A., Xiao, X., Qin, Y., Doughty, R. B., Hooker, K. V., & David Hambright, K. (2017). Continued decrease of open surface water body area in Oklahoma during 1984-2015. *The Science of the Total Environment*, *595*, 451–460. doi:10.1016/j.scitotenv.2017.03.259 PMID:28395260

Chapter 11
Ecological Areas and Challenges of Urbanization: A Stride Towards Environmental Restoration

Garima Toor
Malaviya National Institute of Technology, Jaipur, India

Tarush Chandra
Malaviya National Institute of Technology, Jaipur, India

ABSTRACT

Ecological areas are the network of protected areas that contribute to the ecosystem's productivity and services. With increased human demands, towns and cities are blooming with changes in landuse patterns around their peripheral areas or in the immediate vicinity. Land intensification and disproportionate urbanization have inflicted various challenges such as qualitative and quantitative depletion of natural resources, ecosystem services, and degradation of environmental quality in and around ecological areas. The chapter will focus on the circumstantial elucidation of ecological areas, their recorded challenges caused by urbanization, and the need for their conservation in previous research studies. The authors explore reported challenges encompassing ecological areas by urbanization. This will help understand the various aspects of urban transformation, like physical, social, cultural, and economic change in and around the ecological areas, and develop measures and strategies for ecosystem conservation and environmental restoration in ecological areas.

INTRODUCTION

The global landscape is decked up with a variety of natural habitats. Having rich biodiversity, they are the hub of ecological processes which are silently at work to maintain a balance of the ecosystem through sustainable use of natural resources. Millennium Ecosystem Assessment 2005 has mentioned that the assortment of life available on earth has been altered vitally and irrevocably by human actions. Biodiversity

DOI: 10.4018/978-1-7998-8331-9.ch011

conservation is itself a target articulated by Convention on Biological Diversity (CBD) with several other legal frameworks at a global and national level. The indigenous wildlife and biodiversity inhabit these protected areas and are more vulnerable to the burgeoning threats occurring from anthropogenic activities. This has forced policy planners and governments to restrict activities and even to cordon them as 'protected areas'. Ecological areas provide various services, named ecosystem services, which comprise provisional services like water, food, raw materials, medicinal plants; regulatory services such as climate regulation, control of air and water pollution, management of floods, droughts, and epidemics; habitat or supporting services such as provide habitat areas, maintenance and formation of soil nutrients; and recreational services for cultural empowerment and spiritual wellness. Therefore, as hubs for biological processes and ecological services, these areas are essential for achieving ecological equilibrium.

For the conservation and management of ecological areas, there are various recognized international organizations, conventions, enacted protocols and declarations. These organizations and conventions were established with an intent to promote coordination, cooperation, and interaction among them for future environmental protection, conservation, and management. These organizations are operated as governing authorities having their expertise in strengthening the standards and practices for conservation goals, research studies related to nature, and sustainable future. The recommendations of the Convention on Biological Diversity (CBD) and the Ramsar Convention have been adopted at the global level. In India, the legal framework regarding the environment has covered various laws and legislation, summits, schemes, and projects such as the wildlife protection act, environment protection act, biodiversity act, project tiger, project elephant, project snow leopard, etc. Mostly these are aimed to regulate, manage and protect biodiversity, wildlife, and available natural resources.

Biodiversity is an essential element for many ecosystem services which these protected areas provide for human survival, welfare and livelihoods. Regardless of the vital contributions to human society, ecology and biodiversity are still declining. The record population growth, migration, and urbanization are unpremeditated. With increased human demands, towns and cities are blooming with changes in landuse patterns around their peripheral areas or in the immediate vicinity. This sprawl in urban areas is globally recognized as significant planning and management issue in several major cities. Land intensification and disproportionate urbanization have made our ecological areas victims of uncontrolled urban growth. This has inflicted various challenges such as qualitative and quantitative depletion of natural resources, the devastation of ecosystem services, and degradation of environmental quality in and around ecological areas. These peaceful areas, having biological importance, are now in continuous threats of human intervention as evident from their shrinking sizes, fragmentation, and habitat destruction, forced extinction of select bio-species, reduction in species count, the decline in migratory species visits, especially migratory bird species, destruction of indigenous ecosystems, shrinking and contaminations of rivers and wetlands, deforestation, etc. The aggravated urban sprawl comes with a complex mix of changes such as landuse transformation and environmental disturbances. The increase in urban sprawl and modification in landuse are reported worldwide as responsible for ecological degradation and biodiversity loss. The examples like the establishment of industries, illegal constructions and development work in the vicinity of ecological areas, violation of water bodies and rivers with untreated waste leading to chemical reactions, recent fire disaster in the world's largest rainforests, etc. all substantiate the above statement.

Contrary to this, evidences of nature rejuvenating itself during lockdown due to the recent global pandemic of COVID-19 have been reported from all over the world. This is the time when fear of getting infected captured the humans in their own homes and nature has took this opportunity to flourish and replenish itself at a global level. The changes recorded in nature and ecosystem are clean and healthy river systems (reduction in water pollution), clear air quality (reduction in air pollution and GHG emissions) as this made possible the visibility of Himalayan peak from Punjab state first time in decades, spotted endangered Gangetic dolphins in river Ganges after 30 years, etc. These incidents have proved that development activities and human interferences are responsible for environmental degradation. This owes much to the development strategies and action plans which mostly focus on issues related to infrastructure, urban economy, politics, and environmental pollution.

Urban growth and landuse changes have been measured in many research studies. In this process, geoinformatics have played an eminent role in spatial and environmental planning. The geospatial datasets and their applications are essential for the development of landuse landcover cartography with characterization and quantification and spatial monitoring. Geoinformatics retains key importance in the interpretation and visualization of these remotely sensed datasets for various spatial assessments. It supports various environmental studies and resource management globally and nationwide. For safeguarding ecological areas and their biodiversity, there is a need to comprehend the characteristics and role of various protected area categories known worldwide and nationwide.

ECOLOGICAL AREA CHARACTERISTICS

Worldwide the ecological areas are termed as 'protected areas'. This definition was first given by the International Union for Conservation of Nature (IUCN) in 2008. The definition is broadly acknowledged in legal framework across the world and describes this protected area as a geographical area with clear demarcation, to attain the enduring nature conservation goals with associated cultural and ecosystem service values, recognized and managed through legal or other active sources of the organization (IUCN, 2014). IUCN has considered the protected area in the following six categories, classified according to the management targets. The defined categories for protected areas are strict nature reserves, wilderness areas, national parks, natural features or monuments, habitat or species management areas, and protected landscapes/ seascapes. Following are the categories of 'protected areas' as classified by IUCN. These are recognized worldwide by eminent international associations and national regimes as the global standard for defining the protected areas.

Strict Nature Reserve

This is the first category of protected areas (Category 1a). A defined zone with the highest protection for wildlife consists of native ecology intact with ecosystem processes that can be efficiently refurbished with marginal management and conservation interventions. To safeguard the conservation values, these ecological reserves are not open for tourism purposes as these reserves are act as an essential platform for ecological monitoring and scientific research. The main objective of the protected area is to conserve the ecosystem, species, and geodiversity species regionally, nationally, and globally. Natural forces have developed these ecological attributes which are mostly deteriorated due to human interventions. These reserves played a vital role in safeguarding the earth's ecological richness which is not viable outside of such strictly protected sites. Great Himalayan National Park Conservation Area, India; Western Ghats, India; Air and Ténéré Natural Reserves, Niger; Gough and Inaccessible Islands, U.K; Mount Nimba, Guinea; Sichuan Giant Panda Sanctuaries, China are the few examples that come under the strict nature reserve category of the protected area which is highly rich with endemic flora and fauna species.

Wilderness Areas

This protected area is also in category 1b but their governance and management permit them to retain their natural character by enforcing slight or no modification within them. Wilderness areas are intended to maintain the enduring veracity of undisturbed biodiversity-rich ecological areas from anthropogenic disturbances; therefore, these areas are open at some levels for public access. Indigenous communities are encouraged and facilitated for the maintenance of ethnic traditions and customs in these areas. These areas are defined with the original range with native floral and faunal species in the ecosystem. The ecological areas under this category are generally larger and have access to human visits in contrast to the previous category (1a) of protected areas. Human inhabitation is usually not found in category (1a), whereas in this category (1b) protected areas are home to indigenous communities in many cases. Nanda Devi Biosphere Reserve, India; Great Gulf Wilderness Trailhead, U.S; Marjory stoneman Douglas wilderness, Florida; Zion wilderness, Utah; Death Valley wilderness, California; Hawaii volcanoes wilderness, Hawaii; etc. are a few examples in the category (1b) of protected areas.

National Parks

National parks come under category (2) which covers wide-ranging natural areas. These parks are defined as the ecosystem character of the area along with ecological processes and biodiversity. The main objective for defining the national park category is to promote and safeguard the ecological processes, natural structures, and ecosystem services with the maintenance of biodiversity and ecological balance. This category is open for human visitation fulfilling inspirational, spiritual, recreational, educational, and cultural purposes. These human activities are allowed for a specific level to prevent threats to wildlife and disturbances in biodiversity-rich areas. This category of the protected area plays a key role in developing biological corridors and other connectivity corridors for specific species and provides large-scale conservation opportunities. This category provides more tourism opportunities than category (1b) as they are restricted with limited infrastructure and human visits. Kaziranga National Park, India; Jim Corbett National Park, India; Koli National Park, Finland; Yellowstone National Park, U.S; Tongariro National Park, New Zealand; Northeast Greenland National Park, Greenland are the few protected area examples.

Natural Monument or Features

This category (3) consists of small protected areas with high-ranking tourist's visit places, intended to maintain and safeguard the specific declared natural monument, their outstanding natural features, and associated biodiversity. The protected areas have the following characters:

- Natural geomorphological or geological features for instance waterfalls, caves, dunes, rock formations, valleys, fossil beds, coral formations, marine features, etc.
- Culturally influenced natural features include ancient tracks and cave houses.
- Natural cultural and sacred sites such as mountains, springs, sacred groves, etc.
- Cultural ecological sites include historical/ archaeological spots which are linked to an ecological area.

Natural coral bridge, Andamans, India; National Fossil Wood Park, Thiruvakkarai, India; Cerro Ñielol Natural Monument, Chile; Carvers Creek State Park, North Carolina; Devils Postpile National Monument, California; Balda Canyon Natural Monument, Georgia are the few examples of protected areas in this category.

Habitat/ Species and Management Area

The category (4) of the protected area is intended to conserve and protect the recognized biological species diversities of flora and fauna including resident or migratory species with their habitats. The management and conservation concerns are usually limited to the whole protected areas, but this category is specifically designated a vital role in covering the disparities in planning and management strategies by valuing the key species, their habitats in the ecosystem. The focus of safeguarding species covers critically endangered species, threatened habitats, stepping stones, and breeding sites. This category particularly aims at the identification of target listed species along with their habitats while other categories target for conservation of a fully functional ecosystem. Tadoba Andhari Tiger Reserve, India; Kutch Desert Wildlife Sanctuary; Amana nature reserve, France; Bear river migratory bird refuge, Utha; Bald Point State Park, Florida; Baird Bay Islands Conservation Park, Australia, etc.

Protected Landscapes/ Seascapes

Protected areas in category (5) have considerable ecological, cultural, recreational, and conservational values. This category is defined to protect and sustain the significant landscapes or seascapes with their biological and conservation values. These kinds of protected areas are intended to sustain the fair interaction of ecology and encourage the preservation of agrobiodiversity and aquatic biodiversity. These areas have resulted from biotic, abiotic, and human interaction with specific characteristics such as landscape/ coastal/ island seascape of high and significant habitats, flora, fauna, and associated cultural and biological values. These protected areas provide good opportunities for reactional and tourism activities. Gulf of Mannar Biosphere Reserve, India; Nilgiri Biosphere reserve, India; Azalea State Natural Reserve, California; Bluestone Wildlife Management Area, Virginia; Cairuçu Environmental Protection Area, Rio de Janeiro; Casuarina Coastal Reserve, Australia; the Channel Islands National Marine Sanctuary, California; Chattahoochee River National Recreation Area, Georgia are the few examples in this category.

Protected Areas with Sustainable use of Natural Resources

The protected area in category (6) is aimed to conserve the ecological units and natural habitats along with sustainable management of cultural, traditional, and conservation values for available natural resources. These protected areas have the objective to promote the protection of the natural ecosystem and sustainable utilization of available natural resources considering the social, economic and environmental aspects. The large natural areas are appropriate for this category which includes the tropical forests, desert lands, intricate wetland areas, boreal forests, coastal areas, etc. Hanumansagar Reserve Forest, India; Palani Hills Forest Conservation area, India; Amanã Sustainable Development Reserve, Brazil; Calakmul Biosphere Reserve, Mexico; Cujubim Sustainable Development Reserve, Brazil; Black Jungle Conservation Reserve, Australia; Gulf of Carpentaria Marine Park, Australia are the few examples of this category.

The Protected area definition provided by the International Union for Conservation of Nature (IUCN) has been acknowledged widely in regional and global frameworks and varied in nations with protection and management levels based on nationally endorsed laws and international regulations. These are universally known as 'protected areas' in various government legislations which vary from region to region based on involved nations and global organizations. Although, there is a variation in nomenclature adopted in countries for protected area categories, yet these are ultimately fusing with the IUCN category as shown in Table 1, taking the case of India, Australia and the United States of America.

Table 1. Protected areas in different countries and their corresponding IUCN category

Country	Protected area category	IUCN category
Australia	National parks	II
	National heritage list	III
	Botanical gardens	IV
	Commonwealth marine reserves	V
	Antarctic specially protected areas	I, II, IV
	Antarctic specially managed areas	I, II, IV
United States	National estuarine research reserves	Ia
	National conservation lands	Ib, III
	National park system	II
	National forest	II
	National trails system	II
	National wilderness preservation system	IV
	National wildlife refuge system	IV
	National recreation areas	V
	National wild and scenic river system	V
	National marine sanctuaries	VI
India	National parks	II
	Wildlife sanctuaries	II, IV
	Conservation reserves	III
	Community reserves	VI
	Marine protected area	V

Source: (Compiled by author,2021)

In India, the Wildlife Institute of India (WII) has introduced five categories of protected areas i.e., national park, sanctuary, conservation reserve, community reserve, and marine protected area. The ecological area having a biological, faunal, floral, geomorphological, or zoological association of importance is designated as National Park and Sanctuaries. National parks may find within a wildlife sanctuary. Human activities are allowed in sanctuaries to some extent and are strictly banned by authorities in the case of national parks. These areas are essential natural spots for propagating, developing, and safeguarding the biodiversity and ecosystem. The protected areas are generally acting as buffer zones to migratory corridors or connectivity routes established in between the forests and designated as 'conservation reserves' and 'community reserves'. Conservation reserves are contiguous to sanctuaries or national parks owned by the state government for safeguarding the biodiversity with their habitats in landscape/ seascape. The community-owned land/ private land comes under community reserves for safeguarding the flora and fauna as well as improving the socio-culture and tradition of the indigenous community. For the protection of aquatic flora and fauna, there are marine protected areas that cover the region in the ocean which are restricted for human intervention as regulated similarly in the national park on the land.

LEGAL PROVISIONS: GLOBAL PERSPECTIVE

The protection of the environment has been a major concern, both at the national and global levels. This has forced the intergovernmental organizations to formulate various legal provisions for arising environmental issues. United Nations has initiated various programs to overcome the issue of environmental degradation. IUCN is the oldest and largest of the known organizations in the environment domain. The organization was established in 1948 for tasks related to nature conservation and equitable usage of natural resources. It has to its credit the publication of listed threatened species, known as 'IUCN red list', to recognize the global status of the threatened species. Thereafter, in 1972, United Nation Environment Programme (UNEP), was initiated after the Stockholm conference, as a global environmental authority to meet the Sustainable Development Goals (SDGs). UNEP strengthens the environmental practices and standards and helps in the implementation of environmental obligations at the global and regional levels. UNEP along with World Meteorological Organisation (WMO) has established a dedicated intergovernmental organization recognized as 'Intergovernmental Panel on Climate Change (IPCC)'. This organization is engaged in accessing information regarding impacts of climate change-induced by human disturbances with nature and provides related mitigation and adaptation measures. United Nations has initiated a major assessment of human impacts on the environment in 2001, designated as Millennium Ecosystem Assessment (MEA). It was aimed to review the consequences of ecological transformation for human well-being and development. It has provided the scientific appraisal for ecosystem services and possibilities for the restoration and conservation of ecosystems.

International Agreements for Environmental Protection

To lead the nations for sustainable development and nature conservation, United Nations has initiated various Convention of Parties (COPs) to establish treaties and enact agreements for safeguarding the environment worldwide. One of the international treaties was endorsed in 1971 under the 'Ramsar Convention' also recognized as 'Convention on wetlands of international importance, especially as waterfowl habitat. The treaty is among one of the eighth conventions that emphasize the issues related to biodiversity and exemplify the commitments of the member countries to plan for feasible use of wetlands and their resources to preserve the ecological character of the recognized wetlands under this convention. There are international agreements, specifically focusing on species and biodiversity. 'Convention on international trade in endangered species of wild flora and fauna' also abbreviated as CITES is a global accord ratified in 1975 to prevent the survival risk of wild flora and fauna species engaged in international trade. 'Convention on migratory species of wild animals', also known as 'Bonn convention' is another agreement signed in 1979 at the international level which provides a conservation platform for migratory species and their habitats. This agreement assembles the state authorities to set the legitimate organization at the international level for coordinating the migratory routes and related conservation. 'Convention on biological diversity (CBD)' acts as an international legal mechanism for biodiversity conservation. The convention was endorsed in 1993 which aimed to take actions for fair usage of natural resources and ecological components. These accords are followed by member nations to prevent biodiversity loss and environmental degradation.

LEGAL PROVISIONS: NATIONAL PERSPECTIVE

In addition to the aforementioned international agreements, various environmental laws have been ratified from time to time at the national level also. This however exhibits variations across nations, depending on the level and urgency of environmental issues. The government of India has been active in the creation and enactment of a legislative framework in this regard.

Environmental Laws in India

Wildlife Protection Act is one of the important legislations of Indian environmental laws enacted in 1972 for wildlife protection in protected areas of the country. Constitutional provisions including Article 48A and Article 51A are the basis for framing this wildlife protection act. Indian Constitution has directed the state in Article 48A for protection of the environment and safeguards the forest areas and wildlife, whereas, in Article 51A, Indian citizens are imposed with precise fundamental duties for protection and improvement of the natural environment and enhance solicitude towards all alive creatures. This Act has aided our nation to turn out to be a member of the recognized Convention on International Trade in Endangered Species (CITES) of wild flora and fauna. The protected area system of India was established under this act which categorized the ecological areas in the national park, sanctuary (bird and wildlife sanctuary), conservation reserve, community reserve, and marine protected area. A statuary organization – National Wildlife Board and National Tiger Conservation Authority was established under this Act.

The government was alarmed at rapid deforestation resulting in environmental degradation and decided to ratify an act for safeguarding forest cover. For this purpose, the 'Forest conservation act' was enacted in the year 1980 to strengthen the legislation for forests and their products. This act was aimed to deal with forest conservation in specified categories such as reserve forest, protected forest, private forest, and village forest. Later, in 1986, the Environmental (Protection) Act was endorsed to improve and protect the environment and matters related to it. This bill was endorsed in wake of the world's worst industrial disaster – 'Bhopal Gas tragedy and intended to empower strict actions against activities that harm the environment. The natural resources available within the protected areas are essential to regulate fair access and utilization. For this purpose, the 'Biodiversity Act' was ratified in 2000 to ensure an equitable share of available natural resources and their benefits. The legislation for biodiversity was intended to conserve the rich diversity of species available in protected areas. The bill also aimed to verify the biopiracy and collecting knowledge of biodiversity and its use by foreign organizations. This Indian legislation has led to the formation of biodiversity-related authorities and boards such as Biodiversity Authority at three different governance levels- national level (National Biodiversity Authority), state-level (State Biodiversity Boards), and local level (Biodiversity Management Committees).

The environmental issues related to lawsuits filed in the Indian judiciary are also required to be dealt with separately. The government has passed the National Green Tribunal Act in 2010, which laid the foundation for framing the National Green Tribunal (NGT) to handle the swift resolution of environmental issues related cases. This legislation was ratified under Article 21 of the Indian constitution which ensures the Indian citizens have the right to a healthy and safe environment. National Green Tribunal is an authorized body to solve the lawsuits within six months of the corresponding appeal.

For supplementary lead in achieving the purpose of endorsed environmental laws and with the drive of safeguarding forest, wildlife and environment a good number of schemes and projects like Project Tiger (1973), Indian Crocodile Conservation Project (1975), Project Hangul (1980), Project Elephant (1992), Sea Turtle Project (1999), Project Snow Leopard (2009), Ganges Dolphin (2016), Indian Rhino Vision (2020), etc. have been initiated by the government. Alongside, there are institutions like Wildlife Institute of India Dehradun; The energy and resources institute (TERI) New Delhi; Ashoka Trust for Research in Ecology and Environment (ATREE) Bangalore; Centre for Science and Environment (CSE) New Delhi; Bombay Natural History Society (BNHS) Mumbai; Environmental Information System (ENVIS) Centre on wildlife and protected areas, etc. which are working towards ecosystem conservation, wildlife protection and environmental management.

CHALLENGES OF URBANISATION

Urbanization is a well-known rural to urban transformation process performed in human settlements and followed by more population in human settlements. This has been recorded as one of the key influencers in manipulating the world's cartography (Trzyna, 2007). United Nations (2018) has estimated that the worldwide urban share is rising from around one-third in 1950 to two-third in 2050. India has ranked second in urban population size with 461 million urban dwellers in 2018. Between 2018 and 2050, India is estimated to have a major contribution to urban growth by increasing the size as nearly double its urban population with lead to the addition of urban dwellers by 416 million approximately. An urban agglomeration is a continuous spread of urban area constituting two or more contiguous towns or a town and its adjoining outgrowth. The urban area has defined by the Census of India in 2011 as a place that satisfied the criteria of a minimum of 5000 of the population with the density of 400 persons per square kilometre or higher and at least 75% of the male main working population are affianced in non-agricultural pursuits. In the global context, governments have designated human settlements based on varied criteria as urban and rural areas. Aijaz (2018) has reported the different definitions adopted by nations. In Canada, any area with a population density of 400 persons or more per sq. km and consist of 1,000 or more inhabitants comes under the criteria of an urban area. Areas with 2,000 or more inhabitants are classified as "urban" in Argentina, Austria, Ethiopia and Israel. The criteria followed in Republic of Korea (or South Korea), has 50,000 or more inhabitants for an "urban area". In USA, urban areas are defined as agglomeration of 2500 and more inhabitants with population density of 1000 persons per square mile or more.

Human settlements tend to expand themselves with the increase in population caused urban sprawl in the city periphery. Millions of people have migrated over the last century for earning a livelihood and better quality of life. Due to this process, cities and towns are expanding in an unsustainable manner making them more vulnerable and insecure. Expanded cities have consumed a tremendous amount of natural resources and exhaled a large number of harmful emissions and waste. Natural areas have attained a key role for growth, development, and enhancement in a city or town by fulfilling the resource need and deliver several ecosystem-based services such as recharge groundwater tables; provide drinkable water, raw materials, clean purified air, naturally regulated temperature; cultural and recreational facilities.

Some of the research case studies which reported the urban transformation in ecological areas are reviewed in terms of physical, cultural, social, and economic transformation as shown in table 1. Anderson et al. (2013) has recognized that urban resource consumption patterns have undesirably affected the ecosystem services and natural resources. Seto et al. (2012) has estimated that 1.8% of all the global biodiversity hotspot areas will be covered with new urban expansion by 2030. They have identified five biodiversity hotspots (the Caribbean islands; West African Guinean forests; Japan; Philippines; Western ghats and Sri Lanka) which are predicted to have the highest probability to achieve urbanization and have the largest percentage share of increased population by 2030. Expansion of the cities has resulted in fragmentation of natural habitat and increase the average distance between the fragmented patches which intensify the isolation of habitat areas (Elmqvist & Mcdonald, 2014); (Deb et al., 2014); (Sahu, 2006); (Kumar Ranjan et al., 2016); (Bhardwaj, 2017). Dewan & Yamaguchi (2009); Sukhdev (2010); Rajashekariah et al. (2011); Nagendra et al. (2013); Mallick et al. (2013); H.S. & Nagendra (2013); Everard et al. (2018); Reverte et al. (2020) have reported that uncontrolled urban growth and unsustainable planning of cities have adverse effects on natural resources, ecosystem services, biodiversity, and environmental quality. The reported adverse effects are an encroachment on protected areas, overutilization of available natural resources, deterioration of the soil quality, degradation of biodiversity, exacerbation of urban heat island effects, heightened pollution levels, and various threats to human wellbeing. Sahu, (2006); Kumar Ranjan et al. (2016); Verma (2017); Agarwal et al. (2018); Everard et al. (2018); Reverte et al. (2020) have reported the deforestation practices and human-wildlife conflicts in the protected areas. These ill practices and urban development in the surroundings of ecological areas are accountable for human-wildlife conflicts as these activities are reported to have disturbed the wildlife species, their habitat areas, and breeding sites, their migrating patterns, and deteriorate the floral diversity and other ecosystem services.

ROLE OF GEOSPATIAL TECHNOLOGY

The need for regular monitoring of the eco-sensitive areas has also been reported to detect the potential change in their ecological characters. Remote sensing and Geospatial technology have been found to be of significant help in ecological studies. Remote sensing technology was derived from aerial photo interpretation in 1968 (Franklin, 2001). This technique uses designed sensors which detect electromagnetic energy from various platforms such as ground-based platform, aerial platform, and satellite platform. The basic sensors in remote sensing are passive and active sensors that work on the illumination sources of the system. Passive sensors evaluate the emission and reflection of the natural light source from the earth's surface whereas active sensors emit the energy first, supplied by their energy source, and measure the reflected energy from the earth's surface. The data captured by sensors are then processed into an interpretable form and further endured for image processing for final satellite datasets.

The use of Remote Sensing applications and software tools based on geospatial techniques for quantitative assessment and detection of change in and around the 'protected areas' have been reported for accuracy and ease of comparison over time and space. Applications of remote sensing in the management and monitoring of ecological areas have been categorized as; forest cover types, forest structure estimation, forest cover change detection, and forest modelling (Franklin, 2001). These studies primarily focus on temporal change detection and analysis to estimate loss or encroachment in forest area(s), grasslands, and marshy lands. Use of these tools and techniques have been reported for assessment of the impact of burgeoning urban areas and their populace on ecological reserves, etc. The datasets and various geospatial systems have played a vital role in mapping and monitoring environmental aspects. Research studies of remote sensing in ecology, have reported the need to engage and integrate biodiversity values and conservation strategies at the initial stages in the physical planning process (Pedersen et al., 2004).

Table 2. Review of reported methods to measure transformations in ecological areas

Research citation	Parameters	Analytical structure and method used	Outcome and inferences
		Measurement of physical transformation	
Sahu, 2006	Change in landuse landcover (LULC) as per classification adopted in the study.	Temporal changes of the year 1991,1996 and 2002 were mapped for Dudhwa National Park, Uttar Pradesh using IRS data from NRSA (National Remote Sensing Agency).	• To examine the spatial and temporal patterns in landuse landcover change in a limited area within the region *(MICRO LEVEL ANALYSIS).* • To assess the impact on landuse landcover due to anthropogenic activities generated by urban development in the ecological area. • The landuse landcover assessment provides a base for the quantification of ecosystem services. The following LULC classification has been found: • Built environment (Settlements) • Forest (dense, open) • Agricultural land • Grassland • Fallow land / open/ wasteland • River • Water bodies
Deb et al., 2014		Landuse landcover patterns of year 1975,1990 and 2010 were mapped for Gumti Wildlife Sanctuary (GWLS), Tripura using Landsat images.	
Kumar Ranjan et al., 2016		Landuse landcover patterns of year 2009, 2011, 2013, 2015, and 2016 were mapped for Dalma Wildlife Sanctuary (DWLS) Jamshedpur-Jharkhand using Landsat images.	
Bhanwar Vishvendra & Sen, 2018		Landuse landcover of 1975 and 2015 were mapped for Ranthambhore tiger reserve using Landsat TM and ETM imagery.	
Reverte et al., 2020		Land use maps of year 1986 and 2016 were prepared for Taubaté Basin located in southeastern Brazil using Google earth images.	
Rajashekariah, 2011	Change in Urban growth	• Urban growth measured from Landsat imagery of the year 1973, 1989, 1999, 2010 (Coimbatore) • Urban growth was measured from Landsat imagery of the year 1973, 1989, 1999, 2010 (Kolkata).	• To analyse the Spatio-temporal growth at the city level (MACRO LEVEL ANALYSIS). • To identify landuse expansion and increase in urbanization in urban areas. The following LULC classification has been found: • Urban area • Rural area • Protected area • Buffer around a protected area • Agricultural land • River • Water bodies
Li et al., 2016		Urban growth was quantified using night-time light imagery of the year 2000 and 2010 data by recording digital numbers (DNs) which are used in index development for urban sprawl.	
Hellwig et al., 2019	Change in Landuse Landcover (LULC)	Landuse landcover pattern of year 2000, 2006, and 2012 were analysed for 10 biogeographical regions located across 39 European countries.	

continues on follwoing page

Table 2. Continued

<table>
<tr><th>Research citation</th><th>Parameters</th><th>Analytical structure and method used</th><th>Outcome and inferences</th></tr>
<tr><td colspan="4">Measurement of economic transformation</td></tr>
<tr><td rowspan="2">Deb et al., 2014</td><td>• Increase of open forest area.
• Converted woodland into agricultural and grazing fields.</td><td>Landuse landcover change detection and field investigation.</td><td rowspan="2">To assess the human dependency on ecological areas for their livelihood.</td></tr>
<tr><td>Dependency of indigenous communities (Scheduled Tribes) or urban population for firewood timber etc.</td><td>Census details and field investigation.</td></tr>
<tr><td>Rajashekariah, 2011</td><td>The decrease in the average yield of traditional crops (turmeric, coconut, and bananas)</td><td>• Field investigation
• Focused group discussions with farmers.
• Change in cropping /yield pattern</td><td>To assess the effects of urbanization soil and water quality and its output.</td></tr>
<tr><td>Bhanwar Vishvendra & Sen, 2018</td><td>Extension of agricultural services</td><td>• Field investigation
• Focused group discussions with farmers.</td><td>To find the illegal encroach to public land resources.</td></tr>
<tr><td>Hellwig et al., 2019</td><td>Afforestation</td><td>Spatial modelling using landcover patterns.</td><td>To assess socioeconomic effects of land cover change.</td></tr>
<tr><td>Reverte et al., 2020</td><td>Quantification of ecosystem services</td><td>• Quantify the relationship of land use categories with the related ecosystem services.
• Calculate the relative significance for identified ecosystem services</td><td>To quantify the loss of ecosystem services.</td></tr>
<tr><td colspan="4">Measurement of cultural transformation</td></tr>
<tr><td>Deb et al., 2014</td><td rowspan="2">Human encroachment</td><td>Four blocks were studied for human habitation and occupation (census data).</td><td rowspan="2">Check the dependency of human on forests.</td></tr>
<tr><td>Bhanwar Vishvendra & Sen, 2018</td><td>Landuse landcover change detection.</td></tr>
<tr><td>Rajashekariah, 2011</td><td>Water quality of river/ wetland</td><td>Use data from Central or state pollution control boards.</td><td>To assess the effects of urbanization on river quality and land quality.</td></tr>
<tr><td></td><td>Human-wildlife conflicts</td><td>• Field investigation
• WWF (World Wildlife Fund) reports.
• Discussion with forest department officials.</td><td>To know the impacts of human interference on wildlife.</td></tr>
<tr><td>Hellwig et al., 2019</td><td>Land abandonment in traditional agricultural landscapes</td><td>Spatial modelling using landcover patterns.</td><td>To check the financial unprofitability and cultural changes.</td></tr>
<tr><td colspan="4">Measurement of social transformation</td></tr>
<tr><td>Deb et al., 2014</td><td>• Increase of open forest area.
• Cultivation of specific plant and tree species for traditional accessories making</td><td>• Field investigation
• Discussion with forest department officials</td><td>To check the interaction of indigenous communities with urban areas.</td></tr>
<tr><td>Hellwig et al., 2019</td><td>Demographic factors (accessibility to cities and population density)</td><td>• Use census data
• Spatial modelling using landcover flows</td><td>To check the socio-economic effects on landcover changes</td></tr>
</table>

Source: (Compiled by author,2021)

SELECT ECOLOGICAL STUDIES USING REMOTE SENSING AND GEOSPATIAL TECHNOLOGY

Sahu (2006) explored the impact of landuse landcover changes using remote sensing technology and has proved its applicability towards resource assessments and forest management. Dewan & Yamaguchi (2009) have illustrated the result of LULC change using remote sensing and GIS, which revealed that Dhaka city of Bangladesh has experienced swift urban sprawl and is responsible for environmental deterioration, reduction of water level in water bodies and wetlands and fading of countryside land, cultivated land, vegetation, etc. Tracking the decline in fallow land, dense forests, and water bodies, Deb et al. (2014) identified Eco-Sensitive Zone (ESZ) around wildlife sanctuary of Gumti, Tripura, using image processing, image classification, and change detection analysis with remote sensing and GIS software. Kumar Ranjan et al. (2016) used geospatial technology-based assessment of LULC change and found that forest fire, shifting cultivation and conversion of forest land into cropland, bare land, and settlement encroachments by humans as the main reasons for deforestation of Dalma wildlife sanctuary, Jamshedpur. Studies on ecological areas using geospatial applications have shown the shrinking sizes of protected areas due to sprawling urban growth and development in and around them. Pettorelli et al. (2016) has reported the need for a monitoring system- a global observing system, to track and record biodiversity changes to prevent future biodiversity loss. Choudhary et al. (2020) has reported the increase in settlement areas and farmlands around the protected areas.

Ecological areas are facing major challenges due to urban sprawl also. Research studies (Willis et al., 2012), (Pettorelli et al., 2016), (Reverte et al., 2020) have assessed the spread of urban growth using change detection method with geospatial tools and satellite datasets. Baral et al. (2014) have highlighted the need to increase the connectivity in protected areas and decrease fragmentation by enlarging and protected small forest patches. The decline in ecosystem services such as availability of food, carbon storage, maintenance of soil quality, water storage, and availability of habitat areas due to urbanization was recorded in Li et al., (2016). Hellwig et al. (2019) has reported the high risk of species extinction in protected areas that were close to urban areas. The LULC transformation shows spatio-temporal growth, landuse expansion and an increase in urbanization in and around urban areas.

Geospatial applications also have been effective to record the change in ecological areas (Deb et al., 2014), (Sharma, 2015). Geospatial technology is a specialized platform for various sets of information technologies which include remote sensing, global positioning systems, aerial photography, surveying, etc. These technologies have facilitated the varied analytical applications on satellite datasets to unravel numerous environmental issues. Geospatial technology offers a wide variety of technical applications for environmental monitoring and is utilized by many planning and scrutinizing agencies. These technologies are used in evaluating and recording spatial datasets, monitoring, and mitigating various environmental challenges worldwide. The relevance of geospatial technology and its applications has been appreciated in the sustenance of the environment by various countries (Chigbu & Onukaogu, 2013).

The spatial datasets covering the earth's surface are provided by various global and national earth observation agencies such as NASA (National Aeronautical Space Administration), EROS (Earth Resource Observation and Science) Centre, ISRO (Indian Space Research Organization), etc. Various global and national geospatial programs accentuate the raised requirement for finer landuse landcover spatial information. These programs, such as the United State climate change science program, international geosphere biosphere program, Landcover, and landuse change (LCLUC) program assisted by the Global land project, National Aeronautics and Space Administration (NASA), Global Observation of Forest and Landcover Dynamics (GOFC / GOLD), Group on Earth Observations (GEO) have been in the lead for conducting earth observation-based research studies. There are various remote sensing systems available at national and global scales, such as CARTOSAT, ASTER, EROS, GOSAT, IKONOS, IRS, GeoEye, KOMPSAT, Landsat, MODIS, Quickbird, Worldview, RADARSAT, etc. used for observing and monitoring landuse and landcover. These datasets are analyzed in computer-based decision support systems such as digital image processing systems, Geographical Information Systems (GIS), and Global Positioning Systems (GPSs).

INFERENCES AND RECOMMENDATIONS

The discussions till date have primarily focussed on the circumstantial elucidation of ecological areas, their recorded challenges caused by urbanization, and the need for their conservation. This has helped in encapsulating an in-depth knowledge regarding the ecological areas and also to understand their executive functions for environmental conservation and management, the related legal framework enacted by the government, organizations, and conventions at the global and national scale. Human civilization has achieved many triumphs of development in terms of technological innovations to a smart and sustainable environment. These human accomplishments have been chronicled in the change of slow ecological devastation.

Nature keeps warning humankind from time to time through epidemics of the recent COVID-19 pandemic or other natural catastrophes such as forest fires, cyclones, mass deaths of various avifaunal and wildlife species, etc. The past research also shows that negligence of nature has resulted in ecological disbalance. A solution for this situation would be human acceptance of this devastation and to prioritize ecological sustainability over their development. Ecological areas are the lungs of the Earth system which can be nourished with green forests rather than the man-made concrete forest. These areas should be invigorated for achieving nature conservation goals and ecological sustainability.

An increase in settlement areas and farmlands around the protected areas and shrinking sizes of protected areas due to sprawling urban growth and development in and around them has been reported in past studies. Most of this issue can be attributed to shrinking or vanishing buffers between urban areas and the nearby protected areas. Buffers of protected areas have frequently faced landuse landcover changes that led to threatening the integrity of protected areas. It is essential to have an optimal buffer area and safeguard it from any of the human threats so that the protected areas do remain secured. Taking due care in planning and providing for buffers around protected areas is one of the definite ways forward.

There is a need for regular impact assessment studies related to biodiversity and biomass, water regime, carbon sequestration occurring in and around protected areas mainly due to uncontrolled and non-eco sensitive urban sprawl. There is an emergent need for the formulation of guidelines and regulations for the development and protection of biodiversity in protected areas. The need for a global monitoring system to record changes in biodiversity and to integrate biodiversity values and conservation strategies at the initial stages in the physical planning process to prevent future biodiversity loss using geospatial applications is imperative for consolidated efforts for protecting our ecological areas.

High fragmentation, deforestation, trespassing, etc, have been reported due to anthropogenic disturbances. This is yet another major threat caused by urbanization. Hence, a planning approach for spatially monitoring and management of protected areas is required for urban planners, environmentalists, decision-makers, and policy makers in the preparation of restoration and conservation action plans.

These degraded hotspots in protected areas should be safeguarded and rejuvenated through restoration and conservation action plans.

FUTURE RESEARCH DIRECTIONS

The world is still facing the COVID pandemic for more than a year and the second wave is proved to be worse than the last in Asian countries. In the case of India, the situation is worst as most of the people are getting infecting and facing severe health issues like fever, lung infection, breathing problems, sore throat, heart attack, panic attack, etc. The country is facing a lack of oxygen for both humans and the environment. Building high rise concrete forests are given priority over deforestation activities and low focus on environmental protection resulted in the nastiest scenario for human today. These incidents have proved that human voracity for their development and disregarding nature is responsible for environmental dilapidation. Having underlined the interrelation of urban sprawl in and around ecological areas and the role of remote sensing and geospatial applications in the assessment of urban transformation the two appear to be promising tools to assess and mitigate the challenges in ecological areas. This will help to refine and resolve the various aspects of urban transformation in and around the ecological areas and also lead to developing measures and strategies for ecosystem conservation and environmental management in ecological areas. Researchers, learners (environment planning and management, environmental studies), urban and environmental planners, environmental specialists, decision-makers, law, and policymakers can be expected to take this forward.

REFERENCES

Agarwal, S., Saxena, A. K., Vyas, V., & Shrivastava, S. (2018). *Rethinking REDD+- A Centre for Science and Environment Assessment*. www.cseindia.org

Aijaz, R. (2018). Measuring urbanization in India | ORF. *ORF Issue Brief, 218*. https://www.orfonline.org/research/measuring-urbanisation-india/

Baral, H., Keenan, R. J., Sharma, S. K., Stork, N. E., & Kasel, S. (2014). Spatial assessment and mapping of biodiversity and conservation priorities in a heavily modified and fragmented production landscape in north-central Victoria, Australia. *Ecological Indicators, 36*, 552–562. doi:10.1016/j.ecolind.2013.09.022

Bhanwar Vishvendra, R. S., & Sen, A. (2018). Geo-Spatial Mapping of Land Use and Land Cover Changes in the Core and Periphery Area of Ranthambore Tiger Reserve, Rajasthan, India, 1975–2015. *Annals of Valahia University of Targoviste. Geographical Series, 18*(1), 62–67. doi:10.2478/avutgs-2018-0007

Bhardwaj, D. K. (2017). Diversity and status of terrestrial avifauna in Jamwa Ramgarh Wildlife Sanctuary, Rajasthan, Jaipur. *Life Sciences Leaflets, 108*, 17–29. https://lifesciencesleaflets.petsd.org/

Chigbu, N., & Onukaogu, D. (2013). Role of Geospatial Technology in Environmental Sustainability in Nigeria-An Overview. *Environment For Sustainability*, 6–10.

Choudhary, A., Deval, K., & Joshi, P. K. (2020). Study of habitat quality assessment using geospatial techniques in Keoladeo National Park, India. *Environmental Science and Pollution Research International*. Advance online publication. doi:10.100711356-020-11666-3 PMID:33205275

Deb, S., Ahmed, A., & Datta, D. (2014). An alternative approach for delineating eco-sensitive zones around a wildlife sanctuary applying geospatial techniques. *Environmental Monitoring and Assessment*, *186*(4), 2641–2651. doi:10.100710661-013-3567-7 PMID:24338098

Dewan, A. M., & Yamaguchi, Y. (2009). Using remote sensing and GIS to detect and monitor land use and land cover change in Dhaka Metropolitan of Bangladesh during 1960-2005. *Environmental Monitoring and Assessment*, *150*(1–4), 237–249. doi:10.100710661-008-0226-5 PMID:18317939

Elmqvist, T., & Mcdonald, R. I. (2014). Urbanization, Biodiversity and Ecosystem Services: Challenges and Opportunities. A Global Assessment. Springer. doi:10.1007/978-94-007-7088-1

Everard, M., Sharma, O. P., Vishwakarma, V. K., Khandal, D., Sahu, Y. K., Bhatnagar, R., Singh, J. K., Kumar, R., Nawab, A., Kumar, A., Kumar, V., Kashyap, A., Pandey, D. N., & Pinder, A. C. (2018). Assessing the feasibility of integrating ecosystem-based with engineered water resource governance and management for water security in semi-arid landscapes: A case study in the Banas catchment, Rajasthan, India. *The Science of the Total Environment*, *612*, 1249–1265. doi:10.1016/j.scitotenv.2017.08.308 PMID:28892868

Franklin, S. E. (2001). *Remote Sensing for Sustainable Forest Management*. Lewis Publishers. doi:10.1201/9781420032857

Hellwig, N., Walz, A., & Markovic, D. (2019). Climatic and socioeconomic effects on land cover changes across Europe: Does protected area designation matter? *PLoS One*, *14*(7), 1–20. doi:10.1371/journal.pone.0219374 PMID:31314769

Kumar, V., Singh, S., Shivhare, N., & Rahul, A. K. (2015). Change in Land Cover Pattern in Ramgarh Dam. *National Conference on Open Source GIS : Opportunities and Challenges*, 186-192.

Kumar Ranjan, A., Anand, A., S, V., & Kumar Singh, R. (2016). LU/LC Change Detection and Forest Degradation Analysis in Dalma Wildlife Sanctuary Using 3S Technology: A Case Study in Jamshedpur-India. *AIMS Geosciences*, *2*(4), 273–285. doi:10.3934/geosci.2016.4.273

Li, B., Chen, D., Wu, S., Zhou, S., Wang, T., & Chen, H. (2016). Spatio-temporal assessment of urbanization impacts on ecosystem services: Case study of Nanjing City, China. *Ecological Indicators*, *71*, 416–427. doi:10.1016/j.ecolind.2016.07.017

Mallick, A., Samal, A., & Santra, S. (2013). Integrated Natural Resource Management and Ecosystem Service. *Biodiversity Watch. International Journal on Biodiversity Issues.*, *2*, 88–102.

Mathur, V. B., Sivakumar, K., Singh, B., & R, A. K. (2009). *A Bibliographical Review for Identifying Research Gap Areas- Keoladeo Ghana National Park- – A World Heritage site*. Wildlife Institute of India, Dehradun. http://www.indiaenvironmentportal.org.in/files/bioliographical_keoladeo_ghana_np.pdf

Mehra, S. P., Mehra, S., & Sharma, K. K. (2013). Urban Avifaunal Biodiversity in context of Udaipur, Rajasthan, India. *Biodiversity Watch. International Journal on Biodiversity Issues*, *2*, 120–146.

Nagendra, H., Lucas, R., Honrado, J. P., Jongman, R. H. G., Tarantino, C., Adamo, M., & Mairota, P. (2013). Remote sensing for conservation monitoring: Assessing protected areas, habitat extent, habitat condition, species diversity, and threats. *Ecological Indicators*, *33*, 45–59. doi:10.1016/j.ecolind.2012.09.014

Pedersen, A. O., Nyhuus, S., Blindheim, T., & Krog, O. M. W. (2004). Implementation of a GIS-based management tool for conservation of biodiversity within the municipality of Oslo, Norway. *Landscape and Urban Planning*, *68*(4), 429–438. doi:10.1016/S0169-2046(03)00148-8

Pettorelli, N., Wegmann, M., Skidmore, A., Mücher, S., Dawson, T. P., Fernandez, M., Lucas, R., Schaepman, M. E., Wang, T., O'Connor, B., Jongman, R. H. G., Kempeneers, P., Sonnenschein, R., Leidner, A. K., Böhm, M., He, K. S., Nagendra, H., Dubois, G., Fatoyinbo, T., & Geller, G. N. (2016). Framing the concept of satellite remote sensing essential biodiversity variables: Challenges and future directions. *Remote Sensing in Ecology and Conservation*, *2*(3), 122–131. doi:10.1002/rse2.15

Rajashekariah, K. (2011). Impact of Urbanisation on Biodiversity: Case studies from India. *WWF Report*, (October). Advance online publication. doi:10.13140/2.1.4952.8324

Reddy, C. S., Manaswini, G., Satish, K. V., Singh, S., Jha, C. S., & Dadhwal, V. K. (2016). Conservation priorities of forest ecosystems: Evaluation of deforestation and degradation hotspots using geospatial techniques. Ecological Engineering (Vol. 91). doi:10.1016/j.ecoleng.2016.03.007

Reverte, F. C., & Garcia, M. (2019, December). da G. M., Brilha, J., & Pellejero, A. U. (2020). Assessment of impacts on ecosystem services provided by geodiversity in highly urbanized areas: A case study of the Taubaté Basin, Brazil. *Environmental Science & Policy*, *112*, 91–106. doi:10.1016/j.envsci.2020.05.015

Sahu, L. (2006). Application of remote sensing in forest resource assessment -A case study in and around Dudhwa National Park. *Statistical Accounting of Land and Forestry Resources*, 72–84.

Seto, K. C., Güneralp, B., & Hutyra, L. R. (2012). Global forecasts of urban expansion to 2030 and direct impacts on biodiversity and carbon pools. *Proceedings of the National Academy of Sciences of the United States of America*, *109*(40), 16083–16088. doi:10.1073/pnas.1211658109 PMID:22988086

Sharma, A. (2015). A Geographical Study of Keoladeo National Park, Bharatpur (Rajasthan) with Using Remote Sensing and GIS. *International Journal of Science, Engineering and Technology*, *3*(6), 201–210.

Sudhira, H. S., & Nagendra, H. (2013). Local Assessment of Bangalore: Graying and Greening in Bangalore – Impacts of Urbanization on Ecosystems, Ecosystem Services and Biodiversity. In *Urbanization, Biodiversity and Ecosystem Services: Challenges and Opportunities: A Global Assessment*. Springer. doi:10.1007/978-94-007-7088-1_7

Sukhdev, P. (2010). Mainstreaming the Economics of Nature: A synthesis of the approach, conclusions, and recommendations of TEEB. *De Economist*, *44*(1).

Trzyna, T. (2007). *Global Urbanization and Protected Areas - Challenges and opportunities posed by a major factor of global change and creative ways of responding*. California Institute of Public Affairs. https://www.ncbi.nlm.nih.gov/pubmed/12971682

UN. (2018). *World Urbanization Prospects: The 2018 Revision* (ST/ESA/SER.A/420). United Nations. https://population.un.org/wup/Publications/Files/WUP2018-Report.pdf

Verma, T. (2017). Role of Planners To Address the Threats of National Parks in India. Case Example : Keoladeo Ghana National Park, Bharatpur, Rajasthan, India. *International Journal of Advances in Mechanical and Civil Engineering, 5*(1), 63–65.

Willis, K. J., Jeffers, E. S., Tovar, C., Long, P. R., Caithness, N., Smit, M. G. D., Hagemann, R., Collin-Hansen, C., & Weissenberger, J. (2012). Determining the ecological value of landscapes beyond protected areas. *Biological Conservation, 147*(1), 3–12. doi:10.1016/j.biocon.2011.11.001

KEY TERMS AND DEFINITIONS

Anthropogenic Disturbance: The interruption in ecological areas due to human-induced development activities.

Biodiversity: The diversity of all life forms available on earth ecosystem including the variability within themselves.

CBD: Center for Biological Diversity.

Conservation: The preservation and protection of ecological systems to maintain the perpetuity and durability of natural environments.

Ecological Area: A naturally created biodiversity-rich hub, interconnected with ecological corridors, for operating ecological processes and ecosystem services.

Ecological Connectivity: The connectedness of naturally developed corridors by wildlife species for movement and flow of the ecological process for the sustenance of life on Earth.

Ecological Restoration: The retrieval of an ecological system, its operations, and progression of an authentic ecosystem.

Ecological Sustainability: A fair utilization and management of natural resources at a pace that does not encourage the depreciation of ecosystem services, biodiversity loss, and uninterrupted operation of ecological processes.

Environmental Protection: The preservation of the natural environment from degradation and contamination of natural resources and maintenance of ecological balance.

ESZ: Eco-sensitive zone.

GIS: Geographical information systems.

Habitat Fragmentation: The disintegration of contiguous ecological areas including wildlife species habitats into distinct land parcels due to anthropogenic activities.

IUCN: International Union for Conservation of Nature.

Protected Areas: A specifically defined and dedicated ecological area demarcated for the sustenance of biological diversity and operation of natural processes and ecosystem services.

Remote Sensing: A satellite-based technology of data detection and collection of Earth's surfaces from space in the form of satellite and aerial photographs or raster datasets integrated with information technology.

Transformation: A process of alteration in essential attributes and functions of a natural or man-made system.

Chapter 12
Applications of Remote Sensing and GIS in Hydrological and Hydrogeological Studies:
Integrated Watershed Management

Athar Hussain
https://orcid.org/0000-0002-7187-5977
Netaji Subhas University of Technology (West Campus), India

Richa Madan
Gurukula Kangri (Deemed), Haridwar, India

Vishal Kamboj
Gurukula Kangri (Deemed), Haridwar, India

ABSTRACT

Integrated watershed management is a holistic approach aimed at linking upstream and downstream areas. It encompasses preventing the watershed from being polluted and maintaining the quality of water and prevention of soil erosion. Remote sensing and GIS techniques play a crucial role in assessing watershed conditions through modeling. These techniques are also very helpful in acquiring the spatial information of the land use, land cover, vegetation, water resources, etc. With the available data, characteristics of aquifers can easily be deciphered. Remote sensing aids in identifying and assessing the topographical features and alienates the unnecessary ones. Integration of GIS with remote sensing along with hydrological, hydrogeological, geomorphologic studies will make the practice of integrated watershed management more effective. The primary aim of this chapter is to encompass the remote sensing and GIS technologies currently in use for watershed monitoring and management.

DOI: 10.4018/978-1-7998-8331-9.ch012

INTRODUCTION

A watershed refers to a spatial area where surface water is collected and which drains water to a common point to be used for various purposes. This watershed happens to be a network of small streams that flow through a hydrological unit, biophysical unit and creates a holistic ecosystem (Wang et al. 2016). Watersheds vary greatly in size extending from thousands of square kilometers to tiny regions. They are the main sources of groundwater recharge, help in maintaining food chain and food web of an ecosystem and influence the microclimate of a region. Nowadays, due to a voluminous increase in pollution in soil, water and air, the climate of the surrounding area declines affecting the watershed ecosystem. Hence, to keep a balance of the ecosystem an integrated watershed management approach is crucial which involves maintaining, planning and improving the quality of the watershed overtime. According to the California Department of Conservation (2015), watershed management refers to the assessment of pertinent characteristics of a watershed focused on the feasible circulation of its resources and a way of making and executing plans, projects, and undertakings to maintain and upgrade watershed work that influence the living organisms and the human networks in a watershed. Through the advancement of watershed management, it has become more integrated conspicuous (German et al. 2007). Integrated watershed management incorporates diverse social, technological, and institutional components, along with conservation, social, and economic goals into the core concepts of watershed management. Its purpose is to connect land and water planning considering both ground and surface water, identifying and planning how alteration of landscape can affect the nearby water, plants, animals and human; land use within a watershed's physical borders. One of the fundamental standards of watershed management is the assessment of the interrelationships among land use, water and soil and the linkages among land and downstream regions (Ffolliott et al. 2002). Physical changes in a watershed can bring about an assortment of responses going from present moment occasions, like flooding, avalanches, and point-source contamination, to long haul processes, i.e., degradation of soil and water and non-point-source contamination. Consequently, the advancement and appraisal of watershed management, requires the incorporation of a huge range of spatial and temporal data information (Wang et al. 2016).

REMOTE SENSING AND GEOGRAPHICAL INFORMATION SYSTEM TECHNIQUES

Remote sensing (RS) and geographical information systems (GIS) are widely used techniques over last few decades for hydrological studies. They are modern day applications used for assessing the changing pattern of land use, land cover; mapping and modelling of the water resources; evaluation of point and non-point sources, stream order; and mapping of the watershed. Remote sensing derives information and images about Earth's area and ocean surfaces by analysing electromagnetic radiation reflected or emitted from the planet's surface (Brooks et al. 2003; Brooks et al. 2012). This precise and real-time data makes way for identifying, surveying, characterizing and monitoring the activities within a watershed, including land use/cover, physiography, soil distribution, and drainage features, as well as lake temperature, water level, and algal dynamics (Pandey et al. 2007; Hampton 2013).

Besides, for watershed assessment, RS technique is used with combination of GIS to provide the relevant and accurate spatial input data. GIS is a specific tool for watershed management which is used for understanding the condition through modelling and visualizing the impact due to human intervention for sustaining the watershed (Tim and Mallavaram 2003). The techniques used for visualization and modelling are basic tools to assess the processes and dynamics that create and modify the physical, chemical and biological environment of the watershed. The use of GIS tool into hydrological simulation models provides more spatial detail and prompts the capacity to evaluate the slope, aspects and other hydrological characteristics (Brooks et al. 2013). These modelling and visualization capabilities are critical for comprehending the mechanisms and dynamics that define watersheds' physical, biological, and chemical environments. When GIS is used in hydrological simulation models, it produces more geographic information than most other hydrological models allowing detailed analysis of slope, aspect, and hydrological-response units. "Understanding Your Watershed", a program developed by Michigan State University's Institute of Water Research, is a web-based publicly available GIS tool that can be sued for mapping any watershed.

Table 1. Satellites and sensors used for mapping of water resources

Fields	Purpose	Satellites or Sensors
Water Resources	Snow	Landsat, Cryosat, Aqua MODIS, AVHRR
	Glaciers, Lakes, Wetlands, Reservoirs, Soil Moisture	Landsat, SPOT, ICESat, SRTM
Hydrological fluxes	Precipitation, Evapotranspiration	NEXRAD, MODIS, Landsat
Drought and flooding	Drought and flooding	Landsat, MODIS, GRACE, UAV, ENVISAT

(Source: Wang et al. 2016)

RS AND GIS TECHNIQUES USED FOR HYDROLOGICAL AND HYDROGEOLOGICAL STUDY

Mapping of Water Resources

The concept of mapping of water resources using the RS and GIS contains the study of snow, glaciers, rivers, streams, soil moistures, lake, and wetlands using the satellite and sensors data. Mostly used sensors and satellites namely AVHRR, Landsat, Aqua/Terra MODIS, SSM/1, Cryosat etc. are used for monitoring and mapping the water resources (Wang and Xie 2018). The snow and glaciers are the primary source of fresh water resources. Glacier areas are characterized as the degree in two horizontal dimensions (2D region) in the ice mass equilibrium community (Cogley et al. 2011; Wang and Xie 2018). In high mountains, most glaciers lie on steep inclines, and their actual surface area (3D) might be a lot bigger than the 2D region (Wang and Xie 2018). The mapping and assessment of differences between the glacier's region i.e., 3D and 2D regions is carried out using the data provided by Landsat, ASTER GDEM etc. The collected data is applied in GIS with tools namely TIN 3D model. For example, Wang et al. 2017 collected the ASTER GDEM data and Landsat images for assessing the difference between the 2D and 3D areas of the Muzart Glacier catchment and Central Tianshan Mountains. The results of TIN 3D model showed that the 3D areas of Muzrat glacier catchment were 34.2% greater than the 2D areas while in central Tianshan catchment, the 3D areas were 27.9% greater than the 2D areas.

A key boundary layer between the atmosphere and the ground surface is the surface soil. The amount of water in the soil has an impact on local agriculture, ecology, hydrology, and climate. Soil moisture is one of the most important determinants of vegetation growth and ecosystem restoration. The gravimetric method, the neutron technique, and the Time Domain approach are all examples of moisture measuring technologies. In the mid-latitudes and dry and semiarid areas, groundwater is a major supply of freshwater. Rainfall, rock kinds, geological formations, land use, and other variables all have a role in the hydrological cycle. The cost of site exploration, time, and people are reduced when groundwater prospective locations are found using RS approaches. RS is also beneficial in catering the recharging zone and selecting places for artificial recharge, in addition to site selection.

Rainfall Measurements and Design Storm

Rainfall is an important part of the water cycle and the replenishment of water supplies. Flooding is frequently caused by heavy rains. At the mesoscale, typhoon-related heavy rainfall exhibits distinct temporal and space structures (Kang et al. 2013). Satellite rainfall estimates may aid early-warning systems and disaster management by better delineating the structures of heavy rainfall. The storm hyetograph is a visual representation of the temporal progression of heavy rainfall over a specific area (Yen and Chow 1980; Keifer and Chu 1957). The storm hyetograph (also known as design storm) predicts the peak flooding volume given a total rain depth and duration over a certain return time and is crucial for drainage design in storm water management (SCS, USA 1986).

Runoff Prediction and Flood Forecasting

Climate and hydrologic simulations, water resource management, and flood predictions all rely on hydrological models. Hydrological models have progressed from conceptual and clumped models to physically based, scattered models (Sharif et al. 2017). Thus, urban flooding may be simulated using a single physically-based and distributed model that can account for both local severe storm runoff in urban areas and regional rainfall runoff over a vast river catchment, as well as the integrated urban flooding risk (Lee and Choi, 2017). Numerical simulation and black-box modelling are two types of flood inundation forecasting technologies. Numerical simulation is based on theoretical derivation and frequently has good accuracy, but it requires a lot of computational resources and is difficult to utilize for real-time disaster prevention and rescue response in most situations, such as during a typhoon or flash flood. The black-box model, on the other hand, takes a different approach by treating the rainfall-to-inundation process as a black box and simulating the link between input rainfall and output runoff and inundation (Karlsson et al. 1987).

MAPPING OF WATERSHED USING RS AND GIS

The concept of mapping and modelling of watershed involves the hydrological and hydrographical information regarding the watershed using remote sensing and GIS techniques. In recent years, mostly digital elevation models (DEMs) were used for the extraction of hydrological information from a watershed. By using the DEM data, we can extract the stream orders and networks, flow accumulation, flow direction, and basin area of a watershed. This method provides a lot of knowledge regarding the watershed sustainability. The first stage of watershed analysis is the watershed prioritization. The RS data provides the information which helps to assess the watershed prioritization using some methods mainly Sediment Yield Index (SYI) and universal soil loss equation (Kasaragod and Hegde, 2002). The study of watershed prioritization consists of watershed delineation, geomorphological parameters such as stream order, law of stream number and length, drainage area, surface hydrology and watershed boundary (Narayan et al. 2016).

Basic Steps for Preparation of a Watershed DEM

1. Data Collection and Software Used

For preparation of a watershed of any riverine system, digital elevation model (DEM) is used. The DEM file is obtained by the open source of United State Geological Survey (USGS) or Earth Explorer (EE). After obtaining the DEM file of an area, it is run in the Arc GIS software. *An Example of DEM of a Watershed is Alaknanda River Basin of Uttarakhand using RS and GIS Data*

General Overview or Why We Chose the Alaknanda River Basin

The Alaknanda river is a snow fed river originating from the foothills of two glaciers namely Satopanth and Bhagirath Kharak in the northern Indian state of Uttarakhand (Jain et al. 2020). It encompasses the sister stream of Bhagirathi and accounts a large distance and discharge of water for making the Ganga. Before its confluence with the Bhagirathi River, it covers a large watershed area with various stream orders and drainage areas. So, in the context of watershed delineation using the RS & GIS, we chose the Alaknanda river basin watershed.

2. Preparation of Alaknanda River Basin Watershed

The method of classifying points or set of points of the drainage area is known as watershed delineation. The delineation of watershed is important for planning and execution of watershed management program. It can be demarcated by calculating the direction of flow and using it as a function from DEM. The raster of flow direction is used as watershed function for determining the area. The following steps are depicted in Figure 1.

3. Clipping the Area of Interest

The first step is the clipping of area of interest from the large DEM data. After clipping, further steps are applied. The figure 2 shows the large DEM data of Uttarakhand, India. While figure 3 shows the area of interest i.e., Alaknanda riverine system.

4. Fill

This tool is used to remove the raster depression from the DEM data. In a DEM data, raster depression is normally found in nature and could be generated when DEM data processes further. These topographic or raster depressions are known as sinks. These sinks appear when the low elevation raster is in the neighboring position of high elevation raster. The fill process removes these sinks from the DEM data by enhancing the cell values. Figure 4 shows the fill of Alaknanda river Basin.

Figure 1. Basic steps for watershed delineations
(Created by author, 2021)

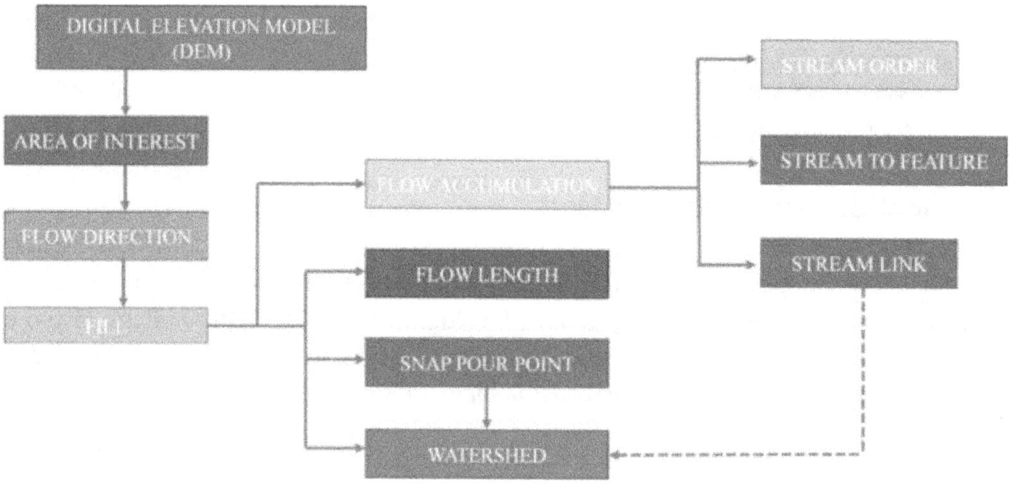

Figure 2. Digital elevation model of Uttarakhand State, India
(Created by author, 2021)

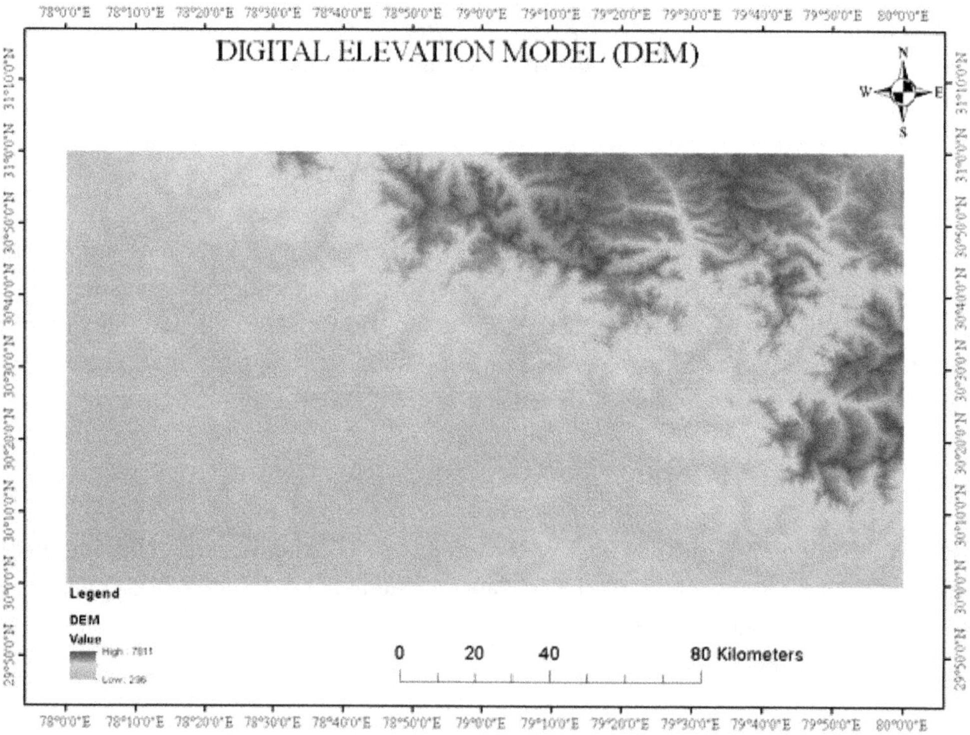

5. Flow Direction

This tool is a vital part of watershed delineation. It is used for preparation of the flow direction in drainage network in a land topography. In this method, the drainage network is determined by a cell elevation in which water can flow. This method is based on the D8 algorithm i.e., 1, 2, 4, 8, 16, 32, 64 and 128. This algorithm number decides the direction of water flow in the cell. For example, if the cell value is 32 then water flows in northwest, if cell value is 4 then water flows southward. In figure 5, the flow direction of Alaknanda river watershed is presented with D8 algorithm methods that define the flow direction and drainage network of Alaknanda river.

Figure 3. Digital Elevation Model of Alaknanda River Basin
(Created by author, 2021)

6. Flow Accumulation

Flow accumulation is an important part of watershed network preparation and also measures the area of contributing runoff cell that makes the watershed network. It is totally dependent on the flow direction. Based on flow direction, it converts all the cell values in two values i.e., 1 and 0. The value 1 shows the higher runoff cell and stream flow while value 0 shows the ridgelines. After flow accumulation process, we can easily define the higher runoff cell and lower runoff cell. In Figure 6, it is clearly depicted that the higher stream flow is in white color while ridgelines are showed in black color of Alaknanda river watershed.

7. Basin

The preparation of basin area is another step of water delineation. Based on flow direction of cells we measure the actual basin areas of any stream network. In a DEM data, there are many basins present shown in Figure 7. So, by the flow directions of cells, we can easily find the actual basin/drainage area of any stream network. In Figure 8, the light color shows the total basin area of Alaknanda river.

Figure 4. Preparation of fill characters of Alaknanda river Basin Watershed
(Created by author, 2021)

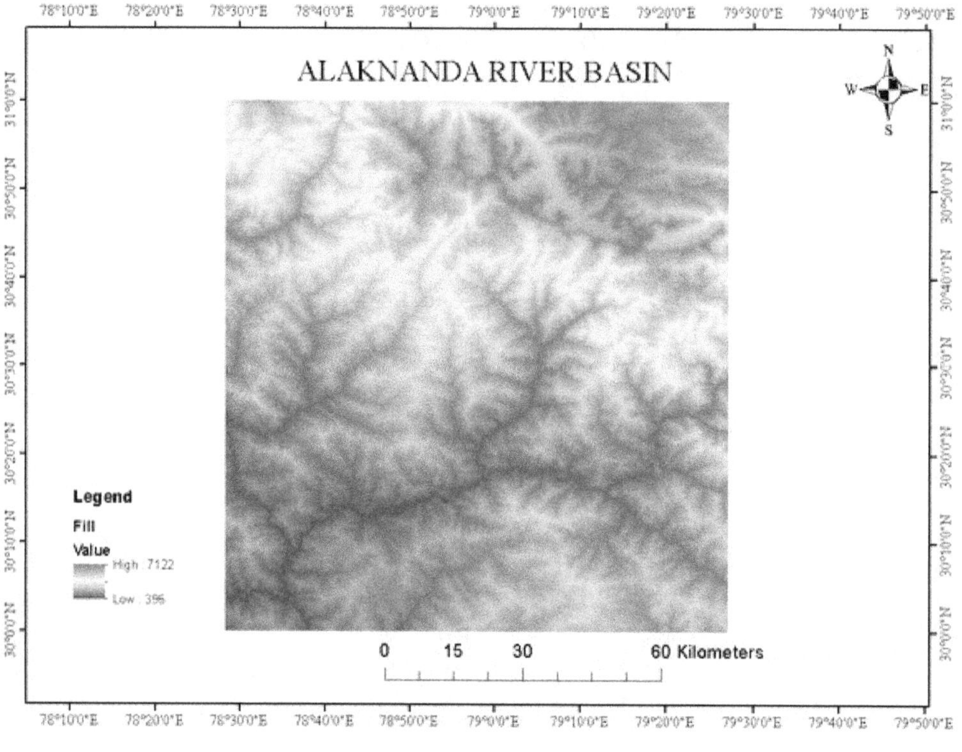

8. Stream Order and Stream to Feather

Stream order is defined as the degree of the relative size of streams. First order stream is the smallest tributary. Headstream is said to be source of river. 80% of Earth's waterways are headwater streams. According to Hack's stream order or classic stream order, number 1 is allotted to a river which drains at the sea. It is the main stem of any river. The tributaries of the main stream are allotted number "2" whereas tributaries discharging water into tributary number "2" are number "3" and so on. The stream order systems are significant for the mapping of a river systematically. This system enables in labelling and ordering of streams in a river system. This method helps in detecting and classifying types of streams on the bases of numbers of tributaries. Some features of streams can be understood by their order.

Figure 5. Preparation of flow direction characters of watershed
(Created by author, 2021)

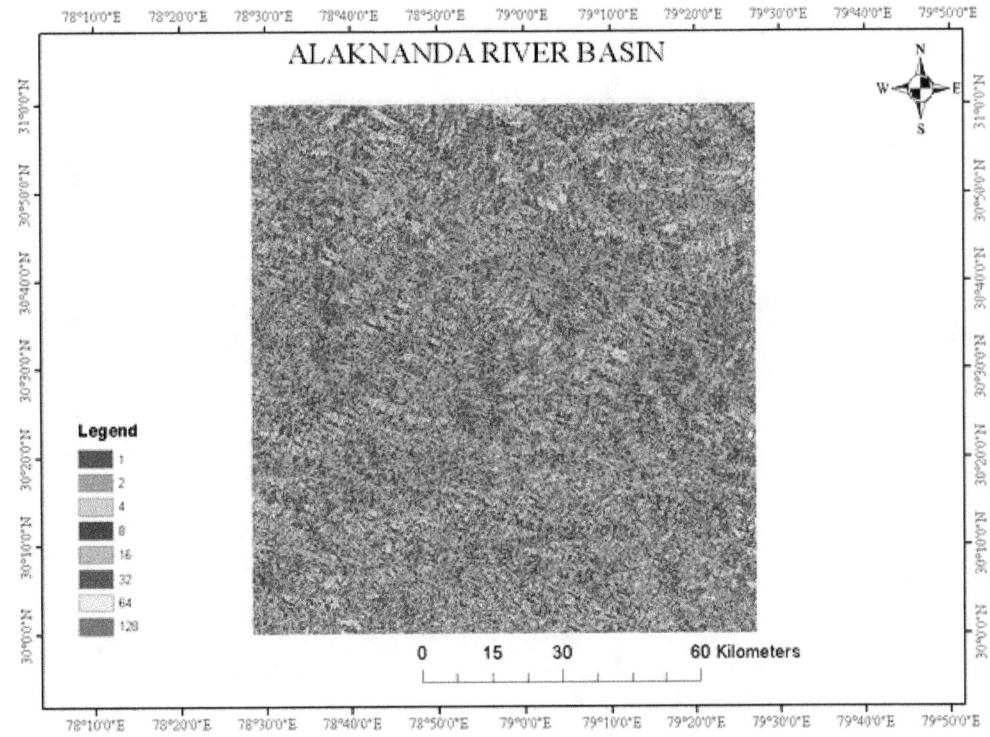

The notion of stream order was given by two scientists, Strahler and Shreve. According to Strahler, in the network of streams the linking of tributaries is known as the stream order. The law states that two same order streams merge and form a new stream, then it is given number one which higher than the previous. Also, when the streams of different orders merge together, the subsequent stream is given the higher number out of the two. It is a commonly used method of stream order. Also, according to the Shreve, outmost tributary is assigned number "1". In this method at a confluence the two numbers are added at a confluence. The similarity of this method with Strahler is that it relies on the accuracy of the sources, but less on scale of the map. Stream which starts from the source are numbered as the first order and all the links are added. Additive orders are called magnitude and demonstrate the number of upstream links. The links combined together form a higher link such as first order added with second order forms third order link and so on.

In the figure 9, the 6-stream order developed in the Alaknanda river watershed is shown. It shows that the Alaknanda river is a combination of various small streams, channels, torrents and other water flow.

9. Watershed

The last step of the watershed delineation is the preparation of watershed. The watershed is the combination of total basin area, stream network, stream order and main river network, and ridgelines. In figure 10, the Alaknanda river watershed with its basin, stream order and network, ridgeline is represented.

Figure 6. Preparation of flow accumulation characters of watershed
(Created by author, 2021)

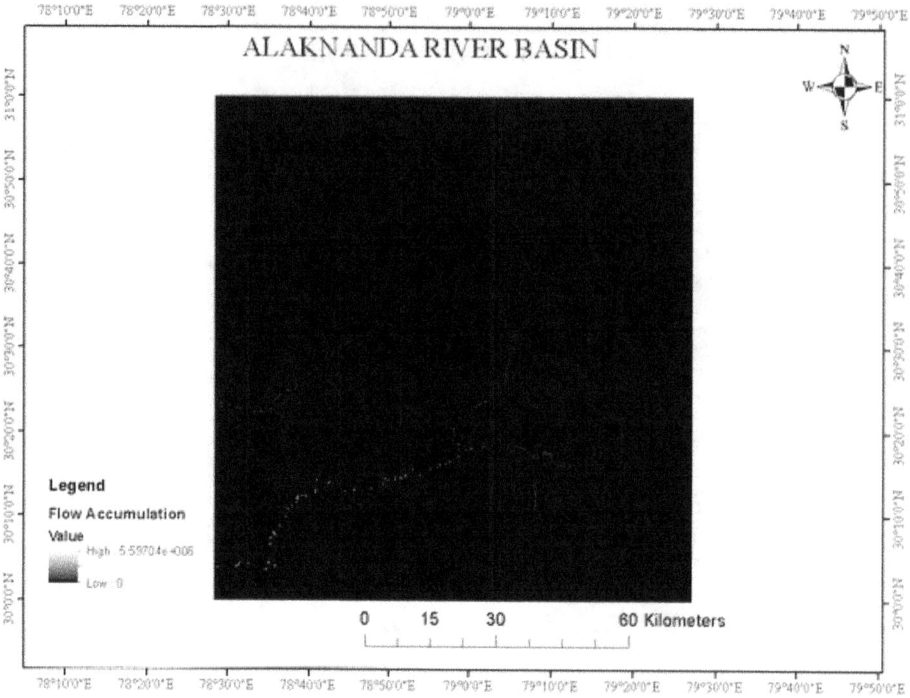

Figure 7. Represent total basin in DEM file
(Created by author, 2021)

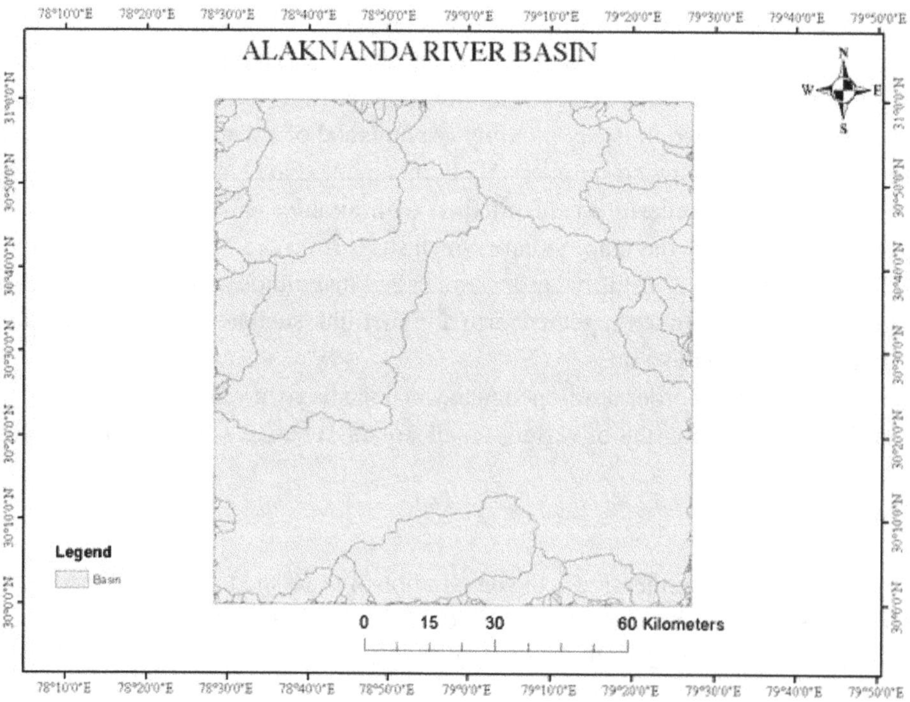

Figure 8. Alaknanda river basin in DEM
(Created by author, 2021)

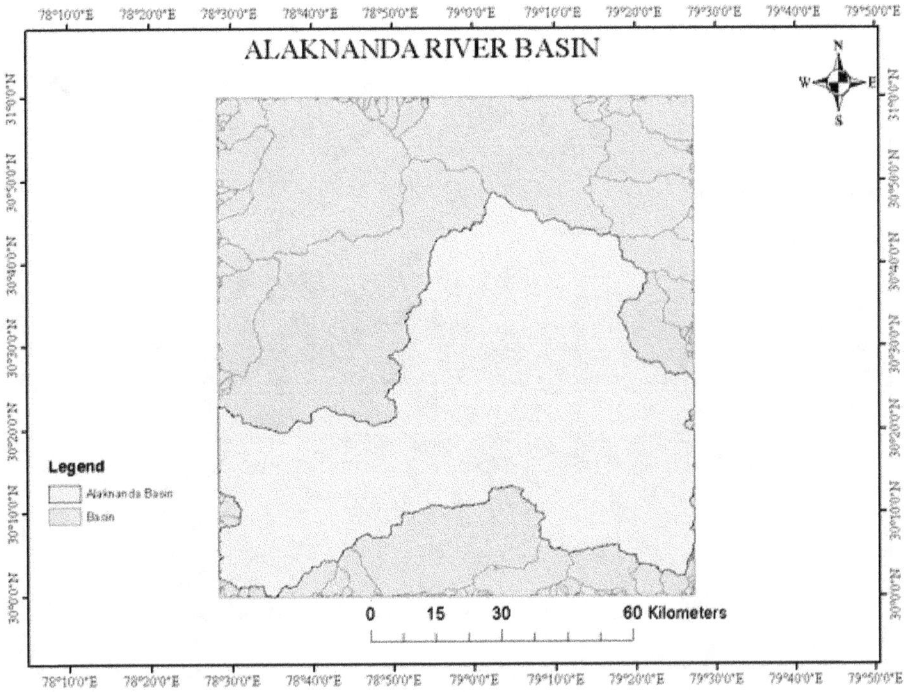

Figure 9. Represents stream order in Alaknanda river basin
(Created by author, 2021)

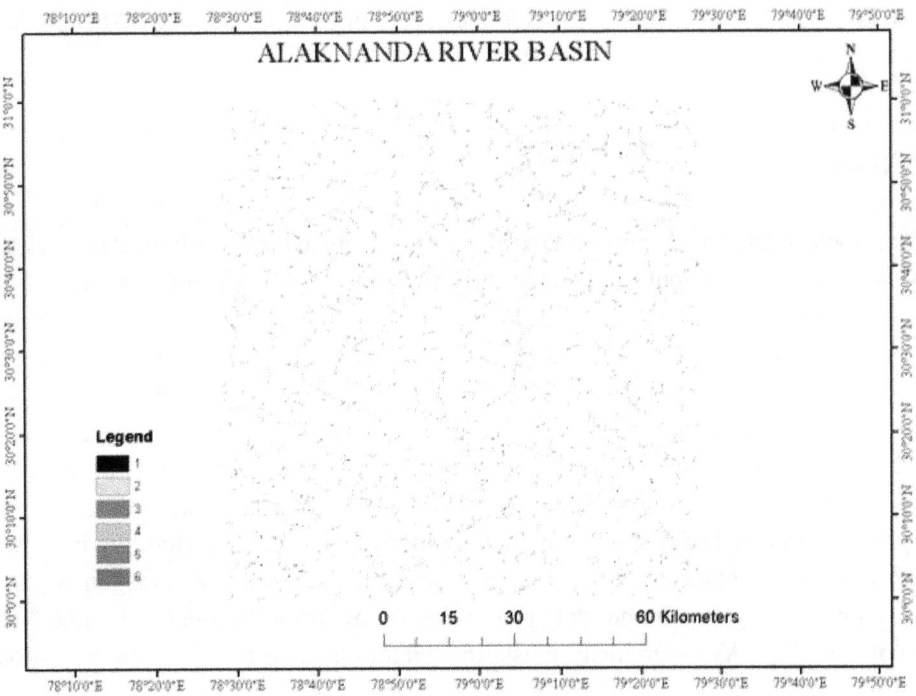

Figure 10. Alaknanda river watershed
(Created by author, 2021)

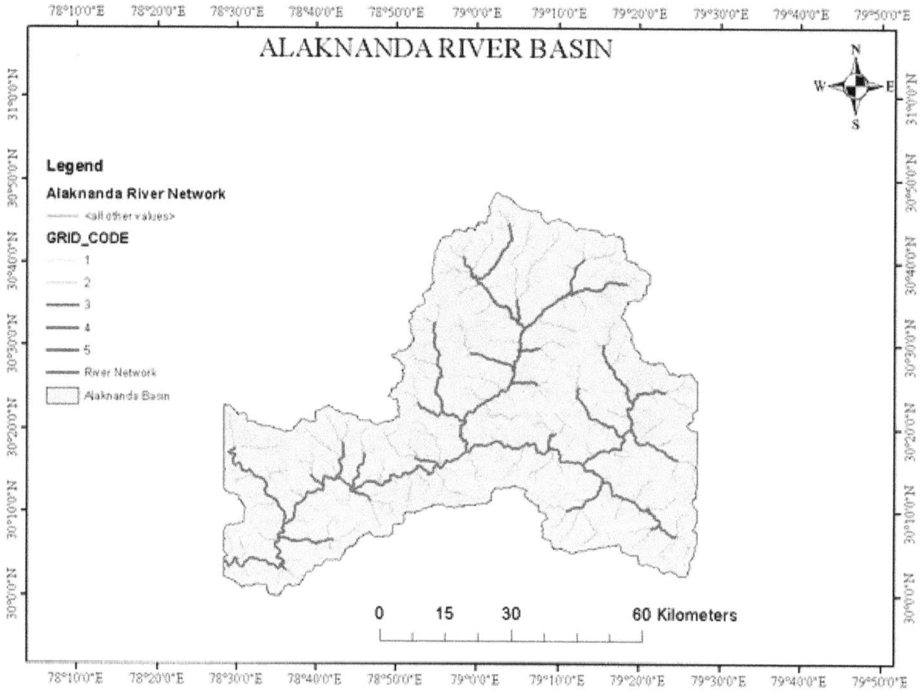

The preparation of a watershed delineation through RS & GIS tool is easy and effective which provides us beneficial knowledge regarding the characteristics, stream orders, basin area and ridgeline. Once we have knowledge about all the characteristics of watershed, we can easily sustain the watershed of any riverine system.

CONCLUSION

In conclusion, remote sensing has become increasingly significant in the hydrological and hydrogeological sector. They can map the geographic and temporal distributions of water resources, quantify hydrologic flow, and monitor the state of hydraulic infrastructures, drought conditions and floods inundation. Together, GIS, statistics, and numerical models can unleash the potential of diverse remote sensing data sets, allowing for better water resource management, drought, and floods catastrophe management. Water resources and flood inundation mapping and risk management rely heavily on remote sensing and geographic information systems (GIS). Remote sensing provides critical data for mapping water resources (snow and glaciers, water bodies, soil moisture, and groundwater), measuring hydrological fluxes (ET, precipitation, and river discharge), and monitoring drought and flooding inundation; meanwhile, GIS provides the best tools for water resource, drought, and flood risk management, as well as the setup, input data processing, and output data processing for hydrologic models. 3D model analysis and visualization of glaciers, UAV video image classification for turfgrass mapping and irrigation planning, ground penetration radar for soil moisture estimation, TRMM and GPM satellite rainfall measurements, storm hyetograph analysis, rainfall runoff and urban flooding simulation, and satellite radar and optical image detection for urban hazard detection are some of the most recent technologies used. GIS is a useful tool, but it prefers to work quietly behind the scenes most of the time. Watershed management has clearly evolved into a more holistic resource management strategy, using integrated and adaptive management techniques to account for biological, physical, and social factors in the environment. This well-rounded strategy has benefited greatly from technological improvements. Knowledge is now more easily exchanged within and between disciplines, data collecting techniques and model building are more accurate, and integrated, multi-level analysis that yields more full information on the watershed system both socially and environmentally is now feasible.

FUNDING DETAILS

There was no funding provided by any government agency or any private organization.

ACKNOWLEDGMENT

Authors are highly grateful to the Department of Civil Engineering, NSUT and Department of Zoology and Environmental Science, Gurukula Kangri (Deemed to be University), Haridwar, India for providing infrastructural facilities.

REFERENCES

Brooks, K. N., Ffolliott, P. F., Gregersen, H. M., & DeBano, L. F. (2003). Hydrology and the Management of Watersheds. Iowa State University Press.

Brooks, K. N., Ffolliott, P. F., & Magner, J. A. (2012). *Hydrology and the Management of Watersheds*. John Wiley & Sons. doi:10.1002/9781118459751

California Department of Conservation. (2015). *Watershed Program*. Author.

Cogley, J. G., Hock, R., Rasmussen, L. A., Arendt, A. A., Bauder, A., Braithwaite, R. J., Jansson, P., Kaser, G., Möller, M., & Nicholson, L. (2011). *Glossary of Glacier Mass Balance and Related Terms; IHP-VII Technical Documents in Hydrology No. 86, IACS Contribution No. 2*. UNESCO-IHP.

Ffolliott, P. F., Baker, M. B., Edminster, C. B., Dillon, M. C., & Mora, K. L. (Eds.). (2012). *Land stewardship through watershed management: perspectives for the 21st century*. Springer Science & Business Media.

German, L., Mansoor, H., Alemu, G., Mazengia, W., Amede, T., & Stroud, A. (2007). Participatory integrated watershed management: Evolution of concepts and methods in an ecoregional program of the eastern African highlands. *Agricultural Systems*, *94*(2), 189–204. doi:10.1016/j.agsy.2006.08.008

Hampton, S. E. (2013). Understanding lakes near and far. *Science, 342*(6160), 815–816. doi:10.1126cience.1244732 PMID:24233716

Kang, M. S., Goo, J. H., Song, I., Chun, J. A., Her, Y. G., Hwang, S. W., & Park, S. W. (2013). Estimating design floods based on the critical storm duration for small watersheds. *Journal of Hydro-environment Research, 7*(3), 209–218. doi:10.1016/j.jher.2013.01.003

Karlsson, M., & Yakowitz, S. (1987). Rainfall-runoff forecasting methods, old and new. *Stochastic Hydrology and Hydraulics, 1*(4), 303–318. doi:10.1007/BF01543102

Kasaragod, G. R., & Hegde, V. S. (2002). Application of Remote Sensing & GIS in Watershed Prioritization, Development Planning & Monitoring. *Conference: Sensitization Workshop on Integrated Watershed Management under Western Ghats Development Programme.*

Keifer, C. J., & Chu, H. H. (1957). Synthetic storm pattern for drainage design. *Journal of the Hydraulics Division, 83*(4), 1332–1. doi:10.1061/JYCEAJ.0000104

Lee, J. S., & Choi, H. I. (2017). Improvements to Runoff Predictions from a Land Surface Model with a Lateral Flow Scheme Using Remote Sensing and In Situ Observations. *Water (Basel), 9*(2), 148. doi:10.3390/w9020148

Michigan State University. (2014). *Understanding your watershed interactive mapping program.* https://msustatewide.msu.edu/Programs/Details/2083

Narayan, K., Singh, S. M., Dikshit, P. K. S., & Dwivedi, S. B. (2016). Application of GIS and Remote Sensing Techniques for Effective Watershed Management. *International Journal of Engineering Research & Technology (Ahmedabad), 4*(03), 1–5.

Pandey, A., Chowdary, V. M., & Mal, B. C. (2007). Identification of critical erosion prone areas in the small agricultural watershed using USLE, GIS and remote sensing. *Water Resources Management, 21*(4), 729–746. doi:10.100711269-006-9061-z

Scs, U. S. (1986). *Urban hydrology for small watersheds, technical release no. 55 (TR-55).* US Department of Agriculture, US Government Printing Office.

Sharif, H. O., Al-Zahrani, M., & Hassan, A. E. (2017). Physically, fully-distributed hydrologic simulations driven by GPM satellite rainfall over an urbanizing arid catchment in Saudi Arabia. *Water (Basel), 9*(3), 163. doi:10.3390/w9030163

Tim, U. S., & Mallavaram, S. (2003). Application of GIS technology in watershed-based management and decision making. *Watershed Update, 1*(5), 1-6.

Wang, G., Mang, S., Cai, H., Liu, S., Zhang, Z., Wang, L., & Innes, J. L. (2016). Integrated watershed management: Evolution, development and emerging trends. *Journal of Forestry Research, 27*(5), 967–994. doi:10.100711676-016-0293-3

Wang, X., Chen, H., & Chen, Y. (2017). Large differences between glaciers 3D surface extents and 2D planar areas in Central Tianshan. *Water (Basel)*, *9*(4), 282. doi:10.3390/w9040282

Wang, X., & Xie, H. (2018). A review on applications of remote sensing and geographic information systems (GIS) in water resources and flood risk management. *Water (Basel)*, *10*(5), 608. doi:10.3390/w10050608

Yen, B. C., & Chow, V. T. (1980). Design hyetographs for small drainage structures. *Journal of the Hydraulics Division, 106*(ASCE 15452).

Chapter 13
The Role of Sustainable Architecture in an Era of Environmental Challenges

Seyda Emekci
https://orcid.org/0000-0002-5470-6485
Ankara Yildirim Beyazit University, Turkey

ABSTRACT

Anthropogenic greenhouse gas emissions have increased since the beginning of the Industrial Revolution. This triggered an unnatural warming and the Earth's temperature which has increased dramatically in a short period of time. This brought many problems such as natural disasters, tropical cyclones, heavy precipitation events, extreme heat events, and drought. It is an urgent need to make all processes "sustainable" in order to protect the environment and human life. Since urban areas are estimated to be the source of approximately 80% of global greenhouse gas (GHG) emissions and a significant proportion of these greenhouse gas emissions in urban areas originate from buildings, buildings should be considered as one of the first stops on the road to sustainability. The purpose of this chapter is to highlight the importance of sustainable architecture in addressing environmental challenges and ensuring sustainable urban development.

INTRODUCTION

The accumulation of greenhouse gases in the atmosphere has been increasing rapidly due to various factors such as the use of fossil fuels, increasing world population, production activities in cities, reduction in forest areas, and misuse of land since the industrial revolution. All these factors intensify the natural greenhouse effect. They cause an increase in the surface temperature of the world and climate change with the contribution of increasing urbanization. Intergovernmental Panel on Climate Change (IPCC) defines climate change as a change of climate that occurs over comparable time periods as a result of natural changes or human activities. However, climate change experienced in the 21st century is also referred to as 'anthropogenic' (human-induced) climate change, because greenhouse gas emissions that cause

DOI: 10.4018/978-1-7998-8331-9.ch013

climate change have arisen as a result of human activities (IPCC, 2018). In order for the efforts carried out by international organizations in combating global warming to succeed, it is necessary to consider and evaluate especially urban areas where human activities that cause climate change are concentrated. There is a two-way relationship between climate change and urban areas; on the one hand, while some urban areas are already suffering from the negative side of climate change; on the other hand, urban activities that cause climate change are the biggest contributors to this problem. According to PlaNYC, urban areas"are at the forefront of both the causes and effects of climate change. Urban areas are estimated to be the source of approximately 80% of global greenhouse gas (GHG) emissions" (PlaNYC, 2011). 56.2% of the world's population lives in urban and this number is constantly increasing; In the last 60 years, the number of people living in urban centers has more than fourfold (D. Satterthwaite, 2020). Therefore, urban areas must be encouraged more inclusive growth; adapted to climate change and action should be taken in these areas to reduce the risks of climate change. The problem is not just only related to the environmental issue. It also has devastating effects on communities, individuals socially and economically. According to a study conducted by the World Health Organization, by 2030, climate change is expected to contribute to approximately 250,000 additional deaths per year due to malnutrition, malaria, diarrhea, and heat stress (WHO, 2018, p. 20). In addition to negatively affecting human health, urban areas will face sea-level rise, tropical cyclones, heavy precipitation events, extreme heat events, drought (UNEP, 2019b).

The Brundtland Report, also known as our "common future", warned the world in 1987 of the urgency to make progress towards sustainable development without making the world an inhabitable place for living things or depleting natural resources, or irreversibly damaging the environment. Starting with this report, important steps have been taken in the international arena. The United Nations Environment Program in 1988 and the World "Intergovernmental Panel on Climate Change (IPCC)" established with the support of the Meteorology Organization, 1992 Rio Summit, 1997, Kyoto Climate Change Protocol, 2002 Johannesburg World Summit on Sustainable Development are some of these initiatives. The initiatives require a planning process so that the measures taken do not conflict with each other. Although the common goal of the initiatives is to focus on restricting emissions of greenhouse gases, a measure taken at a lower scale may conflict with a measure taken at a higher scale. For example, the trend of increasing the density of cities, especially in Europe in recent years (Biesbroek et al., 2010), to reduce GHG from transport (Williams et al., 2000), has revealed many problems such as urban heat islands, increased flood risk, poor health consequences for residents (Dawson, 2011; Holderness et al., 2013; Hunt & Watkiss, 2011). This is where the importance of taking multiple parameters into account and applying them in comprehensive spatial planning to achieve a sustainable future emerges. Spatial planning can facilitate communication between local, regional, inter-regional, and global goals, taking into account various regional dimensions, and ensure consistent decisions to mitigate the impacts of climate change. Once the measures are decided within a plan, the root of the problems can be addressed with a bottom-up approach.

Buildings are a major contributor to climate change. They produced more greenhouse gases (GHGs) than any other sector in the GHGs pie and the sector is responsible for one-third of global greenhouse gas emissions in both developed and developing countries (UNEP, 2009). Therefore, the building sector should be the first stop on the road to achieve sustainable development goals and mitigate the adverse effects of climate change. But top-down policies aimed at combating climate change through restrictions on greenhouse gases are almost certainly not sustainable. This issue should be addressed comprehensively and with a bottom-up approach. Therefore, buildings should be considered as one of the first stops. The purpose of this chapter is to highlight the importance of sustainable architecture in addressing environmental challenges and ensuring sustainable urban development.

CLIMATE CHANGE RISKS FACING URBAN AREAS

Provide broad Environmental changes and natural disasters faced by humanity are deeply intertwined with complex urbanization processes. It's happening at an unprecedented rate and magnitude (O'Brien & Leichenko, 2008). The severity and frequency of natural disasters are increasing. In the last 30 years, the worldwide, rate of natural disasters has almost quadrupled. These natural disasters have caused human and economic losses (UNISDR, 2012).

Urban areas, on the one hand, constitute an important pressure factor on the climatic and environmental system; On the other hand, they are human settlements most vulnerable to the effects of extreme weather and climatic events.

More than half of the world's population lives in urban areas and if it continues to increase at this rate, it is estimated that 68% of the world's population will live in urban areas by 2050 (UN DESA, 2021). The climate of urban areas changes with this urbanization; It is generally more polluted, hotter, rainier, and less windy than the surrounding rural countryside (Givoni, 1998). They are more exposed to extreme weather events such as heavy rainstorms, hurricanes, and cyclones caused by climate change (David Satterthwaite, 2006).

Figure 1. Vulnerability to the risk of climate change
Source: Ebinger & Vandycke, 2015

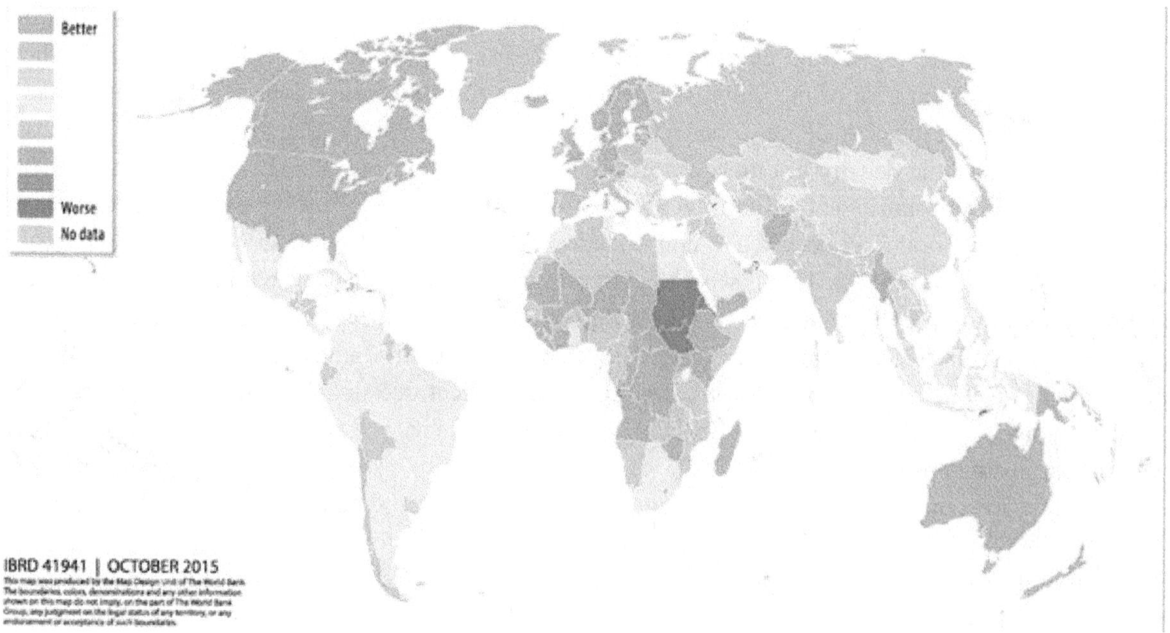

Coastal urban areas are clearly at greater risk due to sea-level rise (Nicholls & Klein, 2005). Rising sea levels can damage building foundations (Lerch, 2017) and increase saltwater inflow to valuable groundwater resources (Abd-Elhamid & Javadi, 2008). In addition, non-coastal urban areas are also very vulnerable to snowmelt due to temperature increase (Adam et al., 2009). Many cities will receive less rainfall due to changing climates and most will be faced with heatwaves and air pollution problems (Givoni, 1998). The possibility for agriculture that cannot adapt quickly enough to the rapidly changing climate will decline in many places and damage the economies (Agovino et al., 2019). Warmer average temperatures will expand the areas where tropical diseases can occur (Patz & Olson, 2006).

THE ROLE OF SPATIAL PLANNING FOR SUSTAINABLE FUTURE

Sustainable development means *"development that meets the needs of present generations without compromising the ability of future generations to meet their own needs"* (WCED, 1987). Based on this generally accepted definition, it is possible to define sustainable development as the efficient use of resources in line with both future needs and current needs. The resources should not be perceived in just one dimension. Since sustainable development requires economic, social, and environmental integration, which are three interconnected dimensions, resources should be addressed in these three important axes. These include efficient use of limited land resources and balanced regional business development and balanced use of resources including natural and landscape resources, land, water, and air (Breuer, 2002). Spatial planning can be used as a tool to coordinate socio-economic development by ensuring the efficient use of these resources. Spatial planning plays an important role in ensuring a sustainable future as it has a long-term perspective (Low et al., 2005; Wheeler, 2004). Using spatial planning as an instrument on the road to sustainability, as it provides a long-term perspective, enables the concepts of development and conservation to be seen as complementary rather than contradictory.

Climate change has been one of the biggest challenges that humanity has faced today as a result of the inefficient use of resources. Solving this problem requires effort at all levels of governance. While national governments sign legally binding commitments to reduce carbon and other greenhouse gases (e.g. Kyoto Protocol), local governments, local community leaders, decision-makers take action in line with this upper decision (e.g. energy efficiency in buildings, renewable energy sources) (Agyeman et al., 1998). All of these should be done within planning to see the general effects of the measures taken at all levels to reduce the effects of climate change. This is where spatial planning can help.

Spatial planning has a critical role in both mitigating the negative effects of climate change and adapting to its inevitable effects (Stern, 2007; Wilson & Piper, 2010). Spatial planning can facilitate communication between local, regional, inter-regional, and global goals by taking into account various regional dimensions. In other words, while producing solutions targeting specific geographic regions, allows these solutions to be integrated with solutions in other large or smaller regions. It helps the public to adopt the measures taken by ensuring the participation of the people in a democratic decision-making process so that various social interests can be weighed and balanced in development decisions (Breuer, 2002).

SCALING DOWN SUSTAINABILITY FROM URBAN AREAS TO BUILDING

Urban planning has been challenged by the rapidly increasing population, the increasing frequency of extreme weather events, and natural disasters. It is also being tested with actions to decarbonize economies to mitigate against changing climate. Currently, 55% of the world's population resides in urban areas. This ratio is predicted to be 68% by 2050 (UN DESA, 2021). This ever-increasing urban population also increases the risk of natural disasters (IPCC (International Panel on Climate Change), 2013).

Figure 2. Changes in global population by degree of urbanisation, 1975-2050
Source: OECD & European Commission, 2020

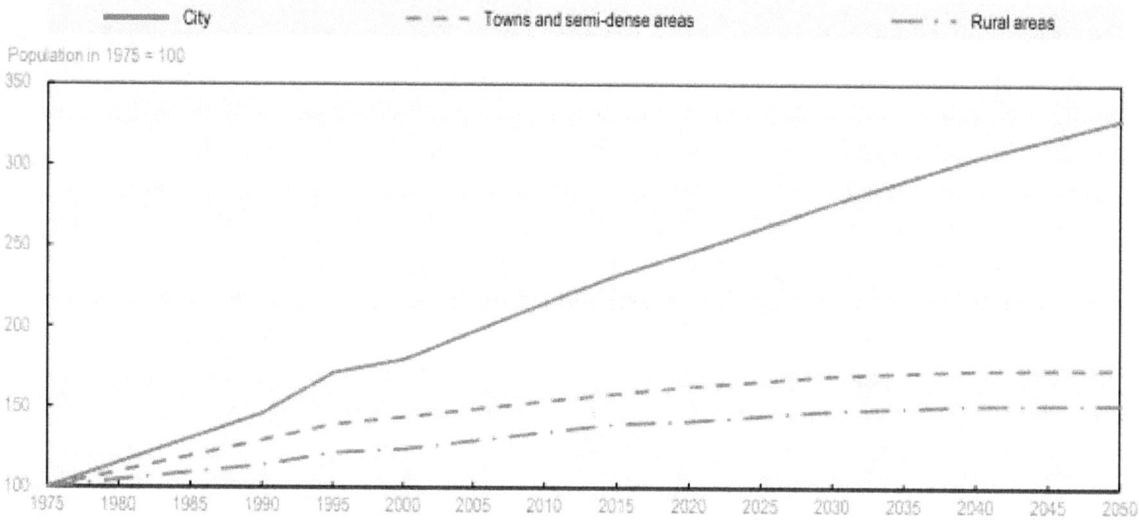

Urban areas produce 71% of global carbon emissions directly or indirectly (IEA (International Energy Agency), 2008, 2008). This fact places them at the top of the areas where the action will be taken in reducing climate change (Reckien et al., 2014). Measures to be taken in urban areas to reduce climate change have the potential to cause conflicts and exchanges (Dawson, 2011; McEvoy et al., 2006). In other words, bona fide intervention implemented in one sector may cause undesirable effects on another sector. For example, in the last decade, the spatial planning of European cities has focused on increasing the density of cities (Biesbroek et al., 2010) in order to reduce GHG from transport (Williams et al., 2000). However, urban concentration poses many problems, such as urban heat islands, increased flood risk, poor health consequences for residents (Dawson, 2011; Holderness et al., 2013; Hunt & Watkiss, 2011). In order to achieve a sustainable future, the importance of considering multiple parameters and applying them within comprehensive spatial planning becomes evident here. The measures should be taken in a way that does not contradict the decisions at the higher scale.

Spatial planning is considered an interdisciplinary field that includes social, engineering, and design sciences. Integrity can be achieved by implementing the decisions made at the upper scale by the actors working at the lower scale in a way that does not allow conflict between decisions. After the upper scale decisions are made, the problems in the lower scales should be handled with the bottom-up approach.

Unlike the top-down approach, the bottom-up approach starts at the specific and moves to the general. The approach depends on specialized knowledge. Since the source of the problem is known, it is easier to take measures and develop a solution for that problem. Since buildings are the biggest source of GHG emission in urban areas, it may be good to reduce the scale, focus on architecture and start from here after planning in a way that will not contradict the upper scale. Because, according to research (e.g (UNEP, 2019a; Yau & Hasbi, 2013), a significant proportion of greenhouse gas emissions in urban areas originate from buildings. Furthermore, the buildings use 50% of the raw material obtained from the natural environment, 40% of the global energy, and 16% of the water resources (King, 2010). In order to mitigate the impacts of climate change, the principle of reducing greenhouse gas emissions from buildings should become the cornerstone of every national climate change strategy (UNEP, 2009). They play a key role in solving the urban-related energy and GHG emissions problems.

Buildings constitute the primary priority and also the primary challenge in this battle, created by changing the climate to reduce the causes and possible effects of climate change. Because people tend to spend most of their time in buildings; In addition, they host the largest part of our social, commercial, cultural, and private activities (Altomonte, 2009). Furthermore, human settlements have been proven to be one of the areas that contributed most to the climate crisis (Steemers, 2010).

Direct emissions from buildings increased by 26% between 1970 and 1990 although this increase did not include emissions from electricity consumed in buildings. When taking into account the electricity required for the operation of mechanical systems and services such as heating and cooling of buildings, the total increase in direct and indirect emissions from the construction sector is 75% (IPCC, 2007). Buildings consume enormous amounts of energy as well as consuming non-renewable resources and emitting greenhouse gases into the atmosphere during their entire life cycle. The energy they consume accounts for approximately half of the total global energy consumed (IEA, 2020). Since this energy source comes mostly from the combustion of oil, coal, and natural gas (Altomonte, 2009), it is among the sectors that make the biggest contribution to climate change with the enormous CO2 emissions they trigger. For this reason, it becomes necessary to develop design strategies to reduce the energy demands of buildings, to ensure that they benefit from clean power sources by reducing their dependence on fossil fuels, and to minimize their waste, to provide efficient use of water and to select the materials used in their construction in accordance with nature.

HOW ARCHITECTURE CAN CONTRIBUTE TO ENVIRONMENTAL SUSTAINABILITY

Humanity has irreversibly changed the relationship between humans and the natural environment with the actions it has taken since the Industrial Revolution. The threat of global warming emerging due to the accumulation of heat-trapping gases in the atmosphere and the resulting changing climate are among the biggest problems facing humanity today. In this context, buildings played a leading role - with the energy they consume and the CO^2 emissions they produce accordingly. In order to mitigate these threats and meet the needs of sustainable development, a new approach to building design and construction that will simultaneously meet the complex requirements of the environment with its limited resources and the needs of contemporary societies is compulsory. In this new approach, new integrated designs that include environmental responsibility, climate awareness, and sustainable production and energy use methods should be brought together.

This section seeks an answer to the question of "How environmental sustainability can be achieved by making the building sector sustainable, which is the biggest contributor of climate change," by emphasizing three main aspects: "Efficient Energy Use", "Water Efficiency and Conservation" and "Material Efficiency".

Efficient Energy Use

More than one-third of the world's total energy consumption is accountable for buildings (Kwok & Lee, 2011; Yoshino et al., 2017). Reducing the energy usage of buildings is, therefore, a global problem. This circumstance highlights the importance of conducting a thorough examination of the link between construction and energy usage (Economidou et al., 2020; Magrini et al., 2020). Through guarantee environmental sustainability, building energy efficiency measures should be addressed from the design phase to the implementation process (Emekci, 2018). Here are some suggestions:

For Energy Savings, Building Design is Crucial

The building design is critical to achieving the objective of decreasing a building's carbon footprint, overall energy consumption, and costs (Chen et al., 2020). The building envelope has the biggest influence on the total energy consumption of the structure (Manioğlu & Yılmaz, 2006). The envelope (foundation, windows, walls, doors, and roof) controls the inside environment and, as a result, the additional energy requirement for heating and cooling. They must be evaluated in conjunction with the climatic parameters of the building's location. The building orientation is another parameter to achieve energy-efficient building. A southern orientation is typically thought to be best for gathering heat in the winter and limiting solar radiation in the summer. In general, the longest wall sections should be directed southward (Mingfang, 2002). The one another factor is compactness. The compactness index is defined as the ratio of the volume to the exterior surface of the building facade. The index is a ratio that indicates how easily a structure can be cooled and heated, as well as how much energy it consumes (Ourghi et al., 2007). One of the most effective ways to save energy in a building is to properly plan its facade. A "double skin façade" is one of the greatest solutions for regulating the interaction between the outside and the inside (Shameri et al., 2011). Traditional building skin facades are recognized to have a variety of issues, including thermal comfort, natural ventilation, and glare, particularly in structures with high glazing skin that are located in hot temperature zones. These issues motivate engineers to seek solutions and improve problems via the use of innovative techniques and technologies such as shading devices, color glass, and tint glass (Haase & Amato, 2006). The double-skin facade systems enable efficient energy usage by minimizing heat loss in cold areas and heat gain in hot climates. In double-skin facade systems, a second façade is built by leaving specific gaps in the facade, and airflow is delivered mechanically or naturally in the area between (Ahmed et al., 2015).

The Selection of Materials is Essential to Reduce Energy Consumption and Emissions

Based on their environmentally friendly characteristics, energy-efficient materials can assist buildings both environmentally and economically. In addition, less energy-consuming materials at the same time generate lower emissions and reduce pollution due to building materials (Yüksek, 2015). Despite the high initial expenditures, the overall cost and energy for building are low in energy-efficient materials and systems, and the long-term economic and environmental benefits are obvious. (Emekci, 2018).

In the Building Sectors, the Percentage of Renewable Resources must be Raised

The energy resources found in nature can be classified as renewable or non-renewable. Non-renewable energy resources are those that are projected to be depleted in the near future and will require a long time to replenish (Sims, 2004). These energy sources include hydrocarbon-containing fossil fuels such as coal, oil, and natural gas. When nonrenewable fossil fuels are used, greenhouse gases such as carbon dioxide (CO2) and methane are produced; climate change is observed as a result of their existence in significant levels in the atmosphere owing to their heat capacity (Owusu & Asumadu-Sarkodie, 2016). Renewable sources are viewed as clean energy sources and their optimum usage minimizes the environmental effect of the renewable energy resources, produces little secondary waste, and is viable, based on present and future social and economic requirements (Panwar et al., 2011).

Move towards Passive Heating/Cooling Systems of Buildings

The application of passive technologies is one of the efficient methods to decreasing cooling and heating energy consumption in residential buildings. These approaches can minimize the energy consumption, peak loads, and variations in air inside, enhance building thermal comfortably and reduce fossil fuel and GHG emissions (Omer, 2008). Green roofs, roof ponds, phase change materials (PCM), thermal energy storage, wind catchers, subterranean structures, earth to air heat exchangers, green walls, and Trombe walls are some examples of passive construction approaches (Goudarzi & Mostafaeipour, 2017).

Energy Efficiency Issues should be Considered in the Urban Design Process

Because urban design deals with the entire design of a particular region and incorporates many elements of different urban development disciplines, integrating efficiency as a key design feature would imply that energy efficiency awareness would become part of every phase of the development process (Gosling & Walker, 1992). In the planning phase, it is possible to discover the variables that impact whether the building is energy-efficient or not, such as building density, building orientation, dominant housing type, building form, and energy savings using these factors effectively (Pacheco et al., 2012).

Water Efficiency and Conservation

Water is one of the principal contributors to the habitability of the globe. The worldwide water loss or unaccounted quantity, in general, has been projected to be more than 30 percent of global production (Center for Neighborhood Technology (CNT), 2013). In fact, about a third of the world's population is now experiencing physical water shortage due to a fast-rising population and water usage per capita in many places (Oki & Kanae, 2006). In the development of fair, stable, and productive societies and ecosystems, water has a major role to play. The United Nations has thus designated water safety as one of 17 (Goal 6) sustainable development objectives (SDGs) (UN, 2019). The world will experience severe drought as a result of climate change in the next years. Water has been used extensively in building construction and operations. Increased urban water usage has resulted in a considerable decrease in water tables, necessitating massive projects that divert resources away from farmland (Roodman & Lenssen, 1995). However, individual efforts combined with sustainable design decisions have a significant influence in decreasing consumption (Akadiri et al., 2012). Here are some solutions.

Recycling and Reusing are Key to Tackling the Water Consumption-Based Problem

The world's freshwater supply is limited and under threat from pollution. Increasing water requirements to provide farming, industry, and cities lead to rivalry in terms of allocating a limited amount of freshwater. Water recycling primarily converts non-potable wastes into useable water, therefore saving both economic and environmental costs associated with the development of new water sources. Water recycling and reuse are the processes of collecting, cleaning, and reusing wastewater, particularly from cities, industry, and agriculture. Recycled water, when properly treated, can be used for agriculture, industry, or domestic purposes (Arena et al., 2020). Water in buildings having high amounts of septic waste is referred to as black water, whereas water utilized in the remaining spaces (bathroom, kitchen, etc.) containing domestic waste but not black water is referred to as gray water (Bergdolt et al., 2011). Depending on the treatment level, gray water can be utilized in landscape irrigation, toilet reservoirs, and washing machines (Stec & Kordana, 2015).

Saving Water through Behavior-Changing

Water-saving using this approach is usually carried out by decreasing or efficient use of water (Wolfe et al., 2014). Water can be saved by repairing leaking taps, recycling gray water, taking a shower instead of a bath, turning off faucets when brushing teeth, and using fully loaded dishwashers and washing machines (Patterson, 2004). However, efficiency necessitates greater human participation, financial commitment, and know-how. Water smart meters, low-volume showerheads, front-loading washing machines, water-efficient dishwashers, double-flush toilets, and automated netting systems are just a few examples (Fielding et al., 2012; Lee et al., 2011).

Rainwater Collection, Storage, and Reuse can Help to Save Water

Cities are facing increasing demands on their water supplies as a result of climate change, economic expansion, and population increase. Rainwater collection and use are critical in terms of water conservation. At the building scale, rainwater is often collected from water falling from the roof. In general, it is the rainwater harvesting system that directly collects rainfall from the rooftops and catchments constructed for various purposes, collects the flux from man-made or natural surface catchments and clogs for household, industrial, agricultural and environmental purposes (Gould, 1999). This approach can be a water source that can fulfill a large portion of a house's water needs in places where there is a lot of rain depending on the building form (Şahin & Manioğlu, 2019).

Climate-Appropriate Plants should be Used

When climate-appropriate plants are used, less irrigation is required, resulting in less water usage (Quesnel et al., 2019). Xeriscape is the usage of plants that have been adapted to the local environment and thus use less water, have a better chance of surviving drought conditions, and are more tolerant of pests and diseases. Using Xeriscape methods whenever possible will save money on water, fertilizer, trimming, upkeep, labor, and overall expenditures. It contributes significantly to water conservation through rules based on ecology in the terrain and environmental considerations (Welsh, 2005).

Material Efficiency

Over the last 50 years, global demand for materials has quadrupled and is now soaring at the fastest rate (Allwood et al., 2011). Because the manufacturing of materials consumes one-quarter of global energy (Hertwich et al., 2019), more effective use of these resources provides substantial potential for reducing greenhouse gas (GHG) emissions. To guarantee environmental sustainability, construction materials that are recyclable and reused must be used effectively and efficiently (Meyer et al., 2007). Criteria like the impacts on the natural environment and human health, durability, and cost should not be overlooked. Some solutions are presented here.

In Building Construction, Recyclable Materials should be Given More Priority

In the field of the construction industry, it is extremely vital to determine the merits of building materials and products in terms of their recycling value. In order to produce building materials and products worldwide more than three billion metric tons of raw materials are used annually. This represents around 40%-50% of the world economy (Calkins, 2009). Construction, operation, and demolition stages consume around 30%-40% of all primary energy used worldwide and about 15% of world freshwater resources, and about 40-50% of global greenhouse gas production (Asif et al., 2007; Gunnell, 2009). Therefore, in building construction, recyclable materials can be used to cause minimal environmental impact.

Materials with Less Embodied Energy should be Highlighted

Embodied energy (EE) is defined as the energy consumed in the production of the material (Koskela, 1992). This includes raw material extraction, processing, and transportation to the building site. Similarly, the emission of energy-related pollutants (such as $CO2$), which is a problem in the context of global warming and climate change, should be considered across their whole life cycle. As a result, the concept of "embodied carbon" (EC) emerges (Hammond & Jones, 2008). The most effective way to decrease the influence of the built environment on climate change is to utilize alternative materials with low EE and EC (Yu et al., 2011).

Properly Sized Buildings can Accomplish Effective and Efficient Material Usage

The building's size is strongly linked to the amount of materials used. Properly sized buildings can minimize excessive energy use and material consumption as well as the environmental harm produced by the construction (Truong, 2020).

CONCLUSION

Since the beginning of the Industrial Revolution, human activity has greatly increased the volume of greenhouse gases released into the atmosphere. Today, human-induced greenhouse gas emissions are higher than ever (IPCC, 2018). Global warming, fed by human-induced greenhouse gas emissions, changes the world's climate. Climate change is one of the most pressing problems of our time. It threatens the lives of billions of people by causing natural disasters, environmental degradation, and extreme weather conditions. Urban areas with their ever-increasing population are held responsible for this emission (PlaNYC, 2011). This also highlights the fact that urban areas have a great potential to mitigate the effects of climate change. Urban areas must be encouraged more inclusive growth; adapted to climate change and action should be taken in these areas to reduce its devastating effects. Buildings have the largest share in the energy consumption of urban areas. They are responsible for 30-40% of the global energy requirements (Huovila et al., 2007), while approximately 40% of total annual anthropogenic GHG emissions come from buildings (UNEP, 2019a). According to UNEP (2019a), in the absence of measures, energy demands for this sector could rise by 50 percent by 2060. Therefore, making urban areas sustainable should start from the building level.

The sustainability of a building is directly related to the efficient use of available resources. Greenhouse gas emissions of urban areas can be reduced as a result of the efficient use of energy, water, and material resources by the building sector. With a bottom-up approach, a strong step can be taken to achieve sustainable development goals.

REFERENCES

Abd-Elhamid, H. F., & Javadi, A. A. (2008). An Investigation into Control of Saltwater Intrusion Considering the Effects of Climate Change and Sea Level Rise. Presented at the *20th Salt Water Intrusion Meeting*, Naples, FL, USA.

Adam, J. C., Hamlet, A. F., & Lettenmaier, D. P. (2009). Implications of global climate change for snowmelt hydrology in the twenty-first century. *Hydrological Processes, 23*(7), 962–972. doi:10.1002/hyp.7201

Agovino, M., Casaccia, M., Ciommi, M., Ferrara, M., & Marchesano, K. (2019). Agriculture, climate change and sustainability: The case of EU-28. *Ecological Indicators, 105*, 525–543. doi:10.1016/j.ecolind.2018.04.064

Agyeman, J., Evans, B., & Kates, R. W. (1998). Greenhouse gases special: Thinking locally in science, practice and policy. *Local Environment, 3*(3), 245–246. doi:10.1080/13549839808725563

Ahmed, M. M. S., Abel-Rahman, A. K., Ali, A. H. H., & Suzuki, M. (2015). Double Skin Façade: The State of Art on Building Energy Efficiency. *Journal of Clean Energy Technologies, 4*(1), 84–89. doi:10.7763/JOCET.2016.V4.258

Akadiri, P. O., Chinyio, E. A., & Olomolaiye, P. O. (2012). Design of A Sustainable Building: A Conceptual Framework for Implementing Sustainability in the Building Sector. *Buildings, 2*(2), 126–152. Advance online publication. doi:10.3390/buildings2020126

Allwood, J. M., Ashby, M. F., Gutowski, T. G., & Worrell, E. (2011). Material efficiency: A white paper. *Resources, Conservation and Recycling*, *55*(3), 362–381. doi:10.1016/j.resconrec.2010.11.002

Altomonte, S. (2009). Climate Change and Architecture: Mitigation and Adaptation Strategies for a Sustainable Development. *Journal of Sustainable Development*, *1*(1), 97. doi:10.5539/jsd.v1n1p97

Arena, C., Genco, M., & Mazzola, M. R. (2020). Environmental Benefits and Economical Sustainability of Urban Wastewater Reuse for Irrigation—A Cost-Benefit Analysis of an Existing Reuse Project in Puglia, Italy. *Water (Basel)*, *12*(10), 2926. Advance online publication. doi:10.3390/w12102926

Asif, M., Muneer, T., & Kelley, R. (2007). Life cycle assessment: A case study of a dwelling home in Scotland. *Building and Environment*, *42*(3), 1391–1394. doi:10.1016/j.buildenv.2005.11.023

Bergdolt, J., Sharvelle, S., & Roesner, L. (2011). *Guidance Manual for Separation of Graywater from Blackwater for Graywater Reuse*. IWA Publishing., doi:10.2166/9781780400082

Biesbroek, G. R., Swart, R. J., Carter, T. R., Cowan, C., Henrichs, T., Mela, H., Morecroft, M. D., & Rey, D. (2010). Europe adapts to climate change: Comparing National Adaptation Strategies. *Governance. Complexity and Resilience*, *20*(3), 440–450. doi:10.1016/j.gloenvcha.2010.03.005

Breuer, D. (2002). *Spatial planning as an instrument for promoting sustainable development in the Nordic countries: action programme for 2001-2004*. Danish Ministry of Environment and Energy.

Calkins, M. (2009). *Materials for Sustainable Sites: A Complete Guide to the Evaluation, Selection, and Use of Sustainable Construction Materials* (Vol. 1). John Wiley.

Center for Neighborhood Technology (CNT). (2013). *The Case for Fixing the Leaks: Protecting People and Saving Water while Supporting Economic Growth in the Great Lakes Region*. Retrieved from https://www.cnt.org/publications/the-case-for-fixing-the-leaks-protecting-people-and-savingwater-while-supporting

Chen, S., Zhang, G., Xia, X., Setunge, S., & Shi, L. (2020). A review of internal and external influencing factors on energy efficiency design of buildings. *Energy and Building*, *216*, 109944. doi:10.1016/j.enbuild.2020.109944

Dawson, R. J. (2011). Potential pitfalls on the transition to more sustainable cities and how they might be avoided. *Carbon Management*, *2*(2), 175–188. doi:10.4155/cmt.11.8

Ebinger, J. O., & Vandycke, N. (2015). *Moving Toward Climate-Resilient Transport : The World Bank's Experience from Building Adaptation into Programs*. World Bank. Retrieved from https://openknowledge.worldbank.org/handle/10986/23685

Economidou, M., Todeschi, V., Bertoldi, P., D'Agostino, D., Zangheri, P., & Castellazzi, L. (2020). Review of 50 years of EU energy efficiency policies for buildings. *Energy and Building*, *225*, 110322. doi:10.1016/j.enbuild.2020.110322

Emekci, Ş. (2018). *A Life cycle costing based decision support tool for cost-optimal energy efficient design and/or refurbishments* (Ph.D Thesis). Middle East Technical University.

Fielding, K., Spinks, A., Russell, S., & Mankad, A. (2012). *Water Demand Management Study: Baseline Survey of Household Water Use*. Urban Water Security Research Alliance Technical Report.

Givoni, B. (1998). *Climate considerations in building and urban design*. John Wiley & Sons, Inc.

Gosling, M.-C., & Walker, A. J. (1992). Energy efficiency within the urban design process. *Cities (London, England)*, *9*(2), 111–116. doi:10.1016/0264-2751(92)90045-7

Goudarzi, H., & Mostafaeipour, A. (2017). Energy saving evaluation of passive systems for residential buildings in hot and dry regions. *Renewable & Sustainable Energy Reviews*, *68*, 432–446. doi:10.1016/j.rser.2016.10.002

Gould, J. (1999). *Contributions Relating to Rainwater Harvesting*. (Paper prepared for the World Commission on Dams Secretariat (WCD). *Thematic Review*, *4*, 3.

Gunnell, K. (2009). *Green Building in South Africa: Emerging Trends*. Department of Environmental Affairs and Tourism (DEAT). Retrieved from http://soer. deat.gov.za/Green Building in South Africa Emerging Trends GyzKE.pdf.file

Haase, M., & Amato, A. (2006). *Ventilated façade design in hot and humid climate*. Academic Press.

Hammond, G. P., & Jones, C. I. (2008). Embodied energy and carbon in construction materials. *Proceedings of the Institution of Civil Engineers - Energy, 161*(2), 87–98. 10.1680/ener.2008.161.2.87

Hertwich, E. G., Ali, S., Ciacci, L., Fishman, T., Heeren, N., Masanet, E., Asghari, F. N., Olivetti, E., Pauliuk, S., Tu, Q., & Wolfram, P. (2019). Material efficiency strategies to reducing greenhouse gas emissions associated with buildings, vehicles, and electronics—A review. *Environmental Research Letters*, *14*(4), 043004. doi:10.1088/1748-9326/ab0fe3

Holderness, T., Barr, S., Dawson, R., & Hall, J. (2013). An evaluation of thermal Earth observation for characterizing urban heatwave event dynamics using the urban heat island intensity metric. *International Journal of Remote Sensing*, *34*(3), 864–884. doi:10.1080/01431161.2012.714505

Hunt, A., & Watkiss, P. (2011). Climate change impacts and adaptation in cities: A review of the literature. *Climatic Change*, *104*(1), 13–49. doi:10.100710584-010-9975-6

Huovila, P., Ala-Juusela, M., Melchert, L., & Pouffary, S. (2007). *Buildings and Climate Change Status, Challenges and Opportunities*. United Nations Environment Programme.

IEA. (2020). *Buildings – Topics*. Retrieved March 15, 2021, from https://www.iea.org/topics/buildings

IEA. (International Energy Agency). (2008). *World energy outlook*. Retrieved April 18, 2021, from https://www.iea.org/topics/world-energy-outlook

IPCC. (2007). *Climate Change 2007: Mitigation. Contribution of Working Group III to the Fourth Assessment Report of the Intergovernmental Panel on Climate Change*. Cambridge University Press.

IPCC. (2018). *Special Report: Global Warming of 1.5 °C*. Retrieved February 4, 2021, from https://www.ipcc.ch/sr15/chapter/glossary/

IPCC (International Panel on Climate Change). (2013). *Climate Change 2013: The Physical Science Basis, Contribution of Working Group I to the Fifth Assessment Report of the Intergovernmental Panel on Climate Change*. Cambridge University Press.

King, B. (2010). Toprak Mimarisinin Yeniden Doğuşu, Kil Kökenli Taze ve Güncellenmiş Bir Bakış. *Mimarlıkta Malzeme Dergisi*, (17).

Koskela, L. (1992). Application of the New Production Philosophy to Construction (CIFE technical report No. 72). Stanford University.

Kwok, S. S. K., & Lee, E. W. M. (2011). A study of the importance of occupancy to building cooling load in prediction by intelligent approach. *Energy Conversion and Management*, *52*(7), 2555–2564. doi:10.1016/j.enconman.2011.02.002

Lee, M., Tansel, B., & Balbin, M. (2011). Influence of residential water use efficiency measures on household water demand: A four year longitudinal study. Resources, Conservation and Recycling. *Resources, Conservation and Recycling*, *56*(1), 1–6. doi:10.1016/j.resconrec.2011.08.006

Lerch, D. (2017). Six Foundations for Building Community Resilience. In D. Lerch (Ed.), *The Community Resilience Reader: Essential Resources for an Era of Upheaval* (pp. 9–42). Island Press/Center for Resource Economics. doi:10.5822/978-1-61091-861-9_2

Low, N. P., Gleeson, B., Green, R. J., & Radovic, D. (2005). *The Green City: Sustainable Homes Sustainable Suburbs*. University of New South Wales Press.

Magrini, A., Lentini, G., Cuman, S., Bodrato, A., & Marenco, L. (2020). From nearly zero energy buildings (NZEB) to positive energy buildings (PEB): The next challenge - The most recent European trends with some notes on the energy analysis of a forerunner PEB example. *Developments in the Built Environment*, *3*, 100019. doi:10.1016/j.dibe.2020.100019

Manioğlu, G., & Yılmaz, Z. (2006). Economic evaluation of the building envelope and operation period of heating system in terms of thermal comfort. *Energy and Building*, *38*(3), 266–272. doi:10.1016/j.enbuild.2005.06.009

McEvoy, D., Lindley, S., & Handley, J. (2006). Adaptation and mitigation in urban areas: Synergies and conflicts. *Proceedings of the Institution of Civil Engineers. Municipal Engineer*, *159*(4), 185–191. doi:10.1680/muen.2006.159.4.185

Meyer, B., Distelkamp, M., & Wolter, M. I. (2007). Material efficiency and economic-environmental sustainability. Results of simulations for Germany with the model PANTA RHEI. *Ecological Economics*, *63*(1), 192–200. doi:10.1016/j.ecolecon.2006.10.017

Mingfang, T. (2002). Solar control for buildings. *Building and Environment*, *37*(7), 659–664. doi:10.1016/S0360-1323(01)00063-4

Nicholls, R. J., & Klein, R. J. T. (2005). Climate change and coastal management on Europe's coast. In J. Vermaat, W. Salomons, L. Bouwer, & K. Turner (Eds.), *Managing European Coasts* (pp. 199–226). Springer Berlin Heidelberg. doi:10.1007/3-540-27150-3_11

O'Brien, K., & Leichenko, R. (2008). *Environmental Change and Globalization: Double Exposures*. Oxford University Press.

OECD & European Commission. (2020). *Cities in the World: A New Perspective on Urbanisation*. OECD. doi:10.1787/d0efcbda-

Oki, T., & Kanae, S. (2006). Global Hydrological Cycles and World Water Resources. *Science, 313*(5790), 1068–1072. doi:10.1126cience.1128845 PMID:16931749

Omer, A. M. (2008). Renewable building energy systems and passive human comfort solutions. *Renewable & Sustainable Energy Reviews, 12*(6), 1562–1587. doi:10.1016/j.rser.2006.07.010

Ourghi, R., Al-Anzi, A., & Krarti, M. (2007). A simplified analysis method to predict the impact of shape on annual energy use for office buildings. *Energy Conversion and Management, 48*(1), 300–305. doi:10.1016/j.enconman.2006.04.011

Owusu, P. A., & Asumadu-Sarkodie, S. (2016). A review of renewable energy sources, sustainability issues and climate change mitigation. *Cogent Engineering, 3*(1), 1167990. doi:10.1080/23311916.2016.1167990

Pacheco, R., Ordóñez, J., & Martínez, G. (2012). Energy efficient design of building: A review. *Renewable & Sustainable Energy Reviews, 16*(6), 3559–3573. doi:10.1016/j.rser.2012.03.045

Panwar, N. L., Kaushik, S. C., & Kothari, S. (2011). Role of renewable energy sources in environmental protection: A review. *Renewable & Sustainable Energy Reviews, 15*(3), 1513–1524. doi:10.1016/j.rser.2010.11.037

Patterson, R. A. (2004). Water Efficiency, Domestic Appliances and Hydraulic Design for On-site Systems. Presented at the *1 st International Conference on Onsite Wastewater Treatment and Recycling*, Perth, Australia.

Patz, J. A., & Olson, S. H. (2006). Malaria risk and temperature: Influences from global climate change and local land use practices. *Proceedings of the National Academy of Sciences of the United States of America, 103*(15), 5635–5636. doi:10.1073/pnas.0601493103 PMID:16595623

Quesnel, K. J., Ajami, N., & Marx, A. (2019). Shifting landscapes: Decoupled urban irrigation and greenness patterns during severe drought. *Environmental Research Letters, 14*(6), 064012. doi:10.1088/1748-9326/ab20d4

Reckien, D., Flacke, J., Dawson, R. J., Heidrich, O., Olazabal, M., Foley, A., Hamann, J. J.-P., Orru, H., Salvia, M., De Gregorio Hurtado, S., Geneletti, D., & Pietrapertosa, F. (2014). Climate change response in Europe: What's the reality? Analysis of adaptation and mitigation plans from 200 urban areas in 11 countries. *Climatic Change, 122*(1), 331–340. doi:10.100710584-013-0989-8

Roodman, D. M., & Lenssen, N. A. (1995). *Building Revolution: How Ecology and Health Concerns are Transforming Construction. Worldwatch Paper 124*. Worldwatch Institute.

Şahin, N. İ., & Manioğlu, G. (2019). Water conservation through rainwater harvesting using different building forms in different climatic regions. Sustainable Cities and Society, 44, 367–377. doi:10.1016/j.scs.2018.10.010

Satterthwaite, D. (2020). *An urbanising world*. Retrieved February 4, 2021, from https://www.iied.org/urbanising-world

Satterthwaite, D. (2006). Climate change and cities. *Sustainable Development Opinion*, 2.

Shameri, M. A., Alghoul, M. A., Sopian, K., Zain, M. F. M., & Elayeb, O. (2011). Perspectives of double skin façade systems in buildings and energy saving. *Renewable & Sustainable Energy Reviews*, *15*(3), 1468–1475. doi:10.1016/j.rser.2010.10.016

Sims, R. E. H. (2004). Renewable energy: a response to climate change. *Solar World Congress 2001*, *76*(1), 9–17. 10.1016/S0038-092X(03)00101-4

Stec, A., & Kordana, S. (2015). Analysis of profitability of rainwater harvesting, gray water recycling and drain water heat recovery systems. *Resources, Conservation and Recycling*, *105*, 84–94. doi:10.1016/j.resconrec.2015.10.006

Steemers, K. (2010). Towards a research agenda for adapting to climate change. *Building Research and Information*, *31*(3-4), 291–301. doi:10.1080/0961321032000097692

Stern, N. (2007). *The Economics of Climate Change: The Stern Review*. Cambridge University Press., doi:10.1017/CBO9780511817434

Truong, Q. (2020). *4. Material Sustainability*. Birkhäuser. doi:10.1515/9783035619522-006

UN. (2019). *The Sustainable Development Goals Report 2019*. United Nations. Retrieved from https://unstats.un.org/sdgs/report/2019/The-Sustainable-Development-Goals-Report-2019.pdf

UN DESA. (2021). *68% of the world population projected to live in urban areas by 2050, says UN | UN DESA | United Nations Department of Economic and Social Affairs*. Retrieved April 18, 2021, from https://www.un.org/development/desa/en/news/population/2018-revision-of-world-urbanization-prospects.html

UNEP. (2009). *Buildings and Climate Change*. Retrieved from http://admin.indiaenvironmentportal.org.in/files/SBCI-BCCSummary.pdf

UNEP. (2019a). *2019 Global Status Report for Buildings and Construction*. UNEP.

UNEP. (2019b, August 29). *Facts about the Climate Emergency*. Retrieved February 8, 2021, from https://www.unenvironment.org/explore-topics/climate-change/facts-about-climate-emergency

UNISDR. (2012). *2011-Disasters in Numbers*. UNISDR, USAID, CRED.

WCED. (1987). *Report of the World Commission on Environment and Development: Our Common Future*. Retrieved July 28, 2016, from http://www.un-documents.net/A372BC4E-993E-4C10-B368-9E75816E59DB/FinalDownload/DownloadId-1990161B7AFC5CB358F7280210D7A48F/A372BC4E-993E-4C10-B368-9E75816E59DB/our-common-future.pdf

Welsh, D. F. (2005). Xeriscape. In *Water Encyclopedia* (pp. 750–752). American Cancer Society. doi:10.1002/047147844X.aw434

Wheeler, S. M. (2004). *Planning for Sustainability: Creating Livable, Equitable and Ecological Communities*. Routledge. doi:10.4324/9780203300565

WHO. (2018). *Climate change and health*. Retrieved February 8, 2021, from https://www.who.int/news-room/fact-sheets/detail/climate-change-and-health

Williams, K., Burton, E., & Jenks, M. (2000). Achieving sustainable urban form: conclusions. *Achieving Sustainable Urban Form*, 347–355.

Wilson, E., & Piper, J. (2010). *Spatial Planning and Climate Change*. Routledge. doi:10.4324/9780203846537

Wolfe, A. K., Malone, E. L., Heerwagen, J. H., & Dion, J. P. (2014). *Behavioral Change and Building Performance: Strategies for Significant, Persistent, and Measurable Institutional Change* (No. PNNL-23264, 1132691). doi:10.2172/1132691

Yau, Y. H., & Hasbi, S. (2013). A review of climate change impacts on commercial buildings and their technical services in the tropics. *Renewable & Sustainable Energy Reviews*, *18*, 430–441. doi:10.1016/j.rser.2012.10.035

Yoshino, H., Hong, T., & Nord, N. (2017). IEA EBC annex 53: Total energy use in buildings—Analysis and evaluation methods. *Energy and Building*, *152*, 124–136. doi:10.1016/j.enbuild.2017.07.038

Yu, D., Tan, H., & Ruan, Y. (2011). A future bamboo-structure residential building prototype in China: Life cycle assessment of energy use and carbon emission. *Energy and Building*, *43*(10), 2638–2646. doi:10.1016/j.enbuild.2011.06.013

Yüksek, İ. (2015). The Evaluation of Building Materials in Terms of Energy Efficiency. *Periodica Polytechnica. Civil Engineering*, *59*(1), 45–58. doi:10.3311/PPci.7050

ADDITIONAL READING

Campbell, H. (2006). Is the Issue of Climate Change too Big for Spatial Planning? *Planning Theory & Practice*, *7*(2), 201–230. doi:10.1080/14649350600681875

Emekci, Ş. (2021). Determination of Sustainable Architectural Design Criteria in Protected Areas: Focus group method. *Journal of Design+Theory*, *17*(33), 229–242. doi:10.14744/tasarimkuram.2020.82687

Kumar, P., & Geneletti, D. (2015). How are climate change concerns addressed by spatial plans? An evaluation framework, and an application to Indian cities. *Land Use Policy*, *42*, 210–226. doi:10.1016/j.landusepol.2014.07.016

Chapter 14 Requisite Sustainable Development Contemplating Buildings: Economic and Environmental Sustainability

Parul Bhyan
Malaviya National Institute of Technology, Jaipur, India

Bhavna Shrivastava
Malaviya National Institute of Technology, Jaipur, India

Nand Kumar
Malaviya National Institute of Technology, Jaipur, India

ABSTRACT

Sustainable development is a requisite for future generation, as increasing urbanization, destruction of natural resources by anthropic activities, degrading ecosystems for the sake of present economic development at cost of environmental exploitation are increasing by each passing day on earth. The goal of this chapter is to provide meaningful insights for policy-makers and decision-makers towards sustainable development in the construction industry. This study is first-of-its-kind study focusing on the Sustainable Developments Goals and sustainability dimensions and their criteria and indicators in one compilation through literature study. The study concludes that there is a need to integrate the construction industry into the Sustainable Development Goals and their targets to test the built environment sustainability and there is a need to develop the most adequate frameworks for commencing the topic. Two possible frameworks suggested for future research recommendation needed for the Indian context to enhance sustainability within construction industry are based on LCSA and MCDM analysis.

DOI: 10.4018/978-1-7998-8331-9.ch014

INTRODUCTION

"Sustainability" and "Sustainable Development" are substantial notions that vary from time to time and region to region, but nature is always the same – a development with very less destruction to the society, economic, environmental ecosystem, and other far-away factors which may occur caused by so-called "development". To manage this destruction by anthropic activities to ecosystem many initiatives have been taken so far by National, International, and building levels. The concept of sustainability became a part of environmental sustainability when the Brundtland Report in 1987, which it described sustainability as "Leaving sufficient resources for future generations to have a quality of life similar to ours". Rio earth summit, sustainable development goals, Agenda 21, etc. These development efforts from the international level down to local municipalities will ideally make every action more effective. At the national level, India has taken so many initiatives as Smart city initiated on 25 June 2015 and the objective was to provide the best infrastructure and quality of life to their cities to promote sustainable and inclusive cities. Considering the third level of sustainable development as building or construction level, as already known that building construction plays a very vital role in the achievement of sustainability as building construction includes various stages design transportation, construction maintenance so there is a correlation between building and sustainability. For improvement of building quality ISO, 9001 was the first step towards sustainable quality of construction and this was provided by the International Bureau of construction. Before notifying the above examples, firstly, this should be noted that:

1. Why it's a need to talk about sustainability and sustainable development/buildings?
2. Why are international and national governments worried?
3. What sustains?

Environmental issues and challenges caused by the construction industry can be decreased via changes in the process. Even the environmental impacts like energy consumption and Co_2 are important concerns for society, Occupants are majorly more concerned about increased electricity bills and increased life cycle costs resulting from energy consumption. The literature study elaborates that governments for sustainability may examine the challenge between the environment and economy, whereas the social dimension (justice and fairness) dimension is generally not examined. Conflicting issues between the triple bottom line concepts, the economic dimension has a direct impact on the environmental dimension, with the social element centralizing this impact of sustainability three-pillar dimensions. The 17 goals and 169 targets are formatted around 5 p's- (people, planet, prosperity, peace, and partnerships) (Moyer and Bohl 2019). The SDGs have emerged in Rio 20+ United Nations Summit in 2012. Literature shows SDGs, out of 169 targets 74 targets of the agenda, 44% of targets found to be dependent on construction and real estate activities out of which 29 targets 17% of targets are directly dependent and 45 targets 27% of targets indirectly dependent. The rise in understanding of the impacts of conventional construction practices and their harmful impacts supported the idea of, Sustainable Development. Human beings are still struggling for an ideal environment in form of shelter and settlement. Shelter being the priory basic need after food and water, humans along with necessity has always shown in developing various tendencies to make settlement in form of a group. Shelter as a kiosk, hut shares protection to human beings, as towns play a vital role in providing essential needs in the form of housing along with roads and lanes, water supply and drainage system, and other essential services. The sustainable housing development is an engendering concept in developing countries like India with lots of ongoing policies, projects, and future up-gradation. But need to be evaluated in terms of the triple bottom line concept to visualize where the reality of sustainability endeavoring in the construction industry. So redefining the relationship between various parameters of buildings and triple bottom line, this chapter will justify which criteria of triple bottom line serve the major role in building construction industry in India and what is causing lackness giving a clear perspective. So it is a need of a human to understand the intergenerational and intragenerational aspects of sustainable development in the construction industry in developed and developing countries. Assuming sustainable development in the construction sector, it is important to assess and identify the progress and lackness. In sustainability there are many sustainability assessment tools and technologies available in literature majorly concerning the environment and social-economic parameter is less considered. Sustainability assessment tools on a life cycle basis are termed life cycle sustainability assessment tools(Guinée 2016). Initially, the thought of integration of three dimensions of sustainability concerning life cycle basis emerged in 1987. After approx. a decade there was a consideration of social and environmental impacts in life cycle analysis in 1996(O"Brien, Doig, and Clift 1996).

SUSTAINABLE DEVELOPMENT

Evolution of Sustainable development over years initiated in 1968 and yet on a verge of up-gradation with increasing problems over the years. Initially, Garret Hardin an ecologist and philosopher wrote an essay discussing individual interests of human activities overusing natural resources is hazardous for community and future generation and spotlighting towards the idea of sustainable development(Hardin 2001). After 4 years, the tragedy of commons was published, 1972 meadows and authors created a computer simulation software in which they simulated the possibility of reduction of natural resources by anthropic activities and after several years the study turns right which again forced a vision of consideration of sustainable development. Till 1972, the complete focus was towards environment sustainability as a thought of degrading resources as 1st UN Conference on the Environment was conducted in United Nations the impact of humans over the environment and its relation to economic development and considered the importance to preserve "Human Environment". In 1980 human development index was introduced concerning the fact the planet earth has its limitation and growth doesn't mean only economic growth across the globe the concept of measuring economic and social aspects. In 1987, "our common future" the Brundtland report initiated the first widely accepted definition for sustainable development "development that meets the need of the present without compromising the ability the ability for future generation to meet their own needs" by the World Commission of Economic Development. In 1992 on 14th June in Rio De Janerio, Brazil the earth summit was conducted called "AGENDA 21" a complete action plan for sustainable development across the world adopted by over 178 countries. In 1994, John Elkington, a founder of a consultancy firm generated the term "Triple bottom line" considering a firm should consider all three parameters of sustainability equally for sustainable development. "The Future We Want" declaration on 20-22th June 2012 after the first declaration in Rio-de Janeiro, the United Nations Conference on sustainable development (UNCSD) included two major themes: green economy and an institutional framework in 41 pages report (UN2012).

Table 1. Sustainable development evolution over years

Sustainable Development Evolution	Definitions	Sources
Tragedy Of Commons,1968	Garret Hardin A ecologist and philosopher wrote an essay discussing if individual interests over natural resources keep growing will lead to destroying the communal interests and responsibility towards natural resources.	(Hardin 2001)
Limits To Growth,1972	(Behrens 2018) created a computer simulation tool to simulate a planet with limited resources and results seem right after many years in concern of environment degradation.	(Behrens 2018)
1st UN Conference On The Environment,1972	First UN conference to discuss the impact of humanity on the environment and its relation to economic development and considered the importance to preserve "human environment".	http://www.un-documents.net/unchedec.htm
TheHuman Development Index,1980	Understanding the fact the planet earth has its limitation and growth doesn't mean only economic growth across the globe the concept of measuring economic and social aspect was initiated called "Human Development Index"	http://hdr.undp.org/en/content/human-development-index-hdi
Brutland Report,1987	The brutland report also called "our common future" gave the first widely accepteddefinition for sustainable development.	(Keeble 1988)
The Earth Summit,1992	On 14th June 1992, In Rio De Janeiro, Brazil, over 178 countries adopted Agenda 21, a comprehensive action plan for sustainable development across the world.	(Kolbasov 1992)
The Triple Bottom Line,1994	In 1994, John Elkington, a founder of a consultancy firm generated the term "triple bottom line" considering a firm should consider all three parameters of sustainability equally for sustainable development.	https://www.ibrc.indiana.edu/ibr/2011/spring/article2.html
The Future We Want	On 20-22th June 2012, In Rio De Janeiro, Brazil, the conference after 20 years of agenda 21 was revived concerning two major themes green economy and an institutional framework 41 pages report.	(UN2012) https://sustainabledevelopment.un.org/content/documents/733FutureWeWant.pdf
United Nations Agenda 2030, 2015	Revised comprehensive action plan for sustainable development across the world.	

(Source: Compiled by Author, 2021)

Need for Sustainable Development

Sustainable development is progressive economic development without environmental exploitation over years. There are various issues to be considered to have safe human life, reduce exploitation of natural resources, and many more. The need for sustainable development worldwide includes many issues as climate change, poverty, human health, living conditions, education, energy, sanitation, inequality, and opportunities. Environment degradation and increasing economic development include various challenges regeneration of renewal resources, need attention towards overuse of natural resources, etc. Sustainable development includes various principles to be followed including integration and balance between economic and environment sustainability, preservation and prevention of natural resources, increase the quality of life and productivity for natural conservation of resources, and decision making. These all issues were considered in the Brutland Report and 17 goals and 169 indicators were unlisted to find the solutions for sustainable development in developed and developing countries. Sustainable development in 1994 considered broad umbrella including sustainability environment, economic and social equity. SDGs in the construction industry plays a vital role if directly related to a set of criteria and indicators for sustainable assessment and can provide a clear perspective about the accuracy of required sustainable development. There are three pillars or dimensions of sustainability that acts as pillars for sustainable development:

- Environment Sustainability
- Social Sustainability
- Economic Sustainability

Environment Sustainability

Figure 1. Various issues of environment sustinability
(Source: Compiled by Author, 2021)

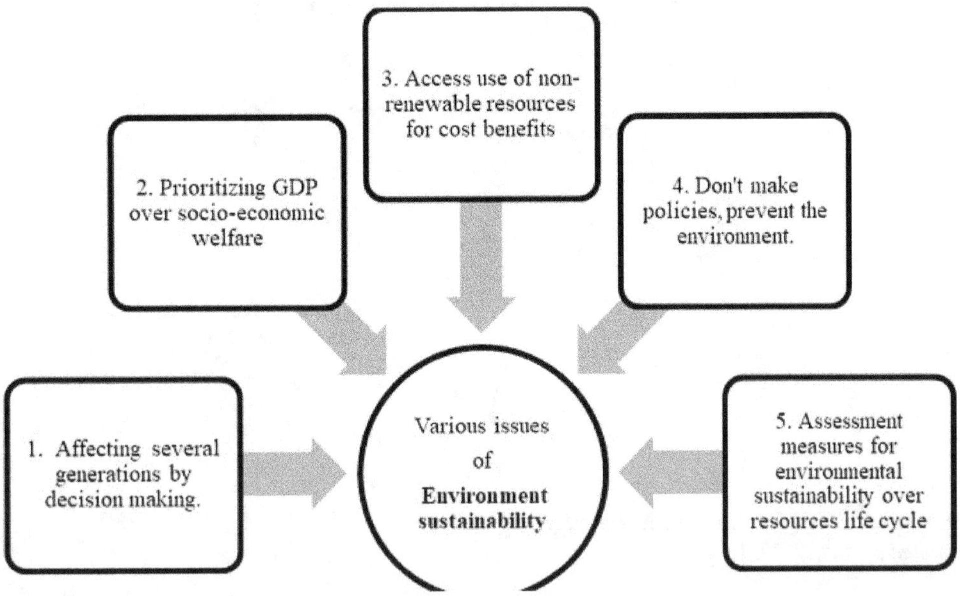

Environment sustainability can be termed as protecting natural resources for the coming generation. It can be also explained as the use of natural ecosystem air, water, and land in a sustainable manner for the upcoming 100 years. Environment sustainability includes various challenges and issues to be solved for enhancing the pace of measures taken. Various issues included in environmental sustainability are shown in figure 1. There are many assessment tools available to assess environmental sustainability like Life cycle assessment LCA and GBRS, etc.

In India, there are three major GBRS used to assess environment sustainability which include India Green Building Council (IGBC), India's first green building assessment system started in 2001. LEED 2011 For INDIA-New construction and GRIHA both started in 2006 .whereas, SVA-GRIHA and GRIHA EB (Existing Building) both started in 2015 and 2017 respectively. GRIHA being the most used green building rating system in the national capital region of India has very fewer projects for residences, single-family houses, and housing projects. Even in its 2019 revised version manual recently added life cycle cost criteria which to some extent includes economic sustainability. Table 2 shows the criterion, eligibility, and rating schemes involved in GBRS in the Indian context.

To improve sustainable development considering SDG13 Climate action, SDG6 Clean Water and Sanitation, SDG7 Affordable and Clean Energy, and SDG15 Life on Land, many environment sustainability indicators are considered to reduce the use of natural resources, less Co_2 emission during construction, etc. Major criteria include Land use, energy, and atmosphere considerations, material and waste management, water efficiency and innovation, and design process considerations shown in Table 3.

Figure 2. GRIHA Criterion
(Source: Compiled by Author, 2021)

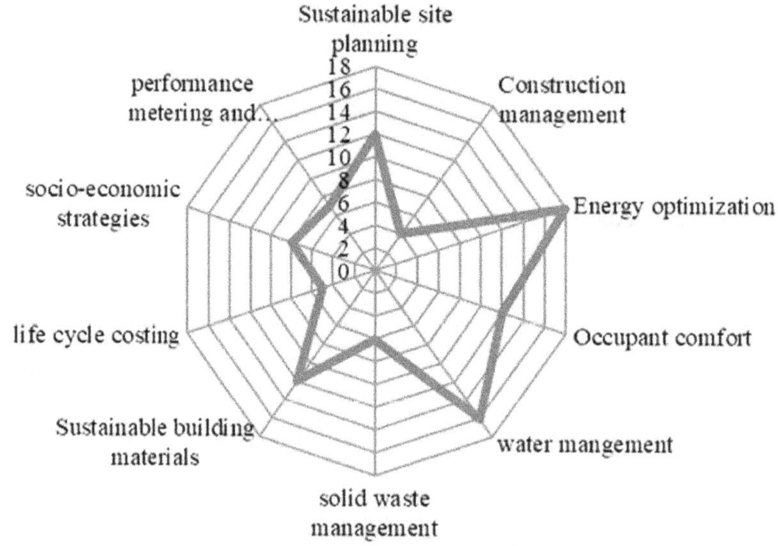

Figure 3. LEED-NC 2011 Criterion weightage
(Source: Compiled by Author, 2021)

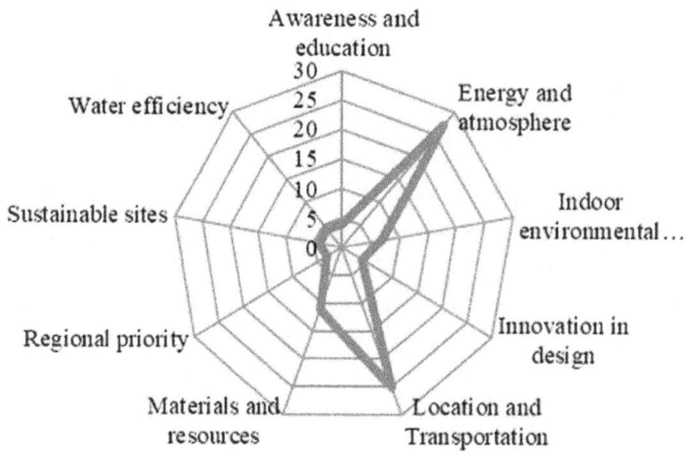

Figure 4. IGBC-2016 Criterion weightage
(Source: Compiled by Author, 2021)

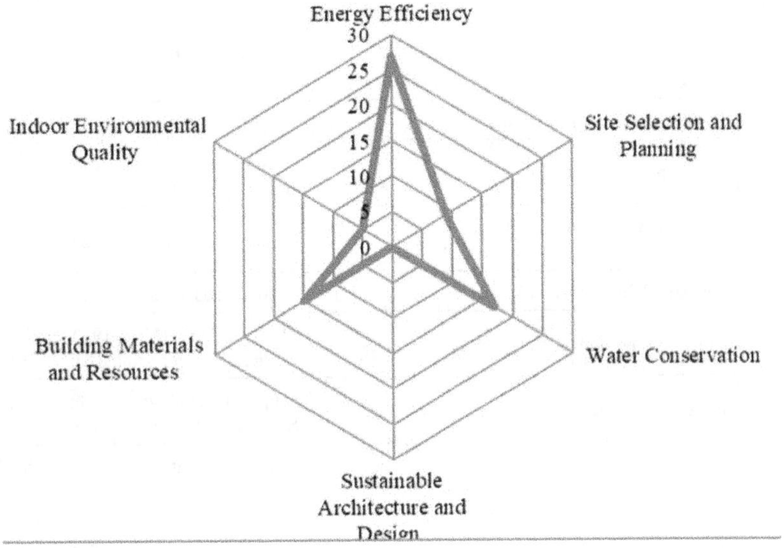

Table 2. Green building rating systems in the Indian context

	India Green Building Council (IGBC)	LEED 2011 For INDIA-NC	GRIHA	SVA-GRIHA	GRIHA EB (Existing Building)
Year	2001	2006	2006	2015	2017
Rating Schemes	Green Homes New Buildings Rating System Green Schools Green Existing Buildings(Non-Residential) Green Factory Buildings Green Townships Green Mass Rapid Transit System Green Landscape	offices, retail & service establishments, institutional buildings (libraries, schools, museums, places of worship, etc.,), hotels, and residential buildings of four or more habitable stories.	Education Health Care Multi-Unit Residential Office As-Built Office Design Office Interiors Retail Centers	residences, commercial offices, motels, dispensaries, schools	offices, retail spaces, institutional buildings, hotels, hospital buildings, healthcare facilities, residences, and multi-family high-rise buildings
Criteria's	Sustainable Architecture And Design Selection Of Species For Landscape Transportation Planning Water Conservation Energy Efficiency Building Material And Resources Indoor Environment Quality Health And Well-Being Health And Hygiene Operation And Maintenance Green Education Innovation And Development	Sustainable Sites Water Efficiency Energy Atmosphere Materials & Resources Indoor Environmental Quality Innovation in Design ™ Regional Priority	Sustainable Site Planning Construction Management Energy Optimization Occupant Comfort Water Management Solid Waste Management Sustainable Building Materials Life Cycle Costing Socio-Economic Strategies Performance Metering And Monitoring Innovation	Landscape Architecture & Energy Water & Waste Materials Lifestyle	Site Selection Maintenance And Housekeeping Energy Water Human Comfort Social Aspects
Eligibility	51% Owner's Ownership	51% Owner's Ownership	Built-Up Area More Than >2500sqm	Built-Up Area Less Than <2500sqm	Built-Up Area More Than >2500sqm

(Source: Compiled by Author, 2021)

Table 3. Environment sustainability criteria and indicators

Criteria	Indicators	Sources
Land Use	Site Selection	(Tupenaite et al., n.d.; Mehta and Porwal 2013; TERI and GRIHA 2019; Vyas and Jha 2016; Mohammed 2019)
	Developing Damaged Area	(Tupenaite et al., n.d.)
	Site Imperviousness Factor	(Lohmeng, Sudasna, and Tondee 2017; "LEED 2011 for India - NC" 2014)
	Preserve And Protect Landscape During Construction	(Doan et al. 2017; Lazar and Chithra 2020)
	Air And Pollution Control	(Rathi, Bansal, and Sharma 2018; Mohammed 2019)
	Construction Management Practices	(TERI and GRIHA 2019; Mehta and Porwal 2013; "LEED 2011 for India - NC" 2014)
	Infrastructure Efficiency	(Tupenaite et al., n.d.; "Integrating and Ranking Sustainability Criteria for Housing," n.d.)
Energy and atmosphere considerations	Energy efficiency	(Ji and Punekar 2018; Ingrao et al. 2018; TERI and GRIHA 2019; Mehta and Porwal 2013; Singh et al. 2016; Mohammed 2019; Lohmeng, Sudasna, and Tondee 2017; Alyami and Rezgui 2012)
	Renewable energy utilization	(Ji and Punekar 2018; Bhatt and Macwan 2012; Ceranic, Beardmore, and Cox 2017; Onat, Kucukvar, and Tatari 2014)
	Zero ozone depletion potential materials	(Ji and Punekar 2018; TERI and GRIHA 2019; Illankoon, Tam, and Le 2017)
	Greenhouse gas emission	(Marrero et al. 2020; Heravi, Rostami, and Kebria 2020; Kaspersen, Lohne, and Bohne 2016; Mehta and Porwal 2013)
Materials and waste management	Construction and demolition waste management planning	(Sebake 2008; Khoshnava et al. 2018; TERI and GRIHA 2019; Mohammed 2019)
	Building Life-cycle impact reduction	(Tupenaite et al., n.d.; M. A. Zanni, Soetanto, and Ruikar 2017; Mohammed 2019)
	Storage and collection of recyclables	
	Treat organic waste on-site	
	Use of regional/local materials	(Khoshnava et al. 2018; Tupenaite et al., n.d.)
Water efficiency	Implementation of alternative water resources	(Lohmeng, Sudasna, and Tondee 2017; "LEED 2011 for India - NC" 2014; TERI and GRIHA 2019; Mahdavinejad et al. 2014)
	Water Demand Reduction	
	Wastewater Treatment	
	Water conservation	
Innovation and design process considerations	Architectural functionality, flexibility and Adaptability Architectural heritage considerations	(Tupenaite et al., n.d.; Mohammed 2019; Sarath Chandra Kumar and Gupta 2014; Alyami and Rezgui 2012)
	Innovation in design	
	Environmental friendly design	("LEED 2011 for India - NC" 2014; TERI and GRIHA 2019)

(Source: Compiled by Author, 2021)

Standards/initiative /policies in India for environment sustainability are EIA Guidelines 2016, Model Building Byelaws 2016, National Energy Policy, National Building Code, Smart Cities, Swachh Bharat Mission, Solid waste Rules 2016, C&D waste Rules 2016, Environmental Clearances, National solar target, Energy efficiency, Passive design, and thermal efficiency, Materials and embodied energies, Green Growth and Building Sector.

Economic Sustainability

Economic sustainability in building construction means the cost of paperwork, stakeholder involvement in the design phase, material, labor, and machine cost for the construction phase, and miscellaneous capital involve in the building when stays useful over time with less maintenance cost and residual cost. Considering the use of capital for less quality work for a minimum period creates a high capital loss in maintenance cost for life cycle cost know as weak economic sustainability. Economic sustainability includes affordability and adaptability of product building having minimum maintenance and long service life. In India is no such rule of adoption of assessment of economic sustainability that plays a major role. Whereas, in India, affordable housing is based on categories LIG and EWS, and by unit area. A meticulous literature review summarized the most important economic sustainability criteria and indicators in Table 4 shown below.

Table 4. Economic sustainability criteria and indicators

Criteria	Indicators	Sources
Capital cost	Land Cost	*(Ahmad and Thaheem 2018b; Kovacic and Zoller 2015; Republic 2019)*
	Contract	
	Miscellaneous	
Life cycle cost	Construction Cost	*(TERI and GRIHA 2019; Yılmaz and Bakış 2015b; Tech 2019; Ahmad and Thaheem 2018b)*
	Operation Cost	
	Maintenance Cost	
	Replacement Cost	
	Residual Cost	
Affordability and economic performance	Project Stakeholder Opinion	*(Ahmad and Thaheem 2018b; Sala, Ciuffo, and Nijkamp 2015)*
	Burden Of Housing Costs	
	Price To Income Ratio	
Manageability aspects of building	Project Stakeholder Opinion	*(Ahmad and Thaheem 2018a; Republic 2019)*
	Complexity Of Building Design	
Adaptability and flexibility	Professional Opinion Regarding Ease Of Major Changes	*(Lazar and Chithra 2020; Zhang et al. 2019; Ahmad and Thaheem 2018a)*
	Professional Opinion Regarding Ease Of Minor Changes	
	No. of Rental Zones	

(Source: Compiled by Author, 2021)

Social Sustainability

Social sustainability is an important dimension of sustainability share security, satisfaction, Health, Comfort, Sense of Place and, safety, etc. In the construction industry, social sustainability shares important criteria's including accessibility, health, and well-being, community consideration shown in Table 5. Social sustainability assessment includes the S-LCA tool majorly some studies identified developing frameworks for respective countries (O"Brien, Doig, and Clift 1996; Lehmann et al. 2013). Literature summarized the most important social criteria and indicators for social sustainability assessment shown in Table 5.

Table 5. Social sustainability criteria and indicators

Criteria	Indicators	Sources
Accessibilities	Access to public transportation	*(Tupenaite et al., n.d.; Bhyan and Shrivastava 2020; Mohammed 2019; Lazar and Chithra 2020)*
	Access to employment opportunities	
	Access to educational institutions	
	Car parking capacity	
	Access to leisure facilities	
	Access to open green public space	
	Access to health care services	
Neighborhood/ community considerations	Safety (crime rate)	*(TERI and GRIHA 2019; Vyas and Jha 2016; Mehta and Porwal 2013)*
	Neighborhood reputation	
	Population density	
	Community cohesion	
	privacy	
Water	Water quality	*("LEED 2011 for India - NC" 2014; Smith 2015)*
Health and well being	Achieving Indoor Comfort Requirements (visual/Thermal/Acoustic)	*(Mehta and Porwal 2013; TERI and GRIHA 2019; Singh et al. 2016; "LEED 2011 for India - NC" 2014)*
	Labour safety and sanitation	
	Maintaining good indoor air quality	
	Dedicated facilities for service staff	
	Increase in environmental awareness	
	Environment tobacco smoke control	
	Enhanced indoor air quality strategies	
	Regional priority: specific credit	
	quality views	
	Winter thermal performance	
	Summer thermal performance	

(Source: Compiled by Author, 2021)

SDGS IN THE CONSTRUCTION INDUSTRY

Within sustainable development, "green buildings" are those which are constructed focusing on environment sustainability, while "sustainable buildings" are those green buildings that share equal importance to all three dimensions of sustainability. Sustainable development is the need to consider the "Three pillars" of sustainability together, i.e. Society, Economy, and the Environment. Environmental protection, social equity, and economic development in the construction industry collectively compose three pillars of the triple bottom line for sustainable housing development. In terms of sustainability three pillars triple bottom line concept, there are two ways to confine that future generation can fulfill their needs. Sustainable developments goals are defined and generated to support the construction industry in terms of reducing environment depletion by using harmful materials in every phase of the life cycle, using cost capital in a sustainable and estimated manner that can reduce the maintenance cost and residual cost can be reimbursed to invest in future by future generation as well as the building should socially acceptable and adaptable. Sustainable development works on these three legs stool, environment and economic sustainability needed to be balanced whereas the social dimension mediates this stool. Balancing the big umbrella of sustainability is still an engendering topic of research as there are a conflict balancing environment and economic sustainability in construction as well other departments affected by anthropic activities.

NCovid-19 Impact On SDGs Globally and Nationally

To achieve sustainable construction goals it is important to study in-depth sustainability concepts within the construction industry. The baseline remains the same for national, international platforms as - people, their habits, and the economy, all are interrelated. The main aim of sustainable development is to ensure that the people can have access to the local amenities by walking and using public transport. Already proven in the ncovid-19 situation, those environments flourish. Himalaya ranges are visible from one of the Punjab districts and are found revving of the ecosystem. So can it be concluded that "deprive in the economic growth in India supported the environment sustainability so far in 2020." Major sources of pollution and CO_2 emission in India –power stations, factories and industries, and transport are majorly under lockdown due to ncovid-19. A major decrease in electricity demand has reduced coal consumption by 6.5 percent in March, as more coal was mined and need is reduced during the economic lockdown, this reduced demand is reported in India after three decades. Similarly, as electricity consumption, a major decrease in Oil consumption by 18 percent in March, is reported. Major closure of transportation via, air road railways has contributed a huge part in carbon emission reduction and may lead to a legacy for its people to use more and more public transport for the cumulative blue-sky vision. The blue sky concept is said to be the clear sky with less pollution in the atmosphere, clean air to breathe with less harmful gases, this concept was devasted from urban areas long 30-40 years back with an increase in urbanization and rapid consumption of natural resources. Indian citizens, usually themselves used to visit rural areas for some closeness to nature which during lockdown was felt in their urban areas. and self-realization is noticed "what has been done with nature by humans?". This pandemic has a huge impact on the economic growth of the country as well a sudden climate change is reported. Economic lockdown –climate change both has a huge impact on biodiversity too, as in day-to-day life it was forgotten to the existence of other species. Now coming to the current scenario. 7 sustainable development goals affect by ncovid-19:

1. SDG2 (Food insecurities)
2. SDG3, which is good health affected at the global level.
3. 1.25 billion Students are affected so SDG goal 4 (Quality education) has a serious challenge.
4. International labor organization (ILO) estimates some 25 million people could lose their jobs.
5. Likewise SDG 6 (Sanitation target),
6. SDG8 (Weak economic growth and absence of decent work)
7. SDG13 Climate action

All above should not down the potential of energy levels. There is a need to search for "Recovery better", and build a healthier, safer, and more prosperous world. Global recovery from it'll impact the longer term of Sustainable development. IISD'S threefold mission is to advance e a stable climate, sustainable resource development, and fair economics. Three measures suggested by IISD:

1. **Resilience is essential** in terms of the human body system anticipate, cope and adopt surrounding situations.
2. **Changes in the environment should be sustained** as a good time for low Carbon investment projects to sustain higher unemployment, CO2 emission should be less.
3. **Inequality is magnified** as poor hit by a virus at the global level in developed countries. Workers in the service economy get less money and maximum at risk. So emergence needs to focus on export and fair trade for enhancing the economy.

NCovid-19 Impact on SDGs in Buildings: A Case Study

Quality of life is the possibility of human satisfaction with the living standard of their lives. Satisfaction happens with less difference between needs and probability of existence. The quality domain includes both the built environment and its neighborhood (Bhyan and Shrivastava 2019). To discuss the NCovid-19 Impact on SDGs in buildings peculiarly in the Indian scenario, case studies have been discussed including housing settlement in Gurgaon being millennium city. The built environment of India includes various region-specific, culture-specific needs of residents. In the Indian scenario, culture shares the importance of living near the ground, and increasing housing projects and trends of living share different views in quality of life in Indian neighborhoods. Taking Into account Gurgaon being India's millennium city and even addressed as the cyber city, covers 48.56% of the area as a residential area. Focussing on one sector of Gurgaon and showing the number of flats in one sector in figure 5. In the ncovid-19 pandemic lockdown, there was a difficult situation for everyone as no shops were opening and there was a complete lockdown in India. In Gurgaon sectors, single-family houses near the market area were more benefited than residents in gated societies with means of the quality of life services i.e. Daily Supplies, Public Spaces, Culture, Health, Elderly, child-care and Entertainment have shown in figure 5. Gated communities need time, not Indian culture and as in ncovid-19, everyone was locked in their houses caused anxiety and a cage for people in the gated community whereas people who live in the market area are still conned with the Indian culture of social interaction. if the living context is not adaptable for residents there is no sense in making buildings and controlling their harmful environmental impacts. So, quality of life depends on three main divisions Living, Pertaining, and Embellishing(Bhyan and Shrivastava 2019). Living includes physical physiological and spiritual dimensions, pertaining include physical, social, and communal dimensions, and embellishing includes practical, leisure, and growth dimensions. This situation can be seen other way around single-family houses in market areas were more prone to the ncovid-19 situation rather than society residents will less social inclusion and health environment. Sustainable development sustains if occupants are satisfied with their quality of life. So it is very important to understand the quality of life needed in the construction sector to select the indicators for assessment of sustainability and promote sustainable development.

Figure 5. Showing different residential settlements in Gurgaon sec 92 and sec 23
(Source: Google images, 2021)

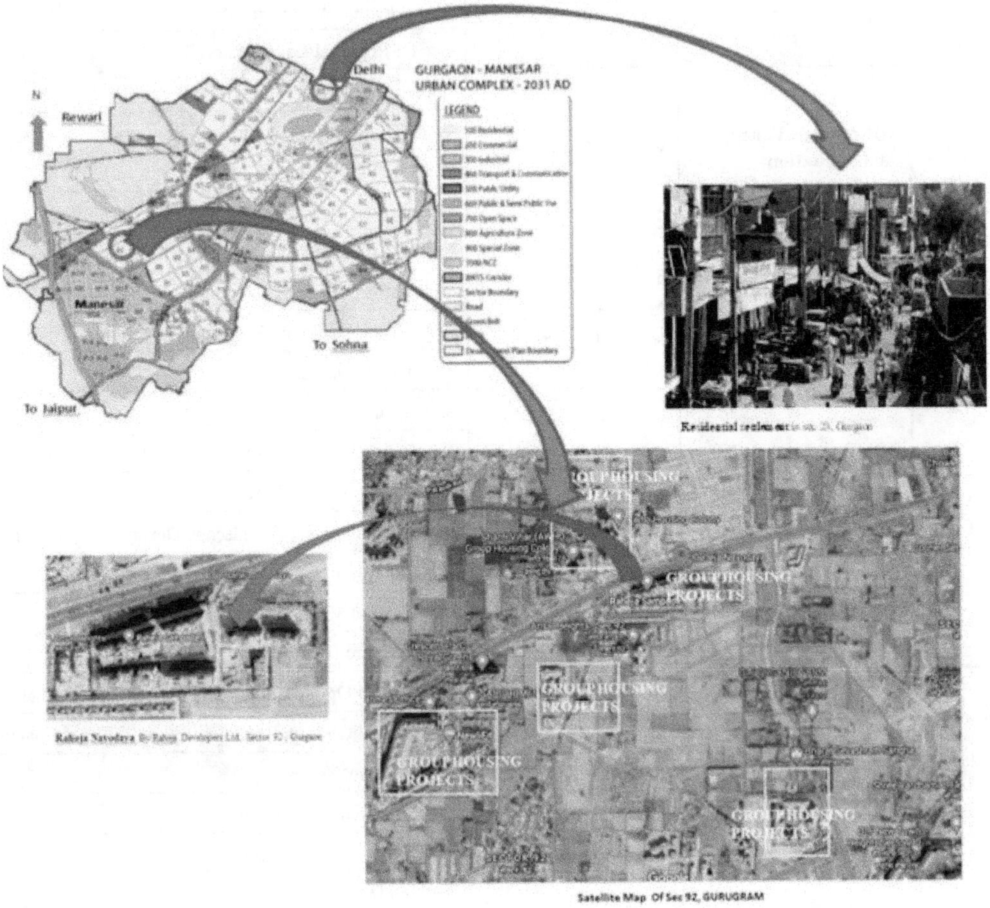

Urbanization Related to Housing Quality

Because of urbanization and the lack of economic opportunities in rural areas, many people move to the cities. They move to the cities that are already dealing with issues of overcrowding, infrastructure, and high cost of living. This forces them to seek shelter in the slums of the urban fringe. United Nation Habitat in 2006 found that 90% of slum residents are in developing countries with struggling economies. In addition, cities were not mean to handle millions of people streaming in when designed.

Table 6. Criteria and indicators of sustainability contemplating SDGs

Sustainability Parameters	Sustainable Development Goals	Criteria	Indicators
Environment	SDG13 Climate Action	Materials And Waste Management	Construction And Demolition Waste Management Planning
			Building Life-Cycle Impact Reduction
			Storage And Collection Of Recyclables
			Utilization Of BIS Recommended Waste Materials In The Building Structure
			Reduction In Embodied Energy Of Building Structure
			Treat Organic Waste On-Site
			Use Of Regional/Local Materials
	SDG6 Clean Water And Sanitation	Water Efficiency	Implementation Of Alternative Water Resources
			Water Demand Reduction
			Wastewater Treatment
			Water Conservation
	SDG7 Affordable And Clean Energy	Energy And Atmosphere Considerations	Energy Efficiency
			Renewable Energy Utilization
			Zero Ozone Depletion Potential Materials
			Greenhouse Gas Emission
	SDG15 Life On Land	Land Use	Site Selection
			Developing Damaged Area
			Site Imperviousness Factor
			Preserve And Protect Landscape During Construction
			Air And Pollution Control
			Construction Management Practices
			Infrastructure Efficiency
Economic	SDG8 Decent Work And Economic Growth	Adaptability And Flexibility	Professional Opinion Regarding Ease Of Major Changes
			Professional Opinion Regarding Ease Of Minor Changes
			No. Of Rental Zones
		Affordability And Economic Performance	Project Stakeholder Opinion
			Burden Of Housing Costs
			Price To Income Ratio
	SDG12 Responsible Growth And Production	Manageability Aspects Of Building	Project Stakeholder Opinion
			Complexity Of Building Design
		Life Cycle Cost	Construction Cost
			Operation Cost
			Maintenance Cost
			Replacement Cost
			Residual Cost

continues on following page

Table 6. Continued

Sustainability Parameters	Sustainable Development Goals	Criteria	Indicators
Social	SDG4 Quality Education	Neighborhood/ Community Considerations	Safety (Crime Rate)
			Neighborhood Reputation
			Population Density
			Community Cohesion
			Privacy
		Accessibility	Access To Public Transportation
			Access To Employment Opportunities
			Access To Educational Institutions
			Car Parking Capacity
			Access To Leisure Facilities
			Access To Open Green Public Space
	SDG3 Good Health And Well Being	Health And Well Being	Achieving Indoor Comfort Requirements (Visual/Thermal/ Acoustic)
			Labour Safety And Sanitation
			Maintaining Good Indoor Air Quality
			Dedicated Facilities For Service Staff
			Increase In Environmental Awareness
			Environment Tobacco Smoke Control
			Enhanced Indoor Air Quality Strategies
			Regional Priority: Specific Credit
			Quality Views
			Winter Thermal Performance
			Summer Thermal Performance

(Source: Compiled by Author, 2021)

Above mentioned SDGs and criteria are results of the literature examined. In India, this is the first kind of study summarizing SDGs regarding buildings and has thoroughly summarized the criteria and sub-indicators, if achieved can make us move towards sustainable development.

CONCLUSION

This study concludes the importance of SDGs in buildings sustainability assessment, initially, the study discusses the theoretical approach of sustainable development then discusses how sustainable development contributes to the construction industry, ncovid-19 impact on sustainable development was also discussed thoroughly, quality of life services and impact of ncovid-19 on two different residential settlements was also discussed. In meticulous literature finding showing results of sustainable development contemplating buildings table 6 summarizes criteria's and indicators with aspects of sustainable development goals majoring, SDG 6, SDG 15, SDG13, SDG 7 In environment dimension of sustainability, SDG 8 and

SDG 12 in the economic dimension of sustainability and SDG 3 and SDG4 in the social dimension of sustainability. The future recommendation suggested the need for frameworks using vastly used in other countries and increasing research interest in India's LCSA framework for inclusion of SDGs with LCSA in the construction industry of India as well as using MCDM analysis for weightage of key performance indicators among the mentioned indicators in Table 6. Figure 6. Shows the graphical representations of the conclusion of this study showing three pillars of sustainability contemplating SDGs.

Figure 6. Sustainable development goals contemplating buildings
(Source: Compiled by Author, 2021)

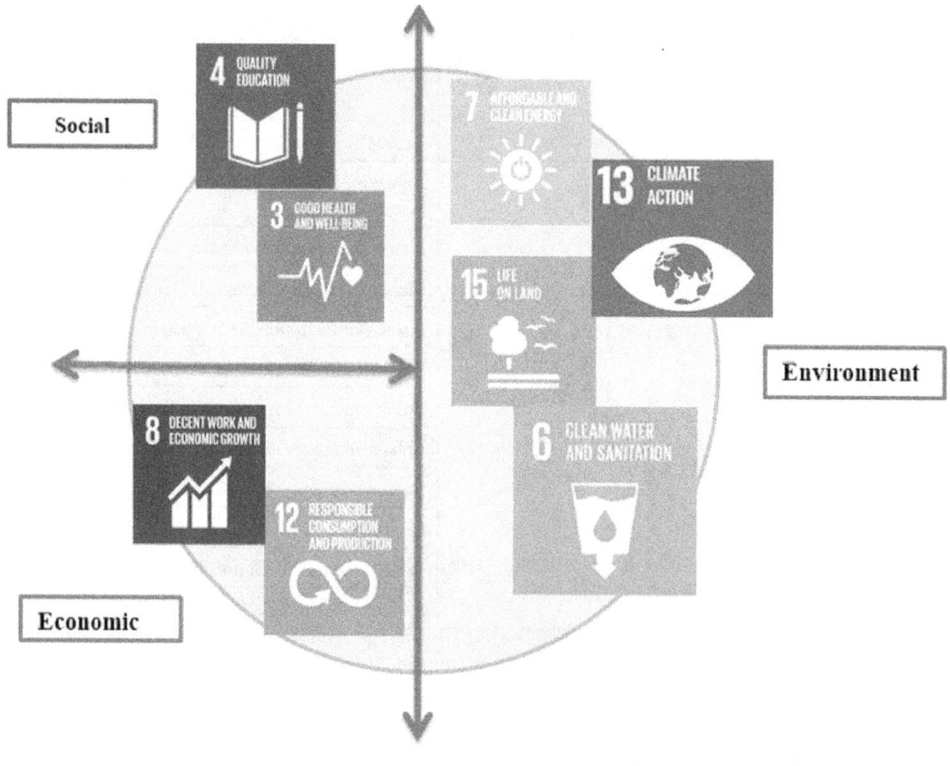

FUTURE RESEARCH DIRECTIONS

To achieve the agenda 2030, there is an excessive need for sustainability assessment of the construction industry directly linked with SDGs. So, there is a tremendous need for a framework that can assess the sustainable development goals in buildings for one score assessment of sustainable development in India. The focus of this chapter is on sustainability assessment, indicators, economic, environment, and social sustainability and concluded criteria's and indicators contemplating building for sustainable development. Showing share of GBRS in environment sustainability, policies /standards /initiatives concerning sustainable development as well as economic and social sustainability dimensions importance and assessment tools over last many years. In India from consideration of sustainable development to actual measuring sustainable development has still a major gap in research. There can be two future research directions in the context of commencing the topic.

1. Sustainable Development Goals evaluation in buildings using life cycle assessment systems
2. Sustainable Development Goals assessment in buildings by MCDM analysis.

Above mentioned two frameworks may help the Indian sustainability assessment system to achieve sustainable development goals peculiarly for the construction industry.

FUNDING

This research received no specific grant from any funding agency in the public, commercial, or not-for-profit sectors.

REFERENCES

Acampa, G., García, J. O., Grasso, M., & Díaz-López, C. (2019). Project sustainability: Criteria to be introduced in BIM. *Valori e Valutazioni, 2019*(23), 119–128.

Ahmad, T., & Thaheem, M. J. (2018). Economic Sustainability Assessment of Residential Buildings : A dedicated assessment framework and implications for BIM Economic sustainability assessment of residential buildings : A dedicated assessment framework and implications for BIM. *Sustainable Cities and Society, 38*(April), 476–491. doi:10.1016/j.scs.2018.01.035

Alejandrino, C., Mercante, I., & Bovea, M. D. (2021). Life cycle sustainability assessment: Lessons learned from case studies. *Environmental Impact Assessment Review, 87*(July), 106517. doi:10.1016/j.eiar.2020.106517

Alwan, Z., Greenwood, D., & Gledson, B. (2015). Rapid LEED evaluation performed with BIM based sustainability analysis on a virtual construction project. *Construction Innovation, 15*(2), 134–150. doi:10.1108/CI-01-2014-0002

Alyami, S. H., & Rezgui, Y. (2012). Sustainable building assessment tool development approach. *Sustainable Cities and Society, 5*(1), 52–62. doi:10.1016/j.scs.2012.05.004

Bahadure, P., & Bahadure, S. (2014). Sustainable Urban Development in India : Challenge & Approaches. *International Conference on Advances in Architecture & Civil Engineering (AARCV 2012) Theme : Architecture, Sub theme : Urban and Regional Planning Sustainable Urban Development in India.*

Behrens, D. (2018). Limits to growth. Encyclopedia of Ecology, 367–369. doi:10.1016/B978-0-444-63768-0.00630-2

Bhyan, A. P., & Shrivastava, B. (2019). *Imperative quality of life in built environment in Indian suburb as per opinion poll.* doi:10.35940/ijrte.C6408.098319

Bhyan, A. P., & Shrivastava, B. (2020). *Cognitive Occupancy Appraisal for Sustainable Housing.* doi:10.35940/ijrte.F9597.059120

Bragança, L., Mateus, R., & Koukkari, H. (2010). Building sustainability assessment. *Sustainability, 2*(7), 2010–2023. doi:10.3390u2072010

Cetiner, I., & Edis, E. (2014). *An environmental and economic sustainability assessment method for the retrofitting of residential buildings*. doi:10.1016/j.enbuild.2014.01.020

Costa, D., Quinteiro, P., & Dias, A. C. (2019). A systematic review of life cycle sustainability assessment: Current state, methodological challenges, and implementation issues. *The Science of the Total Environment, 686*, 774–787. doi:10.1016/j.scitotenv.2019.05.435 PMID:31195285

Fatourehchi, D., & Zarghami, E. (2020). *Social sustainability assessment framework for managing sustainable construction in residential buildings*. doi:10.1016/j.jobe.2020.101761

Gladwin, T. N., Kennelly, J. J., & Krause, T. (1995). Shifting Paradigms for Sustainable for Implications Development : And Theory. *Academy of Management Review, 20*(4), 874–907. doi:10.5465/amr.1995.9512280024

Goubran, S. (2019). *On the Role of Construction in Achieving the SDGs*. Academic Press.

Hardin, G. (2001). *The tragedy of the commons*. Academic Press.

Høgevold, N. M., Svensson, G., Rodriguez, R., & Eriksson, D. (2019). Relative importance and priority of TBL elements on the corporate performance. *Management of Environmental Quality, 30*(3), 609–623. doi:10.1108/MEQ-04-2018-0069

Hossaini, N., Reza, B., Akhtar, S., Sadiq, R., & Hewage, K. (2015). AHP based life cycle sustainability assessment (LCSA) framework: A case study of six storey wood frame and concrete frame buildings in Vancouver. *Journal of Environmental Planning and Management, 58*(7), 1217–1241. doi:10.1080/09640568.2014.920704

Ingrao, C., Messineo, A., Beltramo, R., Yigitcanlar, T., & Ioppolo, G. (2018). How can life cycle thinking support sustainability of buildings? Investigating life cycle assessment applications for energy efficiency and environmental performance. *Journal of Cleaner Production, 201*, 556–569. doi:10.1016/j.jclepro.2018.08.080

Karaca, F., Guney, M., Kumisbek, A., Kaskina, D., & Tokbolat, S. (2020). A new stakeholder opinion-based rapid sustainability assessment method (RSAM) for existing residential buildings. *Sustainable Cities and Society, 60*, 102155. doi:10.1016/j.scs.2020.102155

Karthikvel, S., & Samydoss, C. (2020). *Sustainable development*. Academic Press.

Kolbasov, O. S. (1992). UN Conference on Environment and Development. *Izvestiya - Akademiya Nauk, Seriya Geograficheskaya, 6*(June), 47–54. 10.4135/9781412971867.n128

Kovacic, I., & Zoller, V. (2015). *Building life cycle optimization tools for early design phases*. Academic Press.

Lazar, N., & Chithra, K. (2020). *Comprehensive bibliometric mapping of publication trends in the development of Building Sustainability Assessment Systems*. Academic Press.

Marjaba, G. E. (2020). *Sustainability framework for buildings via data analytics*. Academic Press.

Mehta, P. H. S. (2013). *Green Building Construction for Sustainable Future*. Academic Press.

Moslehi, S., & Reddy, T. A. (2019). *A new quantitative life cycle sustainability assessment framework: Application to integrated energy systems*. Academic Press.

Moyer, J. D. (2019). *Alternative pathways to human development: Assessing trade-offs and synergies in achieving the Sustainable Development Goals*. doi:10.1016/j.futures.2018.10.007

Nama, P. (2019). Review Paper on Recent Advancement in Green Building and its. *Sustainability*, *7*(3), 2365–2368. Advance online publication. doi:10.22214/ijraset.2019.3431

Olakitan Atanda, J. (2019). *Developing a social sustainability assessment framework*. doi:10.1016/j.scs.2018.09.023

Pancharathi, R.K., Sangoju, B., & Chaudhary, S. (2019). *Advances in Sustainable Construction Materials*. Academic Press.

Pandey. (2015). *Impact of Green Building Rating Systems on the Sustainability and Efficacy of Green Buildings Case Analysis of Green Building Index*. Academic Press.

Ragheb, A., El-Shimy, H., & Ragheb, G. (2016). Green Architecture: A Concept of Sustainability. *Procedia - Social and Behavioral Sciences, 216*(October), 778–787. doi:10.1016/j.sbspro.2015.12.075

Rathi, R., Bansal, S., & Sharma, S. (2018). An empirical study on sustainability green building practices in residential projects: With reference to Delhi- NCR region. *International Journal of Mechanical Engineering and Technology*, *9*(1), 276–288.

Rathi, R., Bansal, S., & Sharma, S. (2018). An empirical study on sustainability green building practices in residential projects: With reference to Delhi- NCR region. *International Journal of Mechanical Engineering and Technology*, *9*(1), 276–288.

Rossi, B., Marique, A., Glaumann, M., & Reiter, S. (2012). Life-cycle assessment of residential buildings in three different European locations, basic tool. *Building and Environment*, *51*, 395–401. doi:10.1016/j.buildenv.2011.11.017

Sala, S., Ciuffo, B., & Nijkamp, P. (2015). *A systemic framework for sustainability assessment*. doi:10.1016/j.ecolecon.2015.09.015

Santos, R., Aguiar Costa, A., Silvestre, J. D., & Pyl, L. (2020). Development of a BIM-based Environmental and Economic Life Cycle Assessment tool. *Journal of Cleaner Production*, *265*, 121705. Advance online publication. doi:10.1016/j.jclepro.2020.121705

Singh, S., Lal, R. B., Sridharan, U., & Upadhyay, V. P. (2016). Environmental Sustainability Guidelines for Green Buildings in India: A Review. *Indian Journal of Scientific Research and Technology*, *4*(1), 2321–9262. www.indjsrt.com

Smith, R. M. (2015). *"Green" building in India: A comparative and spatial analysis of the LEED-India and GRIHA rating systems*. doi:10.1080/10225706.2015.1020065

Thanu, H. P., & Jayachamarajendra, S. (n.d.). *Comparative study on Indian building assessment tools and its limitations*. Academic Press.

UN. (2012). *The future we want: Outcome document of the United Nations Conference on Sustainable Development.* https://sustainabledevelopment.un.org/content/documents/733FutureWeWant.pdf

United, T., Conference, N., & Environment, H. (1989). *Declaration of the United Nations Conference on the Human Environment, 1972.* 10.1177/0019556119890340

United Nations. (2015). *The Millennium Development Goals Report.* United Nations. DOI: doi:978-92-1-101320-7

Visser, W., & Brundtland, G. H. (2013). *Our Common Future ('The Brundtland Report').* World Commission on Environment and Development. doi:10.9774/GLEAF.978-1-907643-44-6_12

Vyas, G. S., & Jha, K. N. (2016). *Identification of green building attributes for the development of an assessment tool: A case study in India.* doi:10.1080/10286608.2016.1247832

Wulf, C., Werker, J., Zapp, P., Schreiber, A., Schlör, H., & Kuckshinrichs, W. (2018). *Sustainable Development Goals as a Guideline for Indicator Selection in Life Cycle Sustainability Assessment.* doi:10.1016/j.procir.2017.11.144

Zhang, X., Warner, M. E., & Homsy, G. C. (2017). Environment, Equity, and Economic Development Goals: Understanding Differences in Local Economic Development Strategies. *Economic Development Quarterly, 31*(3), 196–209. doi:10.1177/0891242417712003

Zhang, Y., Wang, H., Gao, W., Wang, F., Zhou, N., Kammen, D. M., & Ying, X. (2019). A survey of the status and challenges of green building development in various countries. *Sustainability (Switzerland), 11*(19), 1–29. doi:10.3390u11195385

KEY TERMS AND DEFINITIONS

E-LCA: Environment life cycle assessment.
GBRS: Green building rating system.
LCA: Life cycle assessment.
MCDM: Multi-criterion decision making.
S-LCA: Social-life cycle assessment.
SDGs: Sustainable Development Goals.

Compilation of References

A salty sore: The dying Sambhar lake. (2019, March 25). Retrieved from India water portal: https://www.indiawaterportal.org/news/salty-sore-dying-sambhar-lake

Abd-Elhamid, H. F., & Javadi, A. A. (2008). An Investigation into Control of Saltwater Intrusion Considering the Effects of Climate Change and Sea Level Rise. Presented at the *20th Salt Water Intrusion Meeting*, Naples, FL, USA.

Abileah, R., Vignudelli, S., & Scozzari, A. (2011). *A completely remote sensing approach to monitoring reservoirs water volume*. Paper published in Fifteenth International Water Technology Conference IWTC 15 2011, Alexandria, Egypt.

Acampa, G., García, J. O., Grasso, M., & Díaz-López, C. (2019). Project sustainability: Criteria to be introduced in BIM. *Valori e Valutazioni*, *2019*(23), 119–128.

Adam, J. C., Hamlet, A. F., & Lettenmaier, D. P. (2009). Implications of global climate change for snowmelt hydrology in the twenty-first century. *Hydrological Processes*, *23*(7), 962–972. doi:10.1002/hyp.7201

Agarwal, S., Saxena, A. K., Vyas, V., & Shrivastava, S. (2018). *Rethinking REDD+- A Centre for Science and Environment Assessment*. www.cseindia.org

Agovino, M., Casaccia, M., Ciommi, M., Ferrara, M., & Marchesano, K. (2019). Agriculture, climate change and sustainability: The case of EU-28. *Ecological Indicators*, *105*, 525–543. doi:10.1016/j.ecolind.2018.04.064

Agyeman, J., Evans, B., & Kates, R. (1998). Greenhouse gases special: Thinking locally in science, practice and policy. *Local Environment*, *3*(3), 245–246. doi:10.1080/13549839808725563

Ahmad, T., & Thaheem, M. J. (2018). Economic Sustainability Assessment of Residential Buildings : A dedicated assessment framework and implications for BIM Economic sustainability assessment of residential buildings : A dedicated assessment framework and implications for BIM. *Sustainable Cities and Society*, *38*(April), 476–491. doi:10.1016/j.scs.2018.01.035

Ahmed, M. M. S., Abel-Rahman, A. K., Ali, A. H. H., & Suzuki, M. (2015). Double Skin Façade: The State of Art on Building Energy Efficiency. *Journal of Clean Energy Technologies*, *4*(1), 84–89. doi:10.7763/JOCET.2016.V4.258

Ahmed, U., Mumtaz, R., Anwar, H., Shah, A. A., Irfan, R., & García-Nieto, J. (2019). Efficient Water Quality Prediction Using Supervised Machine Learning. *Water (Basel)*, *11*(11), 2210. doi:10.3390/w11112210

Aijaz, R. (2018). Measuring urbanization in India | ORF. *ORF Issue Brief, 218*. https://www.orfonline.org/research/measuring-urbanisation-india/

Akadiri, P. O., Chinyio, E. A., & Olomolaiye, P. O. (2012). Design of A Sustainable Building: A Conceptual Framework for Implementing Sustainability in the Building Sector. *Buildings*, *2*(2), 126–152. Advance online publication. doi:10.3390/buildings2020126

Akbari, H., Pomerantz, M., & Taha, H. (2001). Cool Surfaces and Shade Trees to Reduce Energy Use and Improve Air Quality in Urban Areas. *Solar Energy, 70*(3), 295–310. doi:10.1016/S0038-092X(00)00089-X

Akbari, H., & Rose, L. S. (2008). Urban Surfaces and Heat Island Mitigation Potentials. *Human-Living Environment Society English Journal, 11*(2), 85–101. doi:10.1618/jhes.11.85

Alawadi, F. (2010). Detection of surface algal blooms using the newly developed algorithm surface algal bloom Index (SABI). *Remote Sensing of the Ocean, Sea Ice, and Large Water Regions, 2010*, 782506. Advance online publication. doi:10.1117/12.862096

Albrechts, L. (2004). Strategic (spatial) planning reexamined. *Environment and Planning. B, Planning & Design, 31*(5), 743–758. doi:10.1068/b3065

Alejandrino, C., Mercante, I., & Bovea, M. D. (2021). Life cycle sustainability assessment: Lessons learned from case studies. *Environmental Impact Assessment Review, 87*(July), 106517. doi:10.1016/j.eiar.2020.106517

Al-Hamdan, M., Quattrochi, D., Bounoua, L., Lachir, A., & Zhang, P. (2016). Using Landsat, MODIS, and a Biophysical Model to Evaluate LST in Urban Centers. *Remote Sensing, 8*(11), 952. doi:10.3390/rs8110952

Aljoufie, M., Zuidgeest, M., Brussel, M., & van Maarseveen, M. (2013). Spatial-temporal analysis of urban growth and transportation in Jeddah City, Saudi Arabia. *Cities (London, England), 31*, 57–68. doi:10.1016/j.cities.2012.04.008

Allwood, J. M., Ashby, M. F., Gutowski, T. G., & Worrell, E. (2011). Material efficiency: A white paper. *Resources, Conservation and Recycling, 55*(3), 362–381. doi:10.1016/j.resconrec.2010.11.002

Altomonte, S. (2009). Climate Change and Architecture: Mitigation and Adaptation Strategies for a Sustainable Development. *Journal of Sustainable Development, 1*(1), 97. doi:10.5539/jsd.v1n1p97

Alwan, Z., Greenwood, D., & Gledson, B. (2015). Rapid LEED evaluation performed with BIM based sustainability analysis on a virtual construction project. *Construction Innovation, 15*(2), 134–150. doi:10.1108/CI-01-2014-0002

Alyami, S. H., & Rezgui, Y. (2012). Sustainable building assessment tool development approach. *Sustainable Cities and Society, 5*(1), 52–62. doi:10.1016/j.scs.2012.05.004

Amani, M., Ghorbanian, A., Ahmadi, S. A., Kakooei, M., Moghimi, A., Mirmazloumi, S. M., Moghaddam, S. H. A., Mahdavi, S., Ghahremanloo, M., Parsian, S., Wu, Q., & Brisco, B. (2020). Google Earth Engine Cloud Computing Platform for Remote Sensing Big Data Applications: A Comprehensive Review. *IEEE Journal of Selected Topics in Applied Earth Observations and Remote Sensing, 13*, 5326–5350. doi:10.1109/JSTARS.2020.3021052

Aminou, D. M. A. (2002). *MSG's SEVIRI Instrument, MSG Project, ESA Directorate of Earth Observation*. ESTEC.

Aram, F., Higueras García, E., Solgi, E., & Mansournia, S. (2019). Urban green space cooling effect in cities. *Heliyon, 5*(4), e01339. doi:10.1016/j.heliyon.2019.e01339 PMID:31008380

Aram, F., Solgi, E., Garcia, E. H., & Mosavi, A. (2020). Urban heat resilience at the time of global warming : Evaluating the impact of the urban parks on outdoor thermal comfort. *Environmental Sciences Europe, 32*(1), 117. doi:10.118612302-020-00393-8

Arena, C., Genco, M., & Mazzola, M. R. (2020). Environmental Benefits and Economical Sustainability of Urban Wastewater Reuse for Irrigation—A Cost-Benefit Analysis of an Existing Reuse Project in Puglia, Italy. *Water (Basel), 12*(10), 2926. Advance online publication. doi:10.3390/w12102926

Arnfield, A. (2003). Two decades of urban climate research : A review of turbulence, exchanges of energy and water, and the urban heat island. *International Journal of Climatology, 23*(1), 1–26. doi:10.1002/joc.859

Asif, M., Muneer, T., & Kelley, R. (2007). Life cycle assessment: A case study of a dwelling home in Scotland. *Building and Environment, 42*(3), 1391–1394. doi:10.1016/j.buildenv.2005.11.023

Atkins, N. (n.d.). Survey of meteorology - class notes - outline - Lyndon state college. https://apollo.nvu.vsc.edu/classes/met130/notes/

Ayoub, G. M., & Malaeb, L. (2015). Impact of intermittent water supply on water quality in Lebanon. *International Journal of Environment and Pollution, 26*(4), 379. doi:10.1504/IJEP.2006.009328

Azevedo, J. A., Chapman, L., & Muller, C. L. (2016). Quantifying the Daytime and Night-Time Urban Heat Island in Birmingham, UK: A Comparison of Satellite Derived Land Surface Temperature and High Resolution Air Temperature Observations. *Remote Sensing, 8*(2), 153. doi:10.3390/rs8020153

Babin, M., Greb, S. R., Dekker, A. D., Arrigo, K. R., Forget, M. H., & the International Ocean-Colour Coordinating Group. (2015). *Earth Observations in Support of Global Water Quality Monitoring.* International Ocean Colour Coordinating Group. https://repository.oceanbestpractices.org/handle/11329/535

Bagchi, D., & Bussa, R. (2011). Application of Remote Sensing in Water Quality and Water Resources Management – An Overview. *Bhujal News Quaterly.* https://hindi.indiawaterportal.org/content/application-remote-sensing-water-quality-and-water-resources-management-overview/content-type-page/53244

Bahadure, P., & Bahadure, S. (2014). Sustainable Urban Development in India : Challenge & Approaches. *International Conference on Advances in Architecture & Civil Engineering (AARCV 2012) Theme : Architecture, Sub theme : Urban and Regional Planning Sustainable Urban Development in India.*

Balasubramanian, S. V., Pahlevan, N., Smith, B., Binding, C., Schalles, J., Loisel, H., Gurlin, D., Greb, S., Alikas, K., Randla, M., Bunkei, M., Moses, W., Nguyễn, H., Lehmann, M. K., O'Donnell, D., Ondrusek, M., Han, T. H., Fichot, C. G., Moore, T., & Boss, E. (2020). Robust algorithm for estimating total suspended solids (TSS) in inland and nearshore coastal waters. *Remote Sensing of Environment, 246*, 111768. doi:10.1016/j.rse.2020.111768

Baral, H., Keenan, R. J., Sharma, S. K., Stork, N. E., & Kasel, S. (2014). Spatial assessment and mapping of biodiversity and conservation priorities in a heavily modified and fragmented production landscape in north-central Victoria, Australia. *Ecological Indicators, 36*, 552–562. doi:10.1016/j.ecolind.2013.09.022

Barrett, S., & Fudge, C. (1981). *Policy and Action. Essays on the Implementation of Public Policy.* Metheun.

Bartram, J., & Ballance, R. (1998). *Water Quality Monitoring: A Practical Guide to the Design and Implementation of Freshwater Quality Studies and Monitoring Programmes* (1st ed.). CRC Press. https://www.who.int/water_sanitation_health/resourcesquality/wqmchap2.pdf

Batty, M. (1976). Models, Methods and Rationality in Urban and Regional Planning: Developments since 1960. *Area (23rd International Geographical Congress), 8*(2), 93-97.

Behrens, D. (2018). Limits to growth. Encyclopedia of Ecology, 367–369. doi:10.1016/B978-0-444-63768-0.00630-2

Bergdolt, J., Sharvelle, S., & Roesner, L. (2011). *Guidance Manual for Separation of Graywater from Blackwater for Graywater Reuse.* IWA Publishing., doi:10.2166/9781780400082

Berke, P. R. (1995). Natural-hazard reduction and sustainable development: A global assessment. *Journal of Planning Literature, 9*(4), 369–382. doi:10.1177/088541229500900404

Betsill, M., & Bulkeley, H. (2007). Looking back and thinking ahead: A decade of cities and climate change research. *Local Environment, 12*(5), 447–456. doi:10.1080/13549830701659683

Bhagwat, T., Klein, I., Huth, J., & Leinenkugel, P. (2019). Volumetric Analysis of Reservoirs in Drought-Prone Areas Using Remote Sensing Products. *Remote Sensing*, *11*(17), 1974. doi:10.3390/rs11171974

Bhanwar Vishvendra, R. S., & Sen, A. (2018). Geo-Spatial Mapping of Land Use and Land Cover Changes in the Core and Periphery Area of Ranthambore Tiger Reserve, Rajasthan, India, 1975–2015. *Annals of Valahia University of Targoviste. Geographical Series*, *18*(1), 62–67. doi:10.2478/avutgs-2018-0007

Bhardwaj, D. K. (2017). Diversity and status of terrestrial avifauna in Jamwa Ramgarh Wildlife Sanctuary, Rajasthan, Jaipur. *Life Sciences Leaflets*, *108*, 17–29. https://lifesciencesleaflets.petsd.org/

Bhopal Municipal Corporation. (2015). *Slum free city plan for Bhopal metropolitan area*. Author.

Bhyan, A. P., & Shrivastava, B. (2019). *Imperative quality of life in built environment in Indian suburb as per opinion poll*. doi:10.35940/ijrte.C6408.098319

Bhyan, A. P., & Shrivastava, B. (2020). *Cognitive Occupancy Appraisal for Sustainable Housing*. doi:10.35940/ijrte.F9597.059120

Biesbroek, G. R., Swart, R. J., Carter, T. R., Cowan, C., Henrichs, T., Mela, H., Morecroft, M. D., & Rey, D. (2010). Europe adapts to climate change: Comparing National Adaptation Strategies. *Governance. Complexity and Resilience*, *20*(3), 440–450. doi:10.1016/j.gloenvcha.2010.03.005

Binding, C., Greenberg, T., & Bukata, R. (2013). The MERIS Maximum Chlorophyll Index; its merits and limitations for inland water algal bloom monitoring. *Journal of Great Lakes Research*, *39*, 100–107. doi:10.1016/j.jglr.2013.04.005

Bluett, J., Gimson, N., Fisher, G., Heydenrych, C., Freeman, T., & Godfrey, J. (2004). *Good Practice Guide for Atmospheric Dispersion Modelling*. Ministry for the Environment. http://www.mfe.govt.nz/publications/air/atmospheric-dispersion-modelling-jun04/html/page5.html

Bollin, C., Cárdenas, C., Hahn, H., & Vatsa, S. K. (2003). *Disaster Risk Management by Communities and Local Governments*. Secretary, Regional Policy Dialogue, Inter-American Development Bank. https://www.researchgate.net/publication/254421276_Disaster_Risk_Management_by_Communities_and_Local_Governments

Bosanquet, J. L., & Pearson, J. L. (1936). The spread of smoke and gases from chimney. *Transactions of the Faraday Society*, *32*, 1249. Advance online publication. doi:10.1039/tf9363201249

Bounoua, L., Safia, A., Masek, J., Peters-Lidard, C., & Imhoff, M. L. (2009). Impact of Urban Growth on Surface Climate : A Case Study in Oran, Algeria. *Journal of Applied Meteorology and Climatology*, *48*(2), 217–231. doi:10.1175/2008JAMC2044.1

Bounoua, L., Zhang, P., Mostovoy, G., Thome, K., Masek, J., Imhoff, M., Shepherd, M., Quattrochi, D., Santanello, J., Silva, J., Wolfe, R., & Toure, A. M. (2015). Impact of urbanization on US surface climate. *Environmental Research Letters*, *10*(8), 084010. doi:10.1088/1748-9326/10/8/084010

Boyd, C. E. (2015). *Water Quality: An Introduction* (2nd ed.). Springer. https://www.springer.com/us/book/9783319330624

Bragança, L., Mateus, R., & Koukkari, H. (2010). Building sustainability assessment. *Sustainability*, *2*(7), 2010–2023. doi:10.3390u2072010

Breuer, D. (2002). *Spatial planning as an instrument for promoting sustainable development in the Nordic countries: action programme for 2001-2004*. Danish Ministry of Environment and Energy.

Brinkley, C., (2014). *Evaluating the Benefits of Peri-Urban Agriculture*. doi:10.1038/jid.2014.371

Brooks, K. N., Ffolliott, P. F., Gregersen, H. M., & DeBano, L. F. (2003). Hydrology and the Management of Watersheds. Iowa State University Press.

Brooks, K. N., Ffolliott, P. F., & Magner, J. A. (2012). *Hydrology and the Management of Watersheds*. John Wiley & Sons. doi:10.1002/9781118459751

Bruse, M. (2004). *ENVI-met 3.0: Updated Model Overview*. Academic Press.

Bulkeley, H. (2006). 'A Changing Climate for Spatial Planning' in 'Is the Issue of Climate Change too big for Spatial Planning?'. *Planning Theory & Practice*, *7*(2), 201–230. doi:10.1080/14649350600681875

California Department of Conservation. (2015). *Watershed Program*. Author.

Calkins, M. (2009). *Materials for Sustainable Sites: A Complete Guide to the Evaluation, Selection, and Use of Sustainable Construction Materials* (Vol. 1). John Wiley.

Caprara, A., Lima, J. W. de O., Marinho, A. C. P., Calvasina, P. G., Landim, L. P., & Sommerfeld, J. (2009). Irregular water supply, household usage and dengue : A bio-social study in the Brazilian Northeast. *Cadernos de Saude Publica*, *25*(suppl 1), 125–136. doi:10.1590/S0102-311X2009001300012 PMID:19287857

CEGIS. (2013). Final Report on Environmental Impact Assessment of 2x (500-660) MW Coal Based Thermal Power Plant to be Constructed at the Location of Khulna. Center for Environmental and Geographic Information Services (CEGIS).

Census. (2011). *Census of India 2011 Kerala Series-33 Part Xii-B District Census Handbook Ernakulam Village And Town Wise Primary Census Abstract (Pca) Directorate Of Census Operations Kerala*. https://www.censusindia.gov.in/2011census/dchb/3208_PART_B_ERNAKULAM.pdf

Center for Neighborhood Technology (CNT). (2013). *The Case for Fixing the Leaks: Protecting People and Saving Water while Supporting Economic Growth in the Great Lakes Region*. Retrieved from https://www.cnt.org/publications/the-case-for-fixing-the-leaks-protecting-people-and-savingwater-while-supporting

Cervero, R. (1996). Mixed land-uses and commuting: Evidence from the American Housing Survey. *Transportation Research Part A, Policy and Practice*, *30*(5), 361–377. doi:10.1016/0965-8564(95)00033-X

Cervero, R., & Kockelman, K. (1997). Travel demand and the 3ds: Density, diversity, and design. *Transportation Research Part D, Transport and Environment*, *2*(3), 199–219. doi:10.1016/S1361-9209(97)00009-6

Cetiner, I., & Edis, E. (2014). *An environmental and economic sustainability assessment method for the retrofitting of residential buildings*. doi:10.1016/j.enbuild.2014.01.020

Chakravartty, A. (2015, July 4). Choked on salt. *DownToEarth*.

Chaminuka, L., & Nyatsanza, T. D. (2013). An assessment of water shortages and coping mechanisms of Harare residents : A case of Msasa Park and Dzivaresekwa. *Journal of Agriculture and Veterinary Science*, *4*(3), 21–35. doi:10.9790/2380-0432135

Chang, C.-R., Li, M.-H., & Chang, S.-D. (2007). A preliminary study on the local cool-island intensity of Taipei city parks. *Landscape and Urban Planning*, *80*(4), 386–395. doi:10.1016/j.landurbplan.2006.09.005

Chang, S., Jiang, Q., & Zhao, Y. (2018). Integrating CFD and GIS into the Development of Urban Ventilation Corridors : A Case Study in Changchun City, China. *Sustainability*, *10*(6), 1814. doi:10.3390u10061814

Chawla, I., Karthikeyan, L., & Mishra, A. K. (2020). A review of remote sensing applications for water security: Quantity, quality, and extremes. *Journal of Hydrology (Amsterdam)*, *585*, 124826. doi:10.1016/j.jhydrol.2020.124826

Chebud, Y., Naja, G. M., Rivero, R. G., & Melesse, A. M. (2012). Water Quality Monitoring Using Remote Sensing and an Artificial Neural Network. *Water, Air, and Soil Pollution, 223*(8), 4875–4887. doi:10.100711270-012-1243-0

Chen, L., Ng, E., An, X., Ren, C., Lee, M., Wang, U., & He, Z. (2012). Sky view factor analysis of street canyons and its implications for daytime intra-urban air temperature differentials in high-rise, high-density urban areas of Hong Kong : A GIS-based simulation approach. *International Journal of Climatology, 32*(1), 121–136. doi:10.1002/joc.2243

Chen, L., Zhang, G., Xu, Y. J., Chen, S., Wu, Y., Gao, Z., & Yu, H. (2020). Human Activities and Climate Variability Affecting Inland Water Surface Area in a High Latitude River Basin. *Water (Basel), 12*(2), 382. doi:10.3390/w12020382

Chen, S., Zhang, G., Xia, X., Setunge, S., & Shi, L. (2020). A review of internal and external influencing factors on energy efficiency design of buildings. *Energy and Building, 216*, 109944. doi:10.1016/j.enbuild.2020.109944

Chigbu, N., & Onukaogu, D. (2013). Role of Geospatial Technology in Environmental Sustainability in Nigeria-An Overview. *Environment For Sustainability*, 6–10.

Chi, M., Plaza, A., Benediktsson, J. A., Sun, Z., Shen, J., & Zhu, Y. (2016). Big Data for Remote Sensing: Challenges and Opportunities. *Proceedings of the IEEE, 104*(11), 2207–2219. doi:10.1109/JPROC.2016.2598228

Choe, K., Varley, R. C. G., & Bijlani, H. U. (1996). *Coping with Intermittent Water Supply: Problems and Prospects, Dehra Dun, Uttar Pradesh, India*. Academic Press.

Choudhary, A., Deval, K., & Joshi, P. K. (2020). Study of habitat quality assessment using geospatial techniques in Keoladeo National Park, India. *Environmental Science and Pollution Research International*. Advance online publication. doi:10.100711356-020-11666-3 PMID:33205275

Cogley, J. G., Hock, R., Rasmussen, L. A., Arendt, A. A., Bauder, A., Braithwaite, R. J., Jansson, P., Kaser, G., Möller, M., & Nicholson, L. (2011). *Glossary of Glacier Mass Balance and Related Terms; IHP-VII Technical Documents in Hydrology No. 86, IACS Contribution No. 2*. UNESCO-IHP.

Convention on Biological diversity-India. (n.d.). https://www.cbd.int/countries/profile/?country=in

Cook, J., Kimuyu, P., & Whittington, D. (2015). The cost of coping with poor water supply in rural Kenya. *Water Resources Research, 51*, 9127–9140. doi:10.1002/2014WR016259

Costa, D., Quinteiro, P., & Dias, A. C. (2019). A systematic review of life cycle sustainability assessment: Current state, methodological challenges, and implementation issues. *The Science of the Total Environment, 686*, 774–787. doi:10.1016/j.scitotenv.2019.05.435 PMID:31195285

Costanza, R., Costanza, R., Arge, R., Groot, R. De, Farber, S., Grasso, M., & Hannon, B. (1997). *The value of the world' s ecosystem services and natural capital 8009*. doi:10.1016/S0921-8009(98)00020-2

Costanzo, V., Evola, G., & Marletta, L. (2016). Energy savings in buildings or UHI mitigation? Comparison between green roofs and cool roofs. *Energy and Building, 114*, 247–255. doi:10.1016/j.enbuild.2015.04.053

CRED. (2018). Natural Disasters 2018. In *Emergency Events Database, Center for Research on the Epidemiology of Disasters*. https://www.emdat.be/

Datta, D. S. (2017). *Sambhar Salt Lake- Why Conserve It?* academia.edu.

Davidoff, P. (1965). Advocacy and Pluralism in Planning. *Journal of the American Institute of Planners, 31*(4), 331–338. doi:10.1080/01944366508978187

Davidoff, P., & Reiner, T. A. (1962). A Choice Theory of Planning. *Journal of the American Institute of Planners, 28*(2), 103–115. doi:10.1080/01944366208979427

Davies, M., Steadman, P., & Oreszczyn, T. (2008). Strategies for the modification of the urban climate and the consequent impact on building energy use. *Energy Policy, 36*(12), 4548–4551. doi:10.1016/j.enpol.2008.09.013

Dawson, R. J. (2011). Potential pitfalls on the transition to more sustainable cities and how they might be avoided. *Carbon Management, 2*(2), 175–188. doi:10.4155/cmt.11.8

Deb, S., Ahmed, A., & Datta, D. (2014). An alternative approach for delineating eco-sensitive zones around a wildlife sanctuary applying geospatial techniques. *Environmental Monitoring and Assessment, 186*(4), 2641–2651. doi:10.100710661-013-3567-7 PMID:24338098

Deilami, K., Kamruzzaman, M., & Liu, Y. (2018). Urban heat island effect : A systematic review of spatio-temporal factors, data, methods, and mitigation measures. *International Journal of Applied Earth Observation and Geoinformation, 67*, 30–42. doi:10.1016/j.jag.2017.12.009

Dekolo, S., Oduwaye, L., & Nwokoro, I. (2015). Urban sprawl and loss of agricultural land in peri-urban areas of Lagos. *Reg. Stat., 5*(2), 20–33. doi:10.15196/RS05202

Department of Land Resources. (2013). National Land Utilisation Policy. Author.

Department of Planning & Environment. (2014). *A plan for growing Sydney: A strong global city, a great place to live.* Department of Planning & Environment. http://www.strategy.planning.nsw.gov.au/sydney/the-plan/

Devi, G. K., Ganasri, B., & Dwarakish, G. (2015). Applications of Remote Sensing in Satellite Oceanography: A Review. *Aquatic Procedia, 4*, 579–584. doi:10.1016/j.aqpro.2015.02.075

Dewan, A. M., & Yamaguchi, Y. (2009). Using remote sensing and GIS to detect and monitor land use and land cover change in Dhaka Metropolitan of Bangladesh during 1960-2005. *Environmental Monitoring and Assessment, 150*(1–4), 237–249. doi:10.100710661-008-0226-5 PMID:18317939

Dhabriya, S. (1988). *Eco-crisis in the Aravalli hill region.* Environmental Research Publication.

Di Tullio, M., Nocchi, F., Camplani, A., Emanuelli, N., Nascetti, A., & Crespi, M. (2018). Copernicus big data and google earth engine for glacier surface velocity field monitoring: Feasibility demonstration on san rafael and san quintin glaciers. *The International Archives of the Photogrammetry, Remote Sensing and Spatial Information Sciences*, 42(3), 289–294. doi:10.5194/isprs-archives-XLII-3-289-2018

Dogliotti, A., Ruddick, K., Nechad, B., Doxaran, D., & Knaeps, E. (2015). A single algorithm to retrieve turbidity from remotely-sensed data in all coastal and estuarine waters. *Remote Sensing of Environment, 156*, 157–168. doi:10.1016/j.rse.2014.09.020

Du, H., Song, X., Jiang, H., Kan, Z., Wang, Z., & Cai, Y. (2016). Research on the cooling island effects of water body : A case study of Shanghai, China. *Ecological Indicators, 67*, 31–38. doi:10.1016/j.ecolind.2016.02.040

Dutta, V., & Tiwari, A. P. (2003). Cost of services and willingness to pay for reliable urban water supply : A study from Delhi, India. *Water Science and Technology: Water Supply, 5*(6), 135–144. doi:10.2166/ws.2005.0058

Ebinger, J. O., & Vandycke, N. (2015). *Moving Toward Climate-Resilient Transport : The World Bank's Experience from Building Adaptation into Programs.* World Bank. Retrieved from https://openknowledge.worldbank.org/handle/10986/23685

Ecology and Rural Development Society Jodhpur. (2014). *Status Survey of Important Bird Areas of Rajasthan.* Indian Bird Conservation Network.

Economidou, M., Todeschi, V., Bertoldi, P., D'Agostino, D., Zangheri, P., & Castellazzi, L. (2020). Review of 50 years of EU energy efficiency policies for buildings. *Energy and Building, 225*, 110322. doi:10.1016/j.enbuild.2020.110322

Edokpayi, J. N., Odiyo, J. O., & Durowoju, O. S. (2017). *Impact of Wastewater on Surface Water Quality in Developing Countries: A Case Study of South Africa*. Water Quality. doi:10.5772/66561

Eghenter, C. (2000). *Mapping people's forests: The role of mapping in planning community-based management of conservation areas in Indonesia. Academic Press.*

Ehsani, N., Vörösmarty, C. J., Fekete, B. M., & Stakhiv, E. Z. (2017). Reservoir operations under climate change: Storage capacity options to mitigate risk. *Journal of Hydrology (Amsterdam)*, *555*, 435–446. doi:10.1016/j.jhydrol.2017.09.008

Elboshy, B., Kanae, S., Gamaleldin, M., Ayad, H., Osaragi, T., & Elbarki, W. (2018). A framework for pluvial flood risk assessment in Alexandria considering the coping capacity. *Environment Systems & Decisions*, *39*(1), 77–94. doi:10.100710669-018-9684-7

Elmqvist, T., & Mcdonald, R. I. (2014). Urbanization, Biodiversity and Ecosystem Services: Challenges and Opportunities. A Global Assessment. Springer. doi:10.1007/978-94-007-7088-1

Emekcí, Ş. (2018). *A Life cycle costing based decision support tool for cost-optimal energy efficient design and/or refurbishments* (Ph.D Thesis). Middle East Technical University.

Engel, K. (2005). Mitigating global climate change in the United States: A regional approach. *New York University Environmental Law Journal*, *14*, 54–85.

Etzioni, A. (1967). Mixed-Scanning: A "Third" Approach to Decision-Making. *Public Administration Review*, *27*(5), 385–392. doi:10.2307/973394

Everard, M., Sharma, O. P., Vishwakarma, V. K., Khandal, D., Sahu, Y. K., Bhatnagar, R., Singh, J. K., Kumar, R., Nawab, A., Kumar, A., Kumar, V., Kashyap, A., Pandey, D. N., & Pinder, A. C. (2018). Assessing the feasibility of integrating ecosystem-based with engineered water resource governance and management for water security in semi-arid landscapes: A case study in the Banas catchment, Rajasthan, India. *The Science of the Total Environment*, *612*, 1249–1265. doi:10.1016/j.scitotenv.2017.08.308 PMID:28892868

Ewing, R., Bartholomew, K., Winkelman, S., Walters, J., & Che, D. (2008). *Growing cooler: Evidence on urban development and climate change*. ULI.

Ewing, R., Pendall, R., & Chen, D. (2003). Measuring Sprawl and Its Transportation Impacts. *Transportation Research Record: Journal of the Transportation Research Board*, *1831*(1), 175–183. doi:10.3141/1831-20

Faludi, A. (1973). *Planning theory*. Pergamon Press.

FAO. (2012). The State of Food and Agriculture 2012. *The State of Food and Agriculture*. Advance online publication. doi:10.18356/3e6e633a-en

Fatourehchi, D., & Zarghami, E. (2020). *Social sustainability assessment framework for managing sustainable construction in residential buildings*. doi:10.1016/j.jobe.2020.101761

Feng, L., Hu, C., Han, X., Chen, X., & Qi, L. (2014). Long-Term Distribution Patterns of Chlorophyll-a Concentration in China's Largest Freshwater Lake: MERIS Full-Resolution Observations with a Practical Approach. *Remote Sensing*, *7*(1), 275–299. doi:10.3390/rs70100275

Ffolliott, P. F., Baker, M. B., Edminster, C. B., Dillon, M. C., & Mora, K. L. (Eds.). (2012). *Land stewardship through watershed management: perspectives for the 21st century*. Springer Science & Business Media.

Fielding, K., Spinks, A., Russell, S., & Mankad, A. (2012). *Water Demand Management Study: Baseline Survey of Household Water Use*. Urban Water Security Research Alliance Technical Report.

Flores-Anderson, A. I., Griffin, R., Dix, M., Romero-Oliva, C. S., Ochaeta, G., Skinner-Alvarado, J., & Barreno, F. (2020). Hyperspectral satellite remote sensing of water quality in lake atitlán, Guatemala. *Frontiers in Environmental Science*, *8*, 7. Advance online publication. doi:10.3389/fenvs.2020.00007

Flyvbjerg, B. (1996) The dark side of planning: Rationality and 'realrationalitat'. In Explorations in Planning Theory. New Brunswick, NJ: Rutgers.

Franklin, S. E. (2001). *Remote Sensing for Sustainable Forest Management*. Lewis Publishers. doi:10.1201/9781420032857

Frenkiel, F. N. (1956). Atmospheric Pollution and Zoning in an Urban Area. *The Scientific Monthly*, *82*(4), 194–203.

Fritsche, Eppler, Iriarte, Laaks, Wunder, Kaphengst, Wolff, & Heyen. (2015). *Resource-Efficient Land Use – Towards a Global Sustainable Land Use Standard (GLOBALANDS)*. Umweltbundesamt. doi:10.1017/CBO9781107415324.004

Fuller, L. M., & Minnerick, R. J. (2007). *Predicting water quality by relating secchi-disk transparency and chlorophyll a measurements to landsat satellite imagery for Michigan inland lakes, 2001-2006*. https://pubs.usgs.gov/fs/2007/3022/pdf/FS2007-3022.pdf

Gao, H., Birkett, C., & Lettenmaier, D. P. (2012). Global monitoring of large reservoir storage from satellite remote sensing. *Water Resources Research*, *48*(9), 2012WR012063. Advance online publication. doi:10.1029/2012WR012063

German, L., Mansoor, H., Alemu, G., Mazengia, W., Amede, T., & Stroud, A. (2007). Participatory integrated watershed management: Evolution of concepts and methods in an ecoregional program of the eastern African highlands. *Agricultural Systems*, *94*(2), 189–204. doi:10.1016/j.agsy.2006.08.008

Gholizadeh, M., Melesse, A., & Reddi, L. (2016). A Comprehensive Review on Water Quality Parameters Estimation Using Remote Sensing Techniques. *Sensors (Basel)*, *16*(8), 1298. doi:10.339016081298 PMID:27537896

Giri, A., Bharti, V. K., Kalia, S., Arora, A., Balaje, S. S., & Chaurasia, O. P. (2020). A review on water quality and dairy cattle health: A special emphasis on high-altitude region. *Applied Water Science*, *10*(3), 79. Advance online publication. doi:10.100713201-020-1160-0

Girotto, M., & Rodell, M. (2019). Terrestrial water storage. *Extreme Hydroclimatic Events and Multivariate Hazards in a Changing Environment*, 41–64. doi:10.1016/B978-0-12-814899-0.00002-X

Givoni, B. (1998). *Climate considerations in building and urban design*. John Wiley & Sons, Inc.

Gladwin, T. N., Kennelly, J. J., & Krause, T. (1995). Shifting Paradigms for Sustainable for Implications Development : And Theory. *Academy of Management Review*, *20*(4), 874–907. doi:10.5465/amr.1995.9512280024

GoB. (2014). *Budget in Brief 2014-15*. Finance Division, Ministry of Finance, Government of the People's Republic of Bangladesh. http://www.mof.gov.bd/en/index.php?option=com_content& view=article&id=269&Itemid=1

Google. (2021a). *API; Google Earth Engine; Google Developers*. Google. https://developers.google.com/earth-engine/api_docs

Google. (2021b). *Datasets; Google Earth Engine; Google Developers*. Google. https://developers.google.com/earth-engine/datasets

Google. (2021c). *Sentinel 1; Google Earth Engine; Google Developers*. Google. https://developers.google.com/earth-engine/guides/sentinel1

Google. (2021d). *Chlorophyll datasets; Google Earth Engine; Google Developers*. Google. https://developers.google.com/earth-engine/datasets/tags/chlorophyll?hl=en

Google. (2021e). *API; Google Earth Engine; Google Developers*. Google. https://developers.google.com/earth-engine/guides/classification

Gopal, B., & Chauhan, M. (2007). River Yamuna from Source to Delhi: Human Impacts and Approaches to Conservation. In *Restoring River Yamuna: Concepts, Strategies and Socio-Economic Considerations* (pp. 45-69). National Institute of Ecology.

Gorde, S. P., & Jadhav, M. V. (2013). Assessment of Water Quality Parameters: A Review. *International Journal of Engineering Research and Applications, 3*, 2029–2035. https://www.ijera.com/papers/Vol3_issue6/LV3620292035.pdf

Gosling, M.-C., & Walker, A. J. (1992). Energy efficiency within the urban design process. *Cities (London, England), 9*(2), 111–116. doi:10.1016/0264-2751(92)90045-7

Goubran, S. (2019). *On the Role of Construction in Achieving the SDGs*. Academic Press.

Goudarzi, H., & Mostafaeipour, A. (2017). Energy saving evaluation of passive systems for residential buildings in hot and dry regions. *Renewable & Sustainable Energy Reviews, 68*, 432–446. doi:10.1016/j.rser.2016.10.002

Gould, J. (1999). *Contributions Relating to Rainwater Harvesting*. (Paper prepared for the World Commission on Dams Secretariat (WCD). *Thematic Review, 4*, 3.

Government of India. (1991). *Census Digital Library*. Retrieved from Primary Census Abstract: https://censusindia.gov.in/DigitalLibrary/TVDirectory.aspx

Government of India. (2001). *Census Digital Library*. Retrieved from Primary Census Abstract: https://censusindia.gov.in/DigitalLibrary/MFTableSeries.aspx

Government of India. (2011). *Census Digital Library*. Retrieved from Primary Census Abstract: https://censusindia.gov.in/DigitalLibrary/MFTableSeries.aspx

Government of Kerala, & Department of Environment and Climate Change. (2014). *Kerala State Action Plan on Climate Change*. http://www.moef.gov.in/sites/default/files/KERALA STATE ACTION PLAN ON CLIMATE CHANGE.pdf

Graf, L., Asante, S. A., & Kovacs-Győri, A. (2019). *Investigating Conflicts between Gold Mining and Water Protection by Remote Sensing Imagery*. Regional Academy for the United Nations., doi:10.13140/RG.2.2.13344.74244

Greb, S., & Daniela, G. (2018). *Remote sensing of water quality in Wisconsin*. Paper presented at Wisconsin Lakes Convention.

Greenwald, M. J., & Boarnet, M. G. (2001). The Built Environment as a Determinant of Walking Behavior: Analyzing Non-Work Pedestrian Travel in Portland, Oregon. *Transportation Research Record: Journal of the Transportation Research Board, 1780*(1), 33–41. doi:10.3141/1780-05

Griffin, C., McClelland, J., Frey, K., Fiske, G., & Holmes, R. (2018). Quantifying CDOM and DOC in major Arctic rivers during ice-free conditions using Landsat TM and ETM+ data. *Remote Sensing of Environment, 209*, 395–409. doi:10.1016/j.rse.2018.02.060

Gu, K., Fang, Y., Qian, Z., Sun, Z., & Wang, A. (2020). Spatial planning for urban ventilation corridors by urban climatology. *Ecosystem Health and Sustainability, 6*(1), 1747946. doi:10.1080/20964129.2020.1747946

Gu, K., Zhang, Y., & Qiao, J. (2020). Random Forest Ensemble for River Turbidity Measurement From Space Remote Sensing Data. *IEEE Transactions on Instrumentation and Measurement, 69*(11), 9028–9036. doi:10.1109/TIM.2020.2998615

Gunawardena, K. R., Wells, M. J., & Kershaw, T. (2017). Utilising green and bluespace to mitigate urban heat island intensity. *The Science of the Total Environment, 584-585*, 1040–1055. doi:10.1016/j.scitotenv.2017.01.158 PMID:28161043

Gunnell, K. (2009). *Green Building in South Africa: Emerging Trends.* Department of Environmental Affairs and Tourism (DEAT). Retrieved from http://soer. deat.gov.za/Green Building in South Africa Emerging Trends GyzKE.pdf.file

Gupta, R. (2014). The pattern of urban land-use changes: A case study of the indian cities. *Environment & Urbanization Asia*, *5*(1), 83–104. doi:10.1177/0975425314521539

Gwynne, M. D. (1982). The Global Environment Monitoring System (GEMS) of UNEP. *Environmental Conservation*, *9*(1), 35–41. doi:10.1017/S0376892900019469

Haase, M., & Amato, A. (2006). *Ventilated façade design in hot and humid climate.* Academic Press.

Hafeez, S., Wong, M., Ho, H., Nazeer, M., Nichol, J., Abbas, S., Tang, D., Lee, K., & Pun, L. (2019). Comparison of Machine Learning Algorithms for Retrieval of Water Quality Indicators in Case-II Waters: A Case Study of Hong Kong. *Remote Sensing*, *11*(6), 617. doi:10.3390/rs11060617

Hammond, G. P., & Jones, C. I. (2008). Embodied energy and carbon in construction materials. *Proceedings of the Institution of Civil Engineers - Energy*, *161*(2), 87–98. 10.1680/ener.2008.161.2.87

Hampton, S. E. (2013). Understanding lakes near and far. *Science*, *342*(6160), 815–816. doi:10.1126cience.1244732 PMID:24233716

Hanna, S. R., Briggs, G. A., & Hosker, R. P. (1982). *Handbook on Atmospheric Diffusion.* U.S. Department of Energy. doi:10.2172/5591108

Hansen, C.H. (2015). *Google Earth Engine as a Platform for Making Remote Sensing of Water Resources a Reality for Monitoring Inland Waters.* doi:10.13140/RG.2.1.3688.1047

Hardin, G. (2001). *The tragedy of the commons.* Academic Press.

Hart, M. A., & Sailor, D. J. (2009). Quantifying the influence of land-use and surface characteristics on spatial variability in the urban heat island. *Theoretical and Applied Climatology*, *95*(3-4), 397–406. doi:10.100700704-008-0017-5

Heaviside, C., Macintyre, H., & Vardoulakis, S. (2017). The Urban Heat Island : Implications for Health in a Changing Environment. *Current Environmental Health Reports*, *4*(3), 296–305. doi:10.100740572-017-0150-3 PMID:28695487

Hellwig, N., Walz, A., & Markovic, D. (2019). Climatic and socioeconomic effects on land cover changes across Europe: Does protected area designation matter? *PLoS One*, *14*(7), 1–20. doi:10.1371/journal.pone.0219374 PMID:31314769

Henry, W. H. (1969). The air diffusion model as an urban planning tool. *Socio-Economic Planning Sciences*, *3*(4), 329–349. doi:10.1016/0038-0121(69)90056-1

Hertwich, E. G., Ali, S., Ciacci, L., Fishman, T., Heeren, N., Masanet, E., Asghari, F. N., Olivetti, E., Pauliuk, S., Tu, Q., & Wolfram, P. (2019). Material efficiency strategies to reducing greenhouse gas emissions associated with buildings, vehicles, and electronics—A review. *Environmental Research Letters*, *14*(4), 043004. doi:10.1088/1748-9326/ab0fe3

Hirano, Y., & Fujita, T. (2012). Evaluation of the impact of the urban heat island on residential and commercial energy consumption in Tokyo. *Energy*, *37*(1), 371–383. doi:10.1016/j.energy.2011.11.018

Høgevold, N. M., Svensson, G., Rodriguez, R., & Eriksson, D. (2019). Relative importance and priority of TBL elements on the corporate performance. *Management of Environmental Quality*, *30*(3), 609–623. doi:10.1108/MEQ-04-2018-0069

Holderness, T., Barr, S., Dawson, R., & Hall, J. (2013). An evaluation of thermal Earth observation for characterizing urban heatwave event dynamics using the urban heat island intensity metric. *International Journal of Remote Sensing*, *34*(3), 864–884. doi:10.1080/01431161.2012.714505

Hollis, G. E. (1975). The effect of urbanization on floods of different recurrence interval. *Water Resources Research, 11*(3), 431–435. doi:10.1029/WR011i003p00431

Hossaini, N., Reza, B., Akhtar, S., Sadiq, R., & Hewage, K. (2015). AHP based life cycle sustainability assessment (LCSA) framework: A case study of six storey wood frame and concrete frame buildings in Vancouver. *Journal of Environmental Planning and Management, 58*(7), 1217–1241. doi:10.1080/09640568.2014.920704

Howard, L. (1833). *The Climate of London: Deduced from meteorological observations made in the metropolis and at various places around it.* Academic Press.

Hsieh, C.-M., Aramaki, T., & Hanaki, K. (2007). The feedback of heat rejection to air conditioning load during the nighttime in subtropical climate. *Energy and Building, 39*(11), 1175–1182. doi:10.1016/j.enbuild.2006.06.016

Huang, X., & Zhang, L. (2011). A multidirectional and multiscale morphological index for automatic building extraction from multispectral GeoEye-1 imagery. *Photogrammetric Engineering and Remote Sensing, 77*(7), 721–732.

Huang, X., & Zhang, L. (2013). An SVM Ensemble Approach Combining Spectral, Structural, and Semantic Features for the Classification of High-Resolution Remotely Sensed Imagery. *IEEE Transactions on Geoscience and Remote Sensing, 51*(1), 257–272.

Hudson, B. M., Galloway, T. D., & Kaufman, J. L. (1979). Comparison of Current Planning Theories: Counterparts and Contradictions. *Journal of the American Planning Association, 45*(4), 387–398. doi:10.1080/01944367908976980

Humanity Road. (2018). *2018 India Kerala Flooding* (Vol. 501, Issue c). https://reliefweb.int/sites/reliefweb.int/files/resources/Humanity Road Sitrep No 1%2C Kerala India Floods%2C 16Aug2018%281%29.pdf

Hunt, A., & Watkiss, P. (2011). Climate change impacts and adaptation in cities: A review of the literature. *Climatic Change, 104*(1), 13–49. doi:10.100710584-010-9975-6

Huovila, P., Ala-Juusela, M., Melchert, L., & Pouffary, S. (2007). *Buildings and Climate Change Status, Challenges and Opportunities.* United Nations Environment Programme.

Huovinen, P., Ramírez, J., Caputo, L., & Gómez, I. (2019). Mapping of spatial and temporal variation of water characteristics through satellite remote sensing in Lake Panguipulli, Chile. *The Science of the Total Environment, 679*, 196–208. doi:10.1016/j.scitotenv.2019.04.367 PMID:31082593

ICLEI. (2005). *International council for local environmental initiatives.* Retrieved Oct 2005, from http://www.iclei.org/

IEA. (2020). *Buildings – Topics.* Retrieved March 15, 2021, from https://www.iea.org/topics/buildings

IEA. (International Energy Agency). (2008). *World energy outlook.* Retrieved April 18, 2021, from https://www.iea.org/topics/world-energy-outlook

IEEE GRSS Data Fusion Contest, Presented to Image Analysis and Data Fusion Technical Committee. (2014). *IEEE Geoscience and Remote Sensing Society.* GRSS.

Ingrao, C., Messineo, A., Beltramo, R., Yigitcanlar, T., & Ioppolo, G. (2018). How can life cycle thinking support sustainability of buildings? Investigating life cycle assessment applications for energy efficiency and environmental performance. *Journal of Cleaner Production, 201*, 556–569. doi:10.1016/j.jclepro.2018.08.080

Institute for Local Government. (2010) *Understanding the Basics of Land Use and Planning: Guide to Local Planning.* https://www.ca-ilg.org/sites/main/files/file-attachments/2010_-_landuseplanning.pdf?1387495993

IPCC (International Panel on Climate Change). (2013). *Climate Change 2013: The Physical Science Basis, Contribution of Working Group I to the Fifth Assessment Report of the Intergovernmental Panel on Climate Change*. Cambridge University Press.

IPCC. (2007). *Climate Change 2007: Mitigation. Contribution of Working Group III to the Fourth Assessment Report of the Intergovernmental Panel on Climate Change*. Cambridge University Press.

IPCC. (2007a). *Climate change 2007: Impacts, adaptation and vulnerability*. Cambridge University Press.

IPCC. (2007b). *Climate change 2007: Mitigation of climate change*. Cambridge University Press.

IPCC. (2007c). *Climate change 2007: Summary for policymakers of the synthesis report of the IPCC fourth assessment report*. Cambridge University Press.

IPCC. (2007e). *Climate change 2007: The physical science basis*. Cambridge University Press.

IPCC. (2018). *Special Report: Global Warming of 1.5 °C*. Retrieved February 4, 2021, from https://www.ipcc.ch/sr15/chapter/glossary/

IUCN. (2000). *Communities and forest management in South Asia: A regional profile of the working group on community involvement in forest management*. The World Conservation UnioN.

Jacobs, J. (1961). *The death and life of great American cities*. Vintage Books.

Jalal, D., & Ezzedine, T. (2019). Performance analysis of machine learning algorithms for water quality monitoring system. *2019 International Conference on Internet of Things, Embedded Systems and Communications (IINTEC)*. 10.1109/IINTEC48298.2019.9112096

Jarir, S., & Dajani, T. H. (1977). Air pollution control through land-use planning. *Urban Systems*, *2*(1), 35–43. doi:10.1016/0147-8001(77)90004-3

Jellison, R., Williams, W. D., Timms, B., Alcocer, J., & Aladin, N. V. (2008). Salt lakes: values, threats and future. In N. V. Polunin (Ed.), *Aquatic Ecosystems* (pp. 94–110). Cambridge University Press. doi:10.1017/CBO9780511751790.010

Jha, A. K., Bloch, R., & Lamond, J. (2012). *Cities and Flooding A Guide to Integrated Urban Flood Risk Management for the 21st Century*. The World Bank. https://openknowledge.worldbank.org/handle/10986/2241

Jiang, L., Schneider, R., Andersen, O., & Bauer-Gottwein, P. (2017). CryoSat-2 Altimetry Applications over Rivers and Lakes. *Water (Basel)*, *9*(3), 211. doi:10.3390/w9030211

Jinru Xue, J., & Su, B. (2017). Significant Remote Sensing Vegetation Indices: A Review of Developments and Applications. *Journal of Sensors*.

Johnson, G. T., & Watson, I. D. (1984). The Determination of View-Factors in Urban Canyons. *Journal of Applied Meteorology and Climatology*, *23*(2), 329–335. doi:10.1175/1520-0450(1984)023<0329:TDOVFI>2.0.CO;2

Jong, S. M., Meer, F. D., & Clevers, J. G. (2004). Basics of Remote Sensing. In Remote Sensing Image Analysis: Including the Spatial Domain. Springer.

Joseph, J. K., Anand, D., Prajeesh, P., Zacharias, A., Varghese, A. G., Pradeepkumar, A. P., & Baiju, K. R. (2020). Community resilience mechanism in an unexpected extreme weather event: An analysis of the Kerala floods of 2018, India. *International Journal of Disaster Risk Reduction*, *49*, 101741. doi:10.1016/j.ijdrr.2020.101741

Joshi, R. R., Warthe, M., Dwivedi, S., Vijay, R., & Chakrabarti, T. (2011). Monitoring changes in land use land cover of Yamuna riverbed in Delhi: A multi-temporal analysis. *International Journal of Remote Sensing*, *32*(24), 9547–9558. Advance online publication. doi:10.1080/01431161.2011.565377

JPL. (2001). *ASTER Higher Level User Guide, Version 2.0*. https://asterweb.jpl.nasa.gov/content/03_data/04_Documents/ASTERHigherLevelUserGuideVer2May01.pdf

JRC. (2018). *Setting up the modelling exercise*. European Union Joint Research Centre. https://rem.jrc.ec.europa.eu/etex/37.htm

Juhls, B., Overduin, P. P., Hölemann, J., Hieronymi, M., Matsuoka, A., Heim, B., & Fischer, J. (2019). Dissolved organic matter at the fluvial–marine transition in the Laptev Sea using in situ data and ocean colour remote sensing. *Biogeosciences*, *16*(13), 2693–2713. doi:10.5194/bg-16-2693-2019

Jun, C., Ban, Y., & Li, S. (2014). Open access to Earth land-cover map. *Nature*, *514*(7523), 434. doi:10.1038/514434c PMID:25341776

Kahn, M. (1996). New evidence on trends in vehicle emissions. *The RAND Journal of Economics*, *27*(1), 183–196. doi:10.2307/2555798

Kaiser, E. J., & Godschalk, D. R. (1995b). Twentieth century land use planning: A stalwart family tree. *Journal of the American Planning Association*, *61*(3), 365–385. doi:10.1080/01944369508975648

Kalele, A. S. (2019). *Estimation and mapping of turbidity in the lower charles river using Landsat 8 oli satellite imagery*. http://hdl.handle.net/2047/D20323946

Kale, V. S. (2014). *Landscapes and Landforms of India*. Springer. doi:10.1007/978-94-017-8029-2

Kang, M. S., Goo, J. H., Song, I., Chun, J. A., Her, Y. G., Hwang, S. W., & Park, S. W. (2013). Estimating design floods based on the critical storm duration for small watersheds. *Journal of Hydro-environment Research*, *7*(3), 209–218. doi:10.1016/j.jher.2013.01.003

Karaca, F., Guney, M., Kumisbek, A., Kaskina, D., & Tokbolat, S. (2020). A new stakeholder opinion-based rapid sustainability assessment method (RSAM) for existing residential buildings. *Sustainable Cities and Society*, *60*, 102155. doi:10.1016/j.scs.2020.102155

Karlsson, M., & Yakowitz, S. (1987). Rainfall-runoff forecasting methods, old and new. *Stochastic Hydrology and Hydraulics*, *1*(4), 303–318. doi:10.1007/BF01543102

Karnieli, A., Agam, N., & Pinker, R. T. (2010). Use of NDVI and land surface temperature for drought assessment: Merits and limitations. *Journal of Climate*, *23*(3), 618–633.

Karthikvel, S., & Samydoss, C. (2020). *Sustainable development*. Academic Press.

Kasaragod, G. R., & Hegde, V. S. (2002). Application of Remote Sensing & GIS in Watershed Prioritization, Development Planning & Monitoring. *Conference: Sensitization Workshop on Integrated Watershed Management under Western Ghats Development Programme*.

Katuwal, H., & Bohara, A. K. (2011). Coping with poor water supplies : Empirical evidence from Kathmandu Nepal. *Journal of Water and Health*, *9*(1), 143–158. doi:10.2166/wh.2010.151 PMID:21301123

Katyambo, M. M., & Ngigi, M. M. (2017). Spatial Monitoring of Urban Growth Using GIS and Remote Sensing : A Case Study of Nairobi Metropolitan Area, Kenya. *Am. J. Geogr. Inf. Syst.*, *6*, 64–82. doi:10.5923/j.ajgis.20170602.03

KDA. (2011). *Structure Plan, Master Plan and Detailed Area Plan for the Extension of the Khulna Master Plan Area up to Mongla Town. Khulna Development Authority (KDA)*. Ministry of Housing and Public Works, Government of the People's Republic of Bangladesh.

Keifer, C. J., & Chu, H. H. (1957). Synthetic storm pattern for drainage design. *Journal of the Hydraulics Division, 83*(4), 1332–1. doi:10.1061/JYCEAJ.0000104

Khan, S., & Swapan, M. S. H. (2013). From blueprint master plans to democratic planning in South Asian cities: Pursuing good governance agenda against prevalent patron-client networks. *Habitat International, 38*, 183–191.

Kidwell, K. B. (1995). *NOAA Polar Orbiter Data Users Guide : (TIROS-N, NOAA-6, NOAA-7, NOAA-8, NOAA-9, NOAA-10, NOAA-11, NOAA-12, NOAA-13, and NOAA-14)*. National Oceanic and Atmospheric Administration, National Environmental Satellite, Data, and Information Service, National Climatic Data Center, Satellite Data Services Division.

Kimambo, O. N., Chikoore, H., Gumbo, J. R., & Msagati, T. A. (2019). Retrospective analysis of Chlorophyll-a and its correlation with climate and hydrological variations in Mindu Dam, Morogoro, Tanzania. *Heliyon, 5*(11), e02834. doi:10.1016/j.heliyon.2019.e02834 PMID:31763484

Kim, H., Yeh, P. J. F., Oki, T., & Kanae, S. (2009). Role of rivers in the seasonal variations of terrestrial water storage over global basins. *Geophysical Research Letters, 36*(17), L17402. Advance online publication. doi:10.1029/2009GL039006

Kim, Y. H., Im, J., Ha, H. K., Choi, J. K., & Ha, S. (2014). Machine learning approaches to coastal water quality monitoring using GOCI satellite data. *GIScience & Remote Sensing, 51*(2), 158–174. doi:10.1080/15481603.2014.900983

King, B. (2010). Toprak Mimarisinin Yeniden Doğuşu, Kil Kökenli Taze ve Güncellenmiş Bir Bakış. *Mimarlıkta Malzeme Dergisi*, (17).

Kleemeier, E. L. (2010). *Private Operators and Rural Water Supplies: A Desk Review of Experience*. World Bank. http://water.worldbank.org/publications/private-operators-and-rural-water-supplies-desk-review-experience

Kłysik, K., & Fortuniak, K. (1999). Temporal and spatial characteristics of the urban heat island of Łódź, Poland. *Atmospheric Environment, 33*(24), 3885–3895. doi:10.1016/S1352-2310(99)00131-4

Kohli, S., & Arora, S. (2018, September 20). *Degradation of the Aravalis Could Spell Ecological Disaster*. Retrieved from The Times of India: https://weather.com/en-IN/india/news/news/2018-09-20-degradation-aravalis-ecological-disaster

Kolbasov, O. S. (1992). UN Conference on Environment and Development. *Izvestiya - Akademiya Nauk, Seriya Geograficheskaya, 6*(June), 47–54. 10.4135/9781412971867.n128

Koskela, L. (1992). Application of the New Production Philosophy to Construction (CIFE technical report No. 72). Stanford University.

Kovacic, I., & Zoller, V. (2015). *Building life cycle optimization tools for early design phases*. Academic Press.

KPMG. (2018). *Green and Blue Master Plan for Bhopal Smart City*. Author.

Kudat, A., Bell, M. E., Boland, J., Humplick, F., Madanat, S., & Mukherjee, N. (1993). *Reliability of urban water supply in developing countries: The emperor has no clothes*. Author.

Kukla, G., Weiss, H., & Dalfes, H. (1997). *Third Millennium BC Climate Changes and Old World Collapse*. Springer.

Kumar Ranjan, A., Anand, A., S, V., & Kumar Singh, R. (2016). LU/LC Change Detection and Forest Degradation Analysis in Dalma Wildlife Sanctuary Using 3S Technology: A Case Study in Jamshedpur-India. *AIMS Geosciences, 2*(4), 273–285. doi:10.3934/geosci.2016.4.273

Kumar, V., Singh, S., Shivhare, N., & Rahul, A. K. (2015). Change in Land Cover Pattern in Ramgarh Dam. *National Conference on Open Source GIS : Opportunities and Challenges*, 186-192.

Kumpel, E., & Nelson, K. L. (2013). Intermittent Water Supply: Prevalence, Practice, and Microbial Water Quality. *Environmental Science & Technology*, *50*(2), 542–553. doi:10.1021/acs.est.5b03973 PMID:26670120

Kwok, S. S. K., & Lee, E. W. M. (2011). A study of the importance of occupancy to building cooling load in prediction by intelligent approach. *Energy Conversion and Management*, *52*(7), 2555–2564. doi:10.1016/j.enconman.2011.02.002

Lachir, A., Bounoua, L., Zhang, P., Thome, K., & Messouli, M. (2016). Modeling the Urban Impact on Semiarid Surface Climate : A Case Study in Marrakech, Morocco. *Canadian Journal of Remote Sensing*, *42*(4), 379–395. doi:10.1080/07038992.2016.1194746

Laukkonen, J., Blanco, P. K., Lenhart, J., Keiner, M., Cavric, B., & Kinuthia-Njenga, C. (2009). Combining climate change adaptation and mitigation measures at the local level. *Habitat International*, *33*(3), 287–292. doi:10.1016/j.habitatint.2008.10.003

Lazar, N., & Chithra, K. (2020). *Comprehensive bibliometric mapping of publication trends in the development of Building Sustainability Assessment Systems*. Academic Press.

Lee, J. S., & Choi, H. I. (2017). Improvements to Runoff Predictions from a Land Surface Model with a Lateral Flow Scheme Using Remote Sensing and In Situ Observations. *Water (Basel)*, *9*(2), 148. doi:10.3390/w9020148

Lee, M., Tansel, B., & Balbin, M. (2011). Influence of residential water use efficiency measures on household water demand: A four year longitudinal study. Resources, Conservation and Recycling. *Resources, Conservation and Recycling*, *56*(1), 1–6. doi:10.1016/j.resconrec.2011.08.006

Lerch, D. (2017). Six Foundations for Building Community Resilience. In D. Lerch (Ed.), *The Community Resilience Reader: Essential Resources for an Era of Upheaval* (pp. 9–42). Island Press/Center for Resource Economics. doi:10.5822/978-1-61091-861-9_2

Liang, S., Li, X., & Wang, J. (2012). *Advanced remote sensing: terrestrial information extraction and applications*. Academic Press.

Liao, A., Chen, L., Chen, J., He, C., Cao, X., Chen, J., Peng, S., Sun, F., & Gong, P. (2014). High-resolution remote sensing mapping of global land water. *Science China. Earth Sciences*, *57*(10), 2305–2316. doi:10.100711430-014-4918-0

Li, B., Chen, D., Wu, S., Zhou, S., Wang, T., & Chen, H. (2016). Spatio-temporal assessment of urbanization impacts on ecosystem services: Case study of Nanjing City, China. *Ecological Indicators*, *71*, 416–427. doi:10.1016/j.ecolind.2016.07.017

Lindblom, C. E. (1959). The Science of "Muddling Through". *Public Administration Review*, *19*(2), 79–88. doi:10.2307/973677

Lindblom, C. E. (1979). Still Muddling, Not Yet Through. *Public Administration Review*, *39*(6), 517–526. doi:10.2307/976178

Linsky, B. (1954). Air Pollution Planning and Zoning. *Air Repair*, *3*(3), 155–156. doi:10.1080/00966665.1954.10467618

Liu & Wang. (2019). Modelling Reservoir Turbidity Using Landsat 8 Satellite Imagery by Gene Expression Programming. *Water, 11*(7), 1479. doi:10.3390/w11071479

Liu, J., Ma, F., & Li, Y. (2011). The Effect of Anthropogenic Heat on Local Heat Island Intensity and the Performance of Air Conditioning Systems. *Advanced Materials Research*, *250-253*, 2975–2978. . doi:10.4028/www.scientific.net/AMR.250-253.2975

Li, X., Zhou, Y., Yu, S., Jia, G., Li, H., & Li, W. (2019). Urban heat island impacts on building energy consumption : A review of approaches and findings. *Energy*, *174*, 407–419. doi:10.1016/j.energy.2019.02.183

Long, J., & Rice, J. (2018). From sustainable urbanism to climate urbanism. *Urban Studies (Edinburgh, Scotland)*, *56*(5), 004209801877084. doi:10.1177/0042098018770846

Low, N. P., Gleeson, B., Green, R. J., & Radovic, D. (2005). *The Green City: Sustainable Homes Sustainable Suburbs*. University of New South Wales Press.

Lundgren, K., & Kjellstrom, T. (2013). Sustainability Challenges from Climate Change and Air Conditioning Use in Urban Areas. *Sustainability*, *5*(7), 3116–3128. doi:10.3390u5073116

Luo, S., Wang, C., Xi, X., Zeng, H., Li, D., Xia, S., & Wang, P. (2015). Fusion of Airborne Discrete-Return LiDAR and Hyperspectral Data for Land Cover Classification. *Remote Sensing*, *8*(1), 3–12.

Macdonald, R. (2003). Theory and Objectives of Air Dispersion. MME 474A. *Wind Engineering*, 1–27.

Magrini, A., Lentini, G., Cuman, S., Bodrato, A., & Marenco, L. (2020). From nearly zero energy buildings (NZEB) to positive energy buildings (PEB): The next challenge - The most recent European trends with some notes on the energy analysis of a forerunner PEB example. *Developments in the Built Environment*, *3*, 100019. doi:10.1016/j.dibe.2020.100019

Mahlein, A. K., Oerke, E. C., Steiner, U., & Dehne, H. W. (2012). Recent advances in sensing plant diseases for precision crop protection. *European Journal of Plant Pathology*, *133*(1), 197–209.

Majuru, B., Suhrcke, M., & Hunter, P. R. (2016). How Do Households Respond to Unreliable Water Supplies? A Systematic Review. *Environmental Research and Public Health*, *13*(12), 1222. Advance online publication. doi:10.3390/ijerph13121222 PMID:27941695

Malahlela, O. E., Oliphant, T., Tsoeleng, L. T., & Mhangara, P. (2018). Mapping chlorophyll-a concentrations in a cyanobacteria- and algae-impacted Vaal Dam using Landsat 8 OLI data. *South African Journal of Science*, *114*(9/10). Advance online publication. doi:10.17159ajs.2018/4841

Malgorzata, V. W., & Árpád, B. (2010). *Data acquisition and integration* (Vol. 6). Remote Sensing, Nyugat-magyarországi Egyetem.

Mallick, A., Samal, A., & Santra, S. (2013). Integrated Natural Resource Management and Ecosystem Service. *Biodiversity Watch. International Journal on Biodiversity Issues.*, *2*, 88–102.

Manioğlu, G., & Yılmaz, Z. (2006). Economic evaluation of the building envelope and operation period of heating system in terms of thermal comfort. *Energy and Building*, *38*(3), 266–272. doi:10.1016/j.enbuild.2005.06.009

Marjaba, G. E. (2020). *Sustainability framework for buildings via data analytics*. Academic Press.

Ma, S., & Swinton, S. M. (2011). Valuation of ecosystem services from rural landscapes using agricultural land prices. *Ecological Economics*, *70*(9), 1649–1659. doi:10.1016/j.ecolecon.2011.04.004

Masson, V. (2000). A Physically-Based Scheme For The Urban Energy Budget In Atmospheric Models. *Boundary-Layer Meteorology*, *94*(3), 357–397. doi:10.1023/A:1002463829265

Masson, V. (2006). Urban surface modeling and the meso-scale impact of cities. *Theoretical and Applied Climatology*, *84*(1-3), 35–45. doi:10.100700704-005-0142-3

Mathur, V. B., Sivakumar, K., Singh, B., & R, A. K. (2009). *A Bibliographical Review for Identifying Research Gap Areas- Keoladeo Ghana National Park- – A World Heritage site*. Wildlife Institute of India, Dehradun. http://www.indiaenvironmentportal.org.in/files/bioliographical_keoladeo_ghana_np.pdf

Mathur. (2008). Geoscientific Studies and Managing Lakes of Arid and Semi-Arid Regions of Rajasthan. *Proceedings of Taal 2007.*

Matson, M., Mcclain, E. P., McGinnis, D. F. Jr, & Pritchard, J. A. (1978). Satellite Detection of Urban Heat Islands. *Monthly Weather Review, 106*(12), 1725–1734. doi:10.1175/1520-0493(1978)106<1725:SDOUHI>2.0.CO;2

Matthews, M. W. (2011). A current review of empirical procedures of remote sensing in inland and near-coastal transitional waters. *International Journal of Remote Sensing, 32*(21), 6855–6899. doi:10.1080/01431161.2010.512947

McCarthy, M. P., Best, M. J., & Betts, R. A. (2010). Climate change in cities due to global warming and urban effects. *Geophysical Research Letters, 37*(9), n/a. Advance online publication. doi:10.1029/2010GL042845

McEvoy, D., Lindley, S., & Handley, J. (2006). Adaptation and mitigation in urban areas: Synergies and conflicts. *Proceedings of the Institution of Civil Engineers. Municipal Engineer, 159*(4), 185–191. doi:10.1680/muen.2006.159.4.185

McLoughlin, J. B. (1969). *Urban and regional planning; a systems approach.* Faber.

Mehra, S. P., Mehra, S., & Sharma, K. K. (2013). Urban Avifaunal Biodiversity in context of Udaipur, Rajasthan, India. *Biodiversity Watch. International Journal on Biodiversity Issues, 2*, 120–146.

Mehta, P. H. S. (2013). *Green Building Construction for Sustainable Future.* Academic Press.

Melgani, F. (2004). Classification of Hyperspectral Remote Sensing Images with Support Vector Machines. *IEEE Transactions on Geoscience and Remote Sensing, 42*(8), 1778–1790.

Memorandum, A. (2018). *State Relief Commissioner, Disaster Management (Additional Chief Secretary) Government of Kerala. August.*

Meyer, B., Distelkamp, M., & Wolter, M. I. (2007). Material efficiency and economic-environmental sustainability. Results of simulations for Germany with the model PANTA RHEI. *Ecological Economics, 63*(1), 192–200. doi:10.1016/j.ecolecon.2006.10.017

Michigan State University. (2014). *Understanding your watershed interactive mapping program.* https://msustatewide.msu.edu/Programs/Details/2083

Miller, J. D., & Hutchins, M. (2017). The impacts of urbanisation and climate change on urban flooding and urban water quality : A review of the evidence concerning the United Kingdom. *Journal of Hydrology. Regional Studies, 12*, 345–362. doi:10.1016/j.ejrh.2017.06.006

Mingfang, T. (2002). Solar control for buildings. *Building and Environment, 37*(7), 659–664. doi:10.1016/S0360-1323(01)00063-4

Ministry of Agriculture and Farmer Welfare, Government of India. (2018, February 4). *Soil and Land Use Survey of India.* Retrieved from https://slusi.dacnet.nic.in/

Ministry of Statistics and Programme Implementation Government of India. (2013). *Green National Accounts in India A Framework.* Author.

Mirzaei, P. A., & Haghighat, F. (2010). Approaches to study Urban Heat Island – Abilities and limitations. *Building and Environment, 45*(10), 2192–2201. doi:10.1016/j.buildenv.2010.04.001

Mishra, A. K., & Coulibaly, P. (2009). Developments in hydrometric network design: A review. *Reviews of Geophysics, 47*(2), RG2001. Advance online publication. doi:10.1029/2007RG000243

Mishra, S., & Mishra, D. R. (2012). Normalized difference chlorophyll index: A novel model for remote estimation of chlorophyll-a concentration in turbid productive waters. *Remote Sensing of Environment, 117*, 394–406. doi:10.1016/j.rse.2011.10.016

Misra, A. K. (2010). A River about to Die: Yamuna. *Journal of Water Resource and Protection, 02*(05), 489–500. Advance online publication. doi:10.4236/jwarp.2010.25056

Modi, M., Ramachandra, V. P., Ahmed, L. S. K., & Hussain, Z. (2013). A review on theoretical air pollutants dispersion models. *International Journal of Pharmaceutical, Chemical and Biological Sciences, 2013*(3), 1224–1230.

MoEfCC & GIZ. (2014). *The Economics of Ecosystems and Biodiversity TEEB - India Initative*. Work. Doc. 92.

MoEFCC. (2012). *The economics of ecosystems and biodiversity*. TEEB., doi:10.1007/978-90-481-9659-3_80

Mohajerani, A., Bakaric, J., & Jeffrey-Bailey, T. (2017). The urban heat island effect, its causes, and mitigation, with reference to the thermal properties of asphalt concrete. *Journal of Environmental Management, 197*, 522–538. doi:10.1016/j.jenvman.2017.03.095 PMID:28412623

Mohamed, M. (2015). Satellite data and real time stations to improve water quality of Lake Manzalah. *Water Science, 29*(1), 68–76. doi:10.1016/j.wsj.2015.03.002

Mohan, M., Kikegawa, Y., Gurjar, B. R., Bhati, S., & Kolli, N. R. (2013). Assessment of urban heat island effect for different land use–land cover from micrometeorological measurements and remote sensing data for megacity Delhi. *Theoretical and Applied Climatology, 112*(3), 647–658. doi:10.100700704-012-0758-z

Morrison, G. (2016). Regressions with The Discrete Dependent Variables : Effect. *JMR, Journal of Marketing Research, 9*(3), 338–340.

Moslehi, S., & Reddy, T. A. (2019). *A new quantitative life cycle sustainability assessment framework: Application to integrated energy systems*. Academic Press.

Moyer, J. D. (2019). *Alternative pathways to human development: Assessing trade-offs and synergies in achieving the Sustainable Development Goals*. doi:10.1016/j.futures.2018.10.007

Muller, C. L., Chapman, L., Grimmond, C. S. B., Young, D. T., & Cai, X. (2013). Sensors and the city : A review of urban meteorological networks. *International Journal of Climatology, 33*(7), 1585–1600. doi:10.1002/joc.3678

Mundhe, N. N., & Jaybhaye, R. G. (2014). Impact of Urbanization on Land Use/Land Covers Change Using Geo-Spatial Techniques. *International Journal of Geomatics and Geosciences, 5*, 50–60.

Myrup, L. O. (1969). A Numerical Model of the Urban Heat Island. *Journal of Applied Meteorology and Climatology, 8*(6), 908–918. doi:10.1175/1520-0450(1969)008<0908:ANMOTU>2.0.CO;2

Nagendra, H., Lucas, R., Honrado, J. P., Jongman, R. H. G., Tarantino, C., Adamo, M., & Mairota, P. (2013). Remote sensing for conservation monitoring: Assessing protected areas, habitat extent, habitat condition, species diversity, and threats. *Ecological Indicators, 33*, 45–59. doi:10.1016/j.ecolind.2012.09.014

Nama, P. (2019). Review Paper on Recent Advancement in Green Building and its. *Sustainability, 7*(3), 2365–2368. Advance online publication. doi:10.22214/ijraset.2019.3431

Narayan, K., Singh, S. M., Dikshit, P. K. S., & Dwivedi, S. B. (2016). Application of GIS and Remote Sensing Techniques for Effective Watershed Management. *International Journal of Engineering Research & Technology (Ahmedabad), 4*(03), 1–5.

NASA. (2020). *POWER Data Access Viewer*. https://power.larc.nasa.gov/data-access-viewer/

Nas, B., Ekercin, S., Karabörk, H., Berktay, A., & Mulla, D. J. (2010). An Application of Landsat-5TM Image Data for Water Quality Mapping in Lake Beysehir, Turkey. *Water, Air, and Soil Pollution*, *212*(1–4), 183–197. doi:10.100711270-010-0331-2

Nastiti, A., Muntalif, B. S., Roosmini, D., Sudradjat, A., Meijerink, S. V., & Smits, A. J. M. (2017). Coping with poor water supply in peri- urban Bandung, Indonesia : Towards a framework for understanding risks and aversion behaviours. *Environment and Urbanization*, *29*(1), 69–88. doi:10.1177/0956247816686485

National Remote Sensing Center. (2019, February 3). *Bhuvan, Indian Geo Platform of ISRO*. Retrieved from Thematic Map: https://bhuvan.nrsc.gov.in/bhuvan_links.php

National Research Council. (2010). *Acute Exposure Guideline Levels for Selected Airborne Chemicals* (Vol. 8). The National Academies Press. https://www.epa.gov/sites/ production/files/2014-11/documents/sulfurdioxide_final_ volume8_2010.pdf

NDMA. (2010). *National Disaster Management Guidelines: Management of Urban Flooding*. A publication of the National Disaster Management Authority, Government of India.

Nelson, A. C., & Steven, P. (2002). Plan quality and mitigating damage from natural disasters: A case study of the north ridge earthquake with planning policy considerations. *Journal of the American Planning Association*, *68*(2), 194–207. doi:10.1080/01944360208976265

Ng, E., Chen, L., Wang, Y., & Yuan, C. (2012). A study on the cooling effects of greening in a high-density city : An experience from Hong Kong. *Building and Environment*, *47*, 256–271. doi:10.1016/j.buildenv.2011.07.014

Nguy-Robertson, A., May, J., Dartevelle, S., Birkett, C., Lucero, E., Russo, T., Griffin, S., Miller, J., Tetrault, R., & Zentner, M. (2018). Inferring elevation variation of lakes and reservoirs from areal extents: Calibrating with altimeter and in situ data. *Remote Sensing Applications: Society and Environment*, *9*, 116–125. doi:10.1016/j.rsase.2018.01.001

Nicholls, R. J., & Klein, R. J. T. (2005). Climate change and coastal management on Europe's coast. In J. Vermaat, W. Salomons, L. Bouwer, & K. Turner (Eds.), *Managing European Coasts* (pp. 199–226). Springer Berlin Heidelberg. doi:10.1007/3-540-27150-3_11

NOAA. (2015). *Acute Exposure Guideline Levels (AEGLs)*. National Oceanic and Atmospheric Administration (NOAA). https://response.restoration.noaa.gov/oil-and-chemical-spills/chemical-spills/resources/acute-exposure-guideline-levels-aegls.html

O'Brien, K., & Leichenko, R. (2008). *Environmental Change and Globalization: Double Exposures*. Oxford University Press.

Oak Ridge National Laboratory. (2003). *Possible Vulnerabilities of Cochin, India, to Climate Change Impacts and Response Strategies to Increase Resilience*. Author.

OECD & European Commission. (2020). *Cities in the World: A New Perspective on Urbanisation*. OECD. doi:10.1787/d0efcbda-

Oerke, Mahlein, & Steiner. (2014). *Proximal sensing of plant diseases, Detection and Diagnostics of Plant Pathogens*. Academic Press.

Okello, C., Tomasello, B., Greggio, N., Wambiji, N., & Antonellini, M. (2015). Impact of Population Growth and Climate Change on the Freshwater Resources of Lamu Island, Kenya. *Water (Basel)*, *7*(12), 1264–1290. doi:10.3390/w7031264

Oke, T. R. (1981). Canyon geometry and the nocturnal urban heat island : Comparison of scale model and field observations. *Journal of Climatology*, *1*(3), 237–254. doi:10.1002/joc.3370010304

Oke, T. R. (1982). The energetic basis of the urban heat island. *Quarterly Journal of the Royal Meteorological Society, 108*(455), 1–24. doi:10.1002/qj.49710845502

Oke, T. R. (2011). Urban heat island. In *The Handbook of Urban Ecology* (pp. 120–131). Routledge.

Oki, T., & Kanae, S. (2006). Global Hydrological Cycles and World Water Resources. *Science, 313*(5790), 1068–1072. doi:10.1126cience.1128845 PMID:16931749

Olakitan Atanda, J. (2019). *Developing a social sustainability assessment framework.* doi:10.1016/j.scs.2018.09.023

Olayide, O. E., & Ikpi, A. E. (2010). Urban domestic water crisis in Zimbabwe: The case of Kadoma City. *Sustainable Development, 12*(8), 90–99.

Olmanson, L. G., Brezonik, P. L., Finlay, J. C., & Bauer, M. E. (2016). Comparison of Landsat 8 and Landsat 7 for regional measurements of CDOM and water clarity in lakes. *Remote Sensing of Environment, 185*, 119–128. doi:10.1016/j.rse.2016.01.007

Omer, A. M. (2008). Renewable building energy systems and passive human comfort solutions. *Renewable & Sustainable Energy Reviews, 12*(6), 1562–1587. doi:10.1016/j.rser.2006.07.010

Ouma, Y. O., Noor, K., & Herbert, K. (2020). Modelling Reservoir Chlorophyll-a, TSS, and Turbidity Using Sentinel-2A MSI and Landsat-8 OLI Satellite Sensors with Empirical Multivariate Regression. *Journal of Sensors, 2020*, 1–21. doi:10.1155/2020/8858408

Ourghi, R., Al-Anzi, A., & Krarti, M. (2007). A simplified analysis method to predict the impact of shape on annual energy use for office buildings. *Energy Conversion and Management, 48*(1), 300–305. doi:10.1016/j.enconman.2006.04.011

Owusu, P. A., & Asumadu-Sarkodie, S. (2016). A review of renewable energy sources, sustainability issues and climate change mitigation. *Cogent Engineering, 3*(1), 1167990. doi:10.1080/23311916.2016.1167990

Pacheco, R., Ordóñez, J., & Martínez, G. (2012). Energy efficient design of building: A review. *Renewable & Sustainable Energy Reviews, 16*(6), 3559–3573. doi:10.1016/j.rser.2012.03.045

Pancharathi, R.K., Sangoju, B., & Chaudhary, S. (2019). *Advances in Sustainable Construction Materials*. Academic Press.

Pandey. (2015). *Impact of Green Building Rating Systems on the Sustainability and Efficacy of Green Buildings Case Analysis of Green Building Index.* Academic Press.

Pandey, A., Chowdary, V. M., & Mal, B. C. (2007). Identification of critical erosion prone areas in the small agricultural watershed using USLE, GIS and remote sensing. *Water Resources Management, 21*(4), 729–746. doi:10.100711269-006-9061-z

Panwar, N. L., Kaushik, S. C., & Kothari, S. (2011). Role of renewable energy sources in environmental protection: A review. *Renewable & Sustainable Energy Reviews, 15*(3), 1513–1524. doi:10.1016/j.rser.2010.11.037

Park, K., & Lee, M. H. (2019). The development and application of the urban flood risk assessment model for reflecting upon urban planning elements. *Water (Switzerland), 11*(5), 1–17. doi:10.3390/w11050920

Pattanayak & Diwakar. (2013). District-Wise Change Analysis of Land Use-Land Cover In Delhi Territory Using Remote Sensing & GIS. *Journal of Urban and Environmental Engineering, 10*(2), 201-213. doi:10.4090/juee2016.v10n2.201213

Pattanayak, S. K., Yang, J., Whittington, D., & Kumar, K. C. B. (2005). *Coping with unreliable public water supplies : Averting expenditures by households in Kathmandu, Nepal.* doi:10.1029/2003WR002443

Patterson, R. A. (2004). Water Efficiency, Domestic Appliances and Hydraulic Design for On-site Systems. Presented at the *1 st International Conference on Onsite Wastewater Treatment and Recycling*, Perth, Australia.

Patz, J. A., & Olson, S. H. (2006). Malaria risk and temperature: Influences from global climate change and local land use practices. *Proceedings of the National Academy of Sciences of the United States of America, 103*(15), 5635–5636. doi:10.1073/pnas.0601493103 PMID:16595623

Pedersen, A. O., Nyhuus, S., Blindheim, T., & Krog, O. M. W. (2004). Implementation of a GIS-based management tool for conservation of biodiversity within the municipality of Oslo, Norway. *Landscape and Urban Planning, 68*(4), 429–438. doi:10.1016/S0169-2046(03)00148-8

Pekel, J. F., Cottam, A., Gorelick, N., & Belward, A. S. (2016). High-resolution mapping of global surface water and its long-term changes. *Nature, 540*(7633), 418–422. doi:10.1038/nature20584 PMID:27926733

Peng, L. L. H., & Jim, C. Y. (2013). Green-Roof Effects on Neighborhood Microclimate and Human Thermal Sensation. *Energies, 6*(2), 598–618. doi:10.3390/en6020598

Pettorelli, N., Wegmann, M., Skidmore, A., Mücher, S., Dawson, T. P., Fernandez, M., Lucas, R., Schaepman, M. E., Wang, T., O'Connor, B., Jongman, R. H. G., Kempeneers, P., Sonnenschein, R., Leidner, A. K., Böhm, M., He, K. S., Nagendra, H., Dubois, G., Fatoyinbo, T., & Geller, G. N. (2016). Framing the concept of satellite remote sensing essential biodiversity variables: Challenges and future directions. *Remote Sensing in Ecology and Conservation, 2*(3), 122–131. doi:10.1002/rse2.15

Poddar, S., Chacko, N., & Swain, D. (2019). Estimation of chlorophyll-a in northern coastal bay of Bengal using Landsat-8 OLI and sentinel-2 MSI sensors. *Frontiers in Marine Science, 6*, 598. Advance online publication. doi:10.3389/fmars.2019.00598

Pokhrel, Y., Felfelani, F., Satoh, Y., Boulange, J., Burek, P., Gädeke, A., Gerten, D., Gosling, S. N., Grillakis, M., Gudmundsson, L., Hanasaki, N., Kim, H., Koutroulis, A., Liu, J., Papadimitriou, L., Schewe, J., Müller Schmied, H., Stacke, T., Telteu, C. E., ... Wada, Y. (2021). Global terrestrial water storage and drought severity under climate change. *Nature Climate Change, 11*(3), 226–233. doi:10.103841558-020-00972-w

Polehampton, Smith, Ghent, Wooster, Bruniquel, & Dransfeld. (2021). *Copernicus Sentinel-3 SLSTR Land User Handbook*. Academic Press.

Power, A. G. (2010). Ecosystem services and agriculture: Tradeoffs and synergies. *Philosophical Transactions of the Royal Society of London. Series B, Biological Sciences, 365*(1554), 2959–2971. doi:10.1098/rstb.2010.0143 PMID:20713396

Prakash, S., Gairola, R., Papa, F., & Mitra, A. (2014). An assessment of terrestrial water storage, rainfall and river discharge over Northern India from satellite data. *Current Science, 107*(9), 1582–1586. http://www.jstor.org/stable/24107219

Punjab Central Ground Water Board. (2013). Ground water information booklet S.A.S Nagar District, Punjab. Author.

Quesnel, K. J., Ajami, N., & Marx, A. (2019). Shifting landscapes: Decoupled urban irrigation and greenness patterns during severe drought. *Environmental Research Letters, 14*(6), 064012. doi:10.1088/1748-9326/ab20d4

Qureshi, A. S., Qadir, M., Heydari, N., Turral, H., & Javadi, A. (2007). *A Review of Management Strategies for Salt-prone Land and Water Resources in Iran*. International Water Management Institute.

Ragheb, A., El-Shimy, H., & Ragheb, G. (2016). Green Architecture: A Concept of Sustainability. *Procedia - Social and Behavioral Sciences, 216*(October), 778–787. doi:10.1016/j.sbspro.2015.12.075

Rahman, M. A. U. (2016). Urban sustainability through strategic planning: A case of metropolitan planning in Khulna city, Bangladesh. *Journal of Urban Management, 5*(1), 16–22. doi:10.1016/j.jum.2016.06.001

Rai, R. K. (2013). *Temporal Variability of Climatic Parameters of Yamuna River Basin: Spatial Analysis of Persistence, Trend and Periodicity. The Open Hydrology Journal.* doi:10.2174/1874378101004010184

Rajashekariah, K. (2011). Impact of Urbanisation on Biodiversity: Case studies from India. *WWF Report,* (October). Advance online publication. doi:10.13140/2.1.4952.8324

Raparthi, K. (2014). Assessing smart growth strategies in Indian cities: 1441 A grounded theory approach to planning practice. *J. Urban Plann., 1442.* ,05014031 doi:10.1061/(ASCE)UP.1943-5444.0000267

Raparthi, K. (2016). Assessing Climate Change Planning in Indian Cities: Bridging the Gap between Climate Change Research and Practice. *Social Science and Humanities Journal, 1*(3), 160-174.

Raparthi, K. (2015). Analyzing the Relationship between Environmental Planning Policies and Climate Change: Multinomial Logit Regression Model Evaluation of Tarrant County, Texas. *Current Urban Studies, 3*(1), 1–10. doi:10.4236/cus.2015.31001

Raparthi, K. (2016). Assessing the Initiatives in Climate Responsive and Energy Efficient Architecture: Bridging the Gap between Architectural Research and Practice". *International Journal for Science and Advance Research in Technology, 2*(6), 94–101.

Raparthi, K. (2018). Assessing the Role of Urban Planning Policies in Meeting Climate Change Mitigation Goals in Indian Cities. *Journal of Urban Planning and Development, 144*(4), 1–16. doi:10.1061/(ASCE)UP.1943-5444.0000440

Raparthi, K. (2020a). *Investigating Open Space Reservation as the New Urban Commons: A Case of Chennai City* (Vol. 52). NAGARLOK Quarterly Journal of Urban Affairs.

Raparthi, K. (2020b). Ability of Nature Based Solutions Through Urban Green Space Management in Building Urban Resilience. *International Journal of Urban Design., 3*(2), 46–56.

Raparthi, K. (2021). 'Assessing the Relationship between Urban Planning Policies, Gender and Climate Change Mitigation: Regression Model Evaluation of Indian Cities. *Journal of Urban Planning and Development, 147*(2), 05021007. Advance online publication. doi:10.1061/(ASCE)UP.1943-5444.0000677

Rathi, R., Bansal, S., & Sharma, S. (2018). An empirical study on sustainability green building practices in residential projects: With reference to Delhi- NCR region. *International Journal of Mechanical Engineering and Technology, 9*(1), 276–288.

Ray, D., Watts, K., Morris, J., & Handley, P. (2013). *Land Use and Ecosystem Services Project.* Academic Press.

Razzolini, M. T. P., & Günther, W. M. R. (2008). Health impacts due to deficient water access. *Saúde e Sociedade, 17*(1), 21–32. doi:10.1590/S0104-12902008000100003

Reckien, D., Flacke, J., Dawson, R. J., Heidrich, O., Olazabal, M., Foley, A., Hamann, J. J.-P., Orru, H., Salvia, M., De Gregorio Hurtado, S., Geneletti, D., & Pietrapertosa, F. (2014). Climate change response in Europe: What's the reality? Analysis of adaptation and mitigation plans from 200 urban areas in 11 countries. *Climatic Change, 122*(1), 331–340. doi:10.100710584-013-0989-8

Reddy, C. S., Manaswini, G., Satish, K. V., Singh, S., Jha, C. S., & Dadhwal, V. K. (2016). Conservation priorities of forest ecosystems: Evaluation of deforestation and degradation hotspots using geospatial techniques. Ecological Engineering (Vol. 91). doi:10.1016/j.ecoleng.2016.03.007

Ren, C., Ng, E. Y., & Katzschner, L. (2011). Urban climatic map studies : A review. *International Journal of Climatology, 31*(15), 2213–2233. doi:10.1002/joc.2237

Ren, C., Yang, R., Cheng, C., Xing, P., Fang, X., Zhang, S., Wang, H., Shi, Y., Zhang, X., Kwok, Y. T., & Ng, E. (2018). Creating breathing cities by adopting urban ventilation assessment and wind corridor plan – The implementation in Chinese cities. *Journal of Wind Engineering and Industrial Aerodynamics*, *182*, 170–188. doi:10.1016/j.jweia.2018.09.023

Reverte, F. C., & Garcia, M. (2019, December). da G. M., Brilha, J., & Pellejero, A. U. (2020). Assessment of impacts on ecosystem services provided by geodiversity in highly urbanized areas: A case study of the Taubaté Basin, Brazil. *Environmental Science & Policy*, *112*, 91–106. doi:10.1016/j.envsci.2020.05.015

RGoB. (2004). Thimpu Structure Plan. Department of Urban Development and Engineering Services, Ministry of Works and Human Settlements, RGoB, Bhutan.

Rinkesh. (2021). World's 10 Worst Natural Disasters - conserve-energy-future. *Conserve-Energy-Future*. https://www.conserve-energy-future.com/10-worst-natural-disasters.php

Rodriguez-Galiano, V. F., Ghimire, B., Pardo-Igúzquiza, E., Chica-Olmo, M., & Congalton, R. G. (2012). Incorporating the downscaled Landsat TM thermal band in land-cover classification using random forest. *Photogrammetric Engineering and Remote Sensing*, *78*(2), 129–137. doi:10.14358/PERS.78.2.129

Roodman, D. M., & Lenssen, N. A. (1995). *Building Revolution: How Ecology and Health Concerns are Transforming Construction. Worldwatch Paper 124*. Worldwatch Institute.

Rosenzweig, C., Solecki, W., Romero-Lankao, P., Mehrotra, S., Dhakal, S., & Ali Ibrahim, S. (2018). *Second Assessment Report of the Urban Climate Change Research Network*. https://climate-adapt.eea.europa.eu/metadata/publications/second-assessment-report-of-the-urban-climate-change-research-network/uccrn_2015_secondassessmentreport.pdf

Rossi, B., Marique, A., Glaumann, M., & Reiter, S. (2012). Life-cycle assessment of residential buildings in three different European locations, basic tool. *Building and Environment*, *51*, 395–401. doi:10.1016/j.buildenv.2011.11.017

Russo, A., & Zerbe, S. (2016). Quantifying the local-scale ecosystem services provided by urban treed streetscapes in Bolzano, Italy. *AIMS Environmental Science*, *3*(1), 58–76. doi:10.3934/environsci.2016.1.58

Safanelli, J., Poppiel, R., Ruiz, L., Bonfatti, B., Mello, F., Rizzo, R., & Demattê, J. (2020). Terrain Analysis in Google Earth Engine: A Method Adapted for High-Performance Global-Scale Analysis. *ISPRS International Journal of Geo-Information*, *9*(6), 400. doi:10.3390/ijgi9060400

Şahin, N. İ., & Manioğlu, G. (2019). Water conservation through rainwater harvesting using different building forms in different climatic regions. Sustainable Cities and Society, 44, 367–377. doi:10.1016/j.scs.2018.10.010

Sahu, L. (2006). Application of remote sensing in forest resource assessment -A case study in and around Dudhwa National Park. *Statistical Accounting of Land and Forestry Resources*, 72–84.

Sala, S., Ciuffo, B., & Nijkamp, P. (2015). *A systemic framework for sustainability assessment*. doi:10.1016/j.ecolecon.2015.09.015

Salet, W. G. M., & Faludi, A. (2000). *Three approaches to strategic spatial planning. In The revival of strategic spatial planning*. Royal Netherlands Academy of Arts and Sciences.

Sánchez-Aparicio, M., Andrés-Anaya, P., Del Pozo, S., & Lagüela, S. (2020). Retrieving Land Surface Temperature from Satellite Imagery with a Novel Combined Strategy. *Remote Sensing*, *12*(2), 277. doi:10.3390/rs12020277

Santamouris, M. (2013). Using cool pavements as a mitigation strategy to fight urban heat island—A review of the actual developments. *Renewable & Sustainable Energy Reviews*, *26*, 224–240. doi:10.1016/j.rser.2013.05.047

Santos, R., Aguiar Costa, A., Silvestre, J. D., & Pyl, L. (2020). Development of a BIM-based Environmental and Economic Life Cycle Assessment tool. *Journal of Cleaner Production*, *265*, 121705. Advance online publication. doi:10.1016/j.jclepro.2020.121705

Sartorio, F. S. (2005). Strategic Spatial Planning. *disP - The Planning Review, 41*(162), 26-40.

Satterthwaite, D. (2006). Climate change and cities. *Sustainable Development Opinion*, 2.

Satterthwaite, D. (2020). *An urbanising world*. Retrieved February 4, 2021, from https://www.iied.org/urbanising-world

Sazib, N., Bolten, J., & Mladenova, I. (2020). Exploring Spatiotemporal Relations between Soil Moisture, Precipitation, and Streamflow for a Large Set of Watersheds Using Google Earth Engine. *Water (Basel)*, *12*(5), 1371. doi:10.3390/w12051371

Scs, U. S. (1986). *Urban hydrology for small watersheds, technical release no. 55 (TR-55)*. US Department of Agriculture, US Government Printing Office.

Sellers, P. J., Mintz, Y., Sud, Y. C., & Dalcher, A. (1986). A Simple Biosphere Model (SIB) for Use within General Circulation Models. *Journal of the Atmospheric Sciences*, *43*(6), 505–531. doi:10.1175/1520-0469(1986)043<0505:ASBMFU>2.0.CO;2

Sen, S., Roesler, J., Ruddell, B., & Middel, A. (2019). Cool Pavement Strategies for Urban Heat Island Mitigation in Suburban Phoenix, Arizona. *Sustainability*, *11*(16), 4452. doi:10.3390u11164452

Seto, K. C., Güneralp, B., & Hutyra, L. R. (2012). Global forecasts of urban expansion to 2030 and direct impacts on biodiversity and carbon pools. *Proceedings of the National Academy of Sciences of the United States of America*, *109*(40), 16083–16088. doi:10.1073/pnas.1211658109 PMID:22988086

Shakil, S. H., & Ananya, A. H. (2015). Effectiveness of Environmental Impact Assessment (EIA): Bangladesh Perspective. *Bangladesh e-Journal of Sociology*, *12*(1), 115–125.

Shameri, M. A., Alghoul, M. A., Sopian, K., Zain, M. F. M., & Elayeb, O. (2011). Perspectives of double skin façade systems in buildings and energy saving. *Renewable & Sustainable Energy Reviews*, *15*(3), 1468–1475. doi:10.1016/j.rser.2010.10.016

Sharif, H. O., Al-Zahrani, M., & Hassan, A. E. (2017). Physically, fully-distributed hydrologic simulations driven by GPM satellite rainfall over an urbanizing arid catchment in Saudi Arabia. *Water (Basel)*, *9*(3), 163. doi:10.3390/w9030163

Sharma, D., & Kansal, A. (2006). *Current condition of the Yamuna River - An overview of flow, pollution load and human use*. https://fore.yale.edu/files/Current_Condition_of_Yamuna_River.pdf

Sharma, A. (2015). A Geographical Study of Keoladeo National Park, Bharatpur (Rajasthan) with Using Remote Sensing and GIS. *International Journal of Science, Engineering and Technology*, *3*(6), 201–210.

Shirazi, S., Baca, E., McCormick, M., & Litchney, S. (2017). State of California General Plan Guidelines. Governor's Office of Planning and Research, California, USA. https://opr.ca.gov/docs/OPR_COMPLETE_7.31.17.pdf

Sims, R. E. H. (2004). Renewable energy: a response to climate change. *Solar World Congress 2001, 76*(1), 9–17. 10.1016/S0038-092X(03)00101-4

Singh, A., & Raju, A. (2019). Application of Grace Satellite Data for Assessment of Groundwater Resources in Central Ganga Alluvial Plain, Northern India. *Environmental Concerns and Sustainable Development*, 153–162. doi:10.1007/978-981-13-5889-0_7

Singh, R. (2018, January 18). Sambhar: A Lake No More. *Times of India*.

Singh, R. P., Singh, N., Singh, S., & Mukherjee, S. (2016). Normalized Difference Vegetation Index (NDVI) Based Classification to Assess the Change in Land Use/Land Cover (LULC) in Lower Assam, India. *International Journal of Advanced Remote Sensing and GIS*, 1963-1970.

Singh, B. P. (2013). Modern salt (halite) deposits of the Sambhar Lake, Rajasthan and their formative conditions. *Current Science*, 1482–1484.

Singh, D., & Vyas, A. (2014). Planning Strategies for the Development of Peri-Urban Area The Case of Indore Peri urban Area. *International Journal of Scientific and Engineering Research*, *5*, 27–32.

Singh, S., Lal, R. B., Sridharan, U., & Upadhyay, V. P. (2016). Environmental Sustainability Guidelines for Green Buildings in India: A Review. *Indian Journal of Scientific Research and Technology*, *4*(1), 2321–9262. www.indjsrt.com

Smith, R. M. (2015). *"Green" building in India: A comparative and spatial analysis of the LEED-India and GRIHA rating systems*. doi:10.1080/10225706.2015.1020065

Sowmya, K., John, C. M., & Shrivasthava, N. K. (2015). Urban flood vulnerability zoning of Cochin City, southwest coast of India, using remote sensing and GIS. *Natural Hazards*, *75*(2), 1271–1286. doi:10.100711069-014-1372-4

Space Applications Centre ISRO Ahmedabad. (2013). National wetland atlas: Wetlands of international importance under Ramsar convention. Ahmedabad: Ministry of Environment and Forests, Government of India.

Special Correspondent the Hindu. (2020, August). *Kochi Corporation_ Water doesn't drain from Kochi streets _ Kochi News - Times of India*. https://www.thehindu.com/news/cities/Kochi/kochi-suffers-as-drains-overflow/article32299553.ece#

Sproles, E. A., Crumley, R. L., Nolin, A. W., Mar, E., & Moreno, J. I. L. (2018). SnowCloudHydro—A New Framework for Forecasting Streamflow in Snowy, Data-Scarce Regions. *Remote Sensing*, *10*(8), 1276. doi:10.3390/rs10081276

Sripada, R. P., Heiniger, R. W., White, J. G., & Weisz, R. (2005). Aerial color infrared photography for determining late-season nitrogen requirements in corn. *Agronomy Journal*, *97*(5), 1443–1451.

Stec, A., & Kordana, S. (2015). Analysis of profitability of rainwater harvesting, gray water recycling and drain water heat recovery systems. *Resources, Conservation and Recycling*, *105*, 84–94. doi:10.1016/j.resconrec.2015.10.006

Steemers, K. (2010). Towards a research agenda for adapting to climate change. *Building Research and Information*, *31*(3-4), 291–301. doi:10.1080/0961321032000097692

Steiner, F. R., & Butler, K. (Eds.). (2012). *Planning and Urban Design Standards* (2nd ed.). Wiley.

Stern, N. (2007). *The Economics of Climate Change: The Stern Review*. Cambridge University Press., doi:10.1017/CBO9780511817434

Stojče Dimov Ilčev. (2019). *Global Satellite Meteorological Observation (GSMO) Applications* (Vol. 2). Springer International Publishing.

Sudhira, H. S., & Nagendra, H. (2013). Local Assessment of Bangalore: Graying and Greening in Bangalore – Impacts of Urbanization on Ecosystems, Ecosystem Services and Biodiversity. In *Urbanization, Biodiversity and Ecosystem Services: Challenges and Opportunities: A Global Assessment*. Springer. doi:10.1007/978-94-007-7088-1_7

Sukhdev, P. (2010). Mainstreaming the Economics of Nature: A synthesis of the approach, conclusions, and recommendations of TEEB. *De Economist*, *44*(1).

Sun, A. Y., & Scanlon, B. R. (2019). How can Big Data and machine learning benefit environment and water management: A survey of methods, applications, and future directions. *Environmental Research Letters*, *14*(7), 073001. doi:10.1088/1748-9326/ab1b7d

Sunil Kumar, M. K. (2021). *Operation Breakthrough fails to end waterlogging in Kochi _ Kochi News - Times of India.* https://timesofindia.indiatimes.com/city/kochi/op-breakthrough-fails-to-end-waterlogging-in-city/articleshow/82985016.cms

Swinburne, R. G. (1969). Physical Determinism. *Royal Institute of Philosophy Lectures*, *3*, 155–168. doi:10.1017/S0080443600000662

Tang, Q., Gao, H., Yeh, P., Oki, T., Su, F., & Lettenmaier, D. P. (2010). Dynamics of Terrestrial Water Storage Change from Satellite and Surface Observations and Modeling. *Journal of Hydrometeorology*, *11*(1), 156–170. doi:10.1175/2009JHM1152.1

Tang, Z., de Almeida, C., & Fishwick, P. A. (1991). Time series forecasting using neural networks vs. Box- Jenkins methodology. *Simulation*, *57*(5), 303–310. doi:10.1177/003754979105700508

Tan, J., Zheng, Y., Tang, X., Guo, C., Li, L., Song, G., Zhen, X., Yuan, D., Kalkstein, A. J., Li, F., & Chen, H. (2010). The urban heat island and its impact on heat waves and human health in Shanghai. *International Journal of Biometeorology*, *54*(1), 75–84. doi:10.100700484-009-0256-x PMID:19727842

Tannier & Pumain. (2005). Fractals in urban geography: A theoretical outline and an empirical example. *Eur. J. Geogr.*

Taylor, N. (1998). *Urban Planning Theory Since 1945.* SAGE Publications Ltd.

Team, G. C. (n.d.). *Sambhar Lake Area Overview & Issues.* Retrieved from Gram Chetna Kendra: http://www.gck.org.in/aboutus/sambharlake.html

Team, P. (2015). *A Study on Punjab Export Potential and Strategy.* Rakesh Mohan Joshi Ram Singh Indian Institute of Foreign Trade.

TEEB. (2010). *Mainstreaming the economics of nature - A Synthesis of the approach, Conclusions and Recommendations.* TEEB.

TEEB. (2012). *The Economics of Ecosystem & Biodiversity- An Interim Report.* TEEB.

Thanu, H. P., & Jayachamarajendra, S. (n.d.). *Comparative study on Indian building assessment tools and its limitations.* Academic Press.

The National Research Council of National Academics. (2005). *Valuing ecosystem services: Toward better environmental decision-making.* The National Academies Press. doi:10.17226/11139

Tilahun, A. (2015). Application of GIS for Calculate Normalize Difference Vegetation Index (NDVI) using LANDSAT MSS, TM, ETM+ and OLI_TIRS in Kilite Awulalo, Tigray State, Ethiopia. *Journal of Environment and Earth Science*, 27–33.

Tim, U. S., & Mallavaram, S. (2003). Application of GIS technology in watershed-based management and decision making. *Watershed Update*, *1*(5), 1-6.

Tiwari, G. (2001). Traffic flow and safety: Need for new models in heterogenous traffic. In D. Mohan & G. Tiwari (Eds.), *Injury prevention and control* (pp. 71–88). Taylor & Francis.

Tomlinson, C. J., Chapman, L., Thornes, J. E., & Baker, C. J. (2011). Including the urban heat island in spatial heat health risk assessment strategies : A case study for Birmingham, UK. *International Journal of Health Geographics*, *10*(1), 42. doi:10.1186/1476-072X-10-42 PMID:21682872

Toparlar, Y., Blocken, B., Maiheu, B., & van Heijst, G. J. F. (2017). A review on the CFD analysis of urban microclimate. *Renewable & Sustainable Energy Reviews*, *80*, 1613–1640. doi:10.1016/j.rser.2017.05.248

Topp, S. N., Pavelsky, T. M., Jensen, D., Simard, M., & Ross, M. R. V. (2020). Research Trends in the Use of Remote Sensing for Inland Water Quality Science: Moving Towards Multidisciplinary Applications. *Water (Basel)*, *12*(1), 169. doi:10.3390/w12010169

Town and Country Planning GOK. (2009). *Kochi City Development Plan*. https://cochinmunicipalcorporation.kerala.gov.in/web/guest/development-plan

Tripathi, S., & Mahey, K. (2017). Urbanization and economic growth in Punjab (India): An empirical analysis. *Urban Research & Practice*, *10*(4), 379–402. doi:10.1080/17535069.2016.1227875

Truong, Q. (2020). *4. Material Sustainability*. Birkhäuser. doi:10.1515/9783035619522-006

Trzyna, T. (2007). *Global Urbanization and Protected Areas - Challenges and opportunities posed by a major factor of global change and creative ways of responding*. California Institute of Public Affairs. https://www.ncbi.nlm.nih.gov/pubmed/12971682

UDD. (1986). *Rangpur Structure Plan Bangladesh National Physical Planning Project, Urban Development Directorate*. UDD.

UN & UNEP. (2015). *SEEA Experimental Ecosystem Accounting: Technical Recommendations*. Final Draft.

UN DESA. (2021). *68% of the world population projected to live in urban areas by 2050, says UN | UN DESA | United Nations Department of Economic and Social Affairs*. Retrieved April 18, 2021, from https://www.un.org/development/desa/en/news/population/2018-revision-of-world-urbanization-prospects.html

UN Habitat. (2010). *Sustainable Urban Development Through Expanding Equitable Access to Land, Housing, Basic Services and Infrastructure; Dialogue on the Special Theme for the Twenty-Third Session of the Governing Council; Report by the Executive Director*. Rio de Janeiro: UN Habitat.

UN, IIFM, & Ministry of Statistics and Programme Implementation. (2018). *Review of Existing Ecosystem Accounting Initiatives And Literature In India*. Author.

UN. (2012). *The future we want: Outcome document of the United Nations Conference on Sustainable Development*. https://sustainabledevelopment.un.org/content/documents/733FutureWeWant.pdf

UN. (2018). *World Urbanization Prospects: The 2018 Revision* (ST/ESA/SER.A/420). United Nations. https://population.un.org/wup/Publications/Files/WUP2018-Report.pdf

UN. (2019). *The Sustainable Development Goals Report 2019*. United Nations. Retrieved from https://unstats.un.org/sdgs/report/2019/The-Sustainable-Development-Goals-Report-2019.pdf

UNDESA. (2018). *World Urbanization Prospects: The 2018 Revision (ST/ESA/SER.A/420)*. (Vol. 12). United Nations. https://population.un.org/wup/Publications/Files/WUP2018-Report.pdf

UNDRR. (2015). *Sendai Framework for Disaster Risk Reduction 2015-2030* (Issue 1). https://www.preventionweb.net/files/43291_sendaiframeworkfordrren.pdf

UNDRR. (2020). *Hazard Definition & Classification Review: Technical Report.* United Nations Office for Disaster Risk Reduction (UNDRR). https://www.undrr.org/publication/hazard-definition-and-classification-review

UNEP. (2009). *Buildings and Climate Change.* Retrieved from http://admin.indiaenvironmentportal.org.in/files/SBCI-BCCSummary.pdf

UNEP. (2019a). *2019 Global Status Report for Buildings and Construction.* UNEP.

UNEP. (2019b, August 29). *Facts about the Climate Emergency.* Retrieved February 8, 2021, from https://www.unenvironment.org/explore-topics/climate-change/facts-about-climate-emergency

UNEP. (n.d.). Convention on Biological Diversity, in: Convention, The Diversity, Biological Cbd, The Protocol, Cartagena Plan, Strategic Secretariat, The Diversity, Biological Secretary, The Executive. UNEP.

UN-ESCAP. (2019). *The Disaster Riskscape Across Asia-Pacific: Pathways for Resilience, Inclusion and Empowerment.* United Nations. https://www.unescap.org/publications/asia-pacific-disaster-report-2019

Unger, J. (2009). Connection between urban heat island and sky view factor approximated by a software tool on a 3D urban database. *International Journal of Environment and Pollution, 36*(1/2/3), 59. doi:10.1504/IJEP.2009.021817

UNISDR. (2009). Terminology on Disaster Risk Reduction. In *United Nations Office for Disaster Risk Reduction.* UNDRR. doi:10.1021/cen-v064n005.p003

UNISDR. (2012). *2011-Disasters in Numbers.* UNISDR, USAID, CRED.

United Nations Children Fund & World Health Organisation. (2017). *Progress on Drinking Water, Sanitation and Hygiene - Joint Monitoring Programme 2017 Update and SDG Baselines.* WHO. doi:10.1111/tmi.12329

United Nations Developmental Program. (2018). *Kerala: Post Disaster Needs Assessment-Floods and Landslides August 2018.* Author.

United Nations. (2015). *The Millennium Development Goals Report.* United Nations. Doi:978-92-1-101320-7

United Nations. (2017). New urban agenda : H III: Habitat III: Quito 17-20 October 2016. In *United Nations Conference on Housing and Sustainable Urban Development.* United Nations.

United Nations. (2019). *World urbanization prospects 2018 : Highlights.* Author.

United Nations. (2020). *The value of sustainable urbanization.* Author.

United Nations. (n.d.). *System of Environmental Economic Accounting.* https://seea.un.org/

United, T., Conference, N., & Environment, H. (1989). *Declaration of the United Nations Conference on the Human Environment, 1972.* 10.1177/0019556119890340

University of Maryland College Center for Disaster Resilience. (2018). *The Growing Threat of Urban Flooding: A National Challenge.* https://cdr.umd.edu/urban-flooding-report

UN-SPIDER. (2020). *Data application of the month: Water quality monitoring.* UN-SPIDER Knowledge Portal. https://www.un-spider.org/links-and-resources/data-sources/daotm-water-quality

US EPA. (2014). *Learn About Heat Islands* [Overviews and Factsheets]. US EPA. https://www.epa.gov/heatislands/learn-about-heat-islands

Usali, N., & Ismail, M. H. (2010). Use of Remote Sensing and GIS in Monitoring Water Quality. *Journal of Sustainable Development, 3*(3). Advance online publication. doi:10.5539/jsd.v3n3p228

USGS. (2019, March 15). *USGS Science for a Changing World.* Retrieved from Earth Explorer: https://earthexplorer.usgs.gov/

USGS. (2019a). *Landsat 7 (L7) Data Users Handbook.* Author.

USGS. (2019b). *Landsat 8 (L8) Data Users Handbook.* Author.

Varlyguin, D. L., Wright, R., Goetz, S. J., & Prince, S. D. (2001). *Advances in Land Cover Classification for Applications Research: A Case Study from The Mid-Atlantic.* RESAC.

VeneKlasen, L., & Miller, V. (2002). *A new weave of power, people and politics: The action guide for advocacy and citizen participation.* World Neighbors.

Verma, T. (2017). Role of Planners To Address the Threats of National Parks in India. Case Example : Keoladeo Ghana National Park, Bharatpur, Rajasthan, India. *International Journal of Advances in Mechanical and Civil Engineering, 5*(1), 63–65.

Visser, W., & Brundtland, G. H. (2013). *Our Common Future ('The Brundtland Report').* World Commission on Environment and Development. doi:10.9774/GLEAF.978-1-907643-44-6_12

Vogt, M. C., & Vogt, M. E. (2016). Research Article: Near-Remote Sensing of Water Turbidity Using Small Unmanned Aircraft Systems. *Environmental Practice, 18*(1), 18–31. doi:10.1017/S1466046615000459

Vyas, G. S., & Jha, K. N. (2016). *Identification of green building attributes for the development of an assessment tool: A case study in India.* doi:10.1080/10286608.2016.1247832

Wan Guanghua, K.S.S. (2013). *Urbanization in Asia: Governance, Infrastructure and the Environment.* Academic Press.

Wan, Z. (2013). *MODIS Land Surface Temperature Products.* Academic Press.

Wang, C., Jia, M., Chen, N., & Wang, W. (2018). Long-Term Surface Water Dynamics Analysis Based on Landsat Imagery and the Google Earth Engine Platform: A Case Study in the Middle Yangtze River Basin. *Remote Sensing, 10*(10), 1635. doi:10.3390/rs10101635

Wang, G., Mang, S., Cai, H., Liu, S., Zhang, Z., Wang, L., & Innes, J. L. (2016). Integrated watershed management: Evolution, development and emerging trends. *Journal of Forestry Research, 27*(5), 967–994. doi:10.100711676-016-0293-3

Wang, X., Chen, H., & Chen, Y. (2017). Large differences between glaciers 3D surface extents and 2D planar areas in Central Tianshan. *Water (Basel), 9*(4), 282. doi:10.3390/w9040282

Wang, X., & Xie, H. (2018). A review on applications of remote sensing and geographic information systems (GIS) in water resources and flood risk management. *Water (Basel), 10*(5), 608. doi:10.3390/w10050608

WaterAid. (2018). The water gap. *The State of the World's Water.* Advance online publication. doi:10.1029/2010GL044571

WCED. (1987). *Report of the World Commission on Environment and Development: Our Common Future.* Retrieved July 28, 2016, from http://www.un-documents.net/A372BC4E-993E-4C10-B368-9E75816E59DB/FinalDownload/DownloadId-1990161B7AFC5CB358F7280210D7A48F/A372BC4E-993E-4C10-B368-9E75816E59DB/our-common-future.pdf

Welsh, D. F. (2005). Xeriscape. In *Water Encyclopedia* (pp. 750–752). American Cancer Society. doi:10.1002/047147844X.aw434

Wheeler, D., & Hammer, D. (2010). *The Economics of Population Policy for Carbon Emissions Reduction in Developing Countries. CGD Working Paper 229.* Center for Global Development. https://www.cgdev.org/content/publications/detail/1424557

Wheeler, S. M. (2004). *Planning for Sustainability: Creating Livable, Equitable and Ecological Communities*. Routledge. doi:10.4324/9780203300565

WHO. (2018). *Climate change and health*. Retrieved February 8, 2021, from https://www.who.int/news-room/fact-sheets/detail/climate-change-and-health

Wilbanks, T. J. (2003). Integrating climate change and sustainable development in a place-based context. *Climate Policy*, *3*(Supplement 1), S147–S154. doi:10.1016/j.clipol.2003.10.013

Wilbanks, T. J. (2005). Issues in developing a capacity for integrated analysis of mitigation and adaptation. *Environmental Science & Policy*, *8*(6), 541–547. doi:10.1016/j.envsci.2005.06.014

Williams, K., Burton, E., & Jenks, M. (2000). Achieving sustainable urban form: conclusions. *Achieving Sustainable Urban Form*, 347–355.

Willis, K. J., Jeffers, E. S., Tovar, C., Long, P. R., Caithness, N., Smit, M. G. D., Hagemann, R., Collin-Hansen, C., & Weissenberger, J. (2012). Determining the ecological value of landscapes beyond protected areas. *Biological Conservation*, *147*(1), 3–12. doi:10.1016/j.biocon.2011.11.001

Wilson, E., & Piper, J. (2010). *Spatial Planning and Climate Change*. Routledge. doi:10.4324/9780203846537

Wolfe, A. K., Malone, E. L., Heerwagen, J. H., & Dion, J. P. (2014). *Behavioral Change and Building Performance: Strategies for Significant, Persistent, and Measurable Institutional Change* (No. PNNL-23264, 1132691). doi:10.2172/1132691

Wong, M. S., Nichol, J. E., Ng, E. Y. Y., Guilbert, E., Kwok, K. H., To, P. H., & Wang, J. Z. (2010). Gis Techniques For Mapping Urban Ventilation. *Using Frontal Area Index And Least Cost Path Analysis.*, *38*, 6.

Wronski, E. W., Anderson, E. W., Berry, A. E., Bernhart, A. P., & Belyea, H. A. (1966). Air Pollution Considerations in the Planning and Zoning of a Large Rapidly Growing Municipality. *Journal of the Air Pollution Control Association*, *63*(3), 157–158. doi:10.1080/00022470.1966.10468458 PMID:5903516

Wu, Q. R., Lane, C., Li, X., Zhao, K., Zhou, Y., Clinton, N., DeVries, B. E., Golden, H., & Lang, M. W. (2019). Mapping wetland hydrological dynamics using Google Earth Engine (Version 1). figshare. https://github.com/giswqs/Wetland-Hydro-GEE doi:10.6084/m9.figshare.8864921.v1

Wulf, C., Werker, J., Zapp, P., Schreiber, A., Schlör, H., & Kuckshinrichs, W. (2018). *Sustainable Development Goals as a Guideline for Indicator Selection in Life Cycle Sustainability Assessment*. doi:10.1016/j.procir.2017.11.144

Wu, Q., Lane, C. R., Li, X., Zhao, K., Zhou, Y., Clinton, N., DeVries, B., Golden, H. E., & Lang, M. W. (2019). Integrating LiDAR data and multi-temporal aerial imagery to map wetland inundation dynamics using Google Earth Engine. *Remote Sensing of Environment*, *228*, 1–13. doi:10.1016/j.rse.2019.04.015 PMID:33776151

WWF. (2020). *About Agricultural Biodiversity*. http://www.environment.gen.tr/biodiversity/161-about-agricultural-biodiversity.html

Xavier, G. D. O., Almeida, T. D., Oliveira, C. M. M. D., Oliveira, P. D. S. D., Costa, V. H. B., & Granado, L. M. A. (2020). Estimate and evaluation of reservoir metrics in Serra da Mesa dam (GO) using the Google Earth Engine platform. Ambiente e Agua - An Interdisciplinary. *The Journal of Applied Science*, *15*(5), 1. doi:10.4136/ambi-agua.2584

Xia, H., Zhao, J., Qin, Y., Yang, J., Cui, Y., Song, H., Ma, L., Jin, N., & Meng, Q. (2019). Changes in Water Surface Area during 1989–2017 in the Huai River Basin using Landsat Data and Google Earth Engine. *Remote Sensing*, *11*(15), 1824. doi:10.3390/rs11151824

Xiao, X. D., Dong, L., Yan, H., Yang, N., & Xiong, Y. (2018). The influence of the spatial characteristics of urban green space on the urban heat island effect in Suzhou Industrial Park. *Sustainable Cities and Society*, *40*, 428–439. doi:10.1016/j.scs.2018.04.002

Xu, G., Shi, Y., Sun, X., & Shen, W. (2019). Internet of Things in Marine Environment Monitoring: A Review. *Sensors (Basel)*, *19*(7), 1711. doi:10.339019071711 PMID:30974791

Yang, X., Pavelsky, T. M., Allen, G. H., & Donchyts, G. (2020). RivWidthCloud: An Automated Google Earth Engine Algorithm for River Width Extraction From Remotely Sensed Imagery. *IEEE Geoscience and Remote Sensing Letters*, *17*(2), 217–221. doi:10.1109/LGRS.2019.2920225

Yaroshenko, I., Kirsanov, D., Marjanovic, M., Lieberzeit, P. A., Korostynska, O., Mason, A., Frau, I., & Legin, A. (2020). Real-Time Water Quality Monitoring with Chemical Sensors. *Sensors (Basel)*, *20*(12), 3432. doi:10.339020123432 PMID:32560552

Yau, Y. H., & Hasbi, S. (2013). A review of climate change impacts on commercial buildings and their technical services in the tropics. *Renewable & Sustainable Energy Reviews*, *18*, 430–441. doi:10.1016/j.rser.2012.10.035

Yen, B. C., & Chow, V. T. (1980). Design hyetographs for small drainage structures. *Journal of the Hydraulics Division*, *106*(ASCE 15452).

Yigit Avdan, Z., Kaplan, G., Goncu, S., & Avdan, U. (2019). Monitoring the Water Quality of Small Water Bodies Using High-Resolution Remote Sensing Data. *ISPRS International Journal of Geo-Information*, *8*(12), 553. doi:10.3390/ijgi8120553

Yoshino, H., Hong, T., & Nord, N. (2017). IEA EBC annex 53: Total energy use in buildings—Analysis and evaluation methods. *Energy and Building*, *152*, 124–136. doi:10.1016/j.enbuild.2017.07.038

Yuan, C. (2010). Mitigating urban heat island effects of high density cities : A study at Hong Kong. *Architectural Journal*, 120-123.

Yu, D., Tan, H., & Ruan, Y. (2011). A future bamboo-structure residential building prototype in China: Life cycle assessment of energy use and carbon emission. *Energy and Building*, *43*(10), 2638–2646. doi:10.1016/j.enbuild.2011.06.013

Yüksek, İ. (2015). The Evaluation of Building Materials in Terms of Energy Efficiency. *Periodica Polytechnica. Civil Engineering*, *59*(1), 45–58. doi:10.3311/PPci.7050

Zaitunah, A., Samsuri, A. G., & Safitri, R. A. (2018). Normalized difference vegetation index (ndvi) analysis for land cover types using landsat 8 oli in besitang watershed, Indonesia. *Earth and Environmental Science*, 1-9.

Zerah, M.-H. (2000). Household strategies for coping with unreliable water supplies: The case of Delhi. *Habitat International*, *24*(3), 295–307. doi:10.1016/S0197-3975(99)00045-4

Zhang, P., Bounoua, L., Thome, K., Wolfe, R., & Imhoff, M. (2015). Modeling Surface Climate in U.S. Cities Using Simple Biosphere Model SiB2. *Canadian Journal of Remote Sensing*, *41*(6), 525–535. doi:10.1080/07038992.2015.1110013

Zhang, X., Cui, J., Wang, W., & Lin, C. (2017). A Study for Texture Feature Extraction of High-Resolution Satellite Images Based on a Direction Measure and Gray Level Co-Occurrence Matrix Fusion Algorithm. *Sensors (Basel)*, *17*.

Zhang, X., Warner, M. E., & Homsy, G. C. (2017). Environment, Equity, and Economic Development Goals: Understanding Differences in Local Economic Development Strategies. *Economic Development Quarterly*, *31*(3), 196–209. doi:10.1177/0891242417712003

Zhang, Y., Wang, H., Gao, W., Wang, F., Zhou, N., Kammen, D. M., & Ying, X. (2019). A survey of the status and challenges of green building development in various countries. *Sustainability (Switzerland)*, *11*(19), 1–29. doi:10.3390u11195385

Zhao, D., Xing, X., Liu, Y., Yang, J., & Wang, L. (2010). The relation of chlorophyll-a concentration with the reflectance peak near 700 nm in algae-dominated waters and sensitivity of fluorescence algorithms for detecting algal bloom. *International Journal of Remote Sensing*, *31*(1), 39–48. doi:10.1080/01431160902882512

Zhou, D., Xiao, J., Bonafoni, S., Berger, C., Deilami, K., Zhou, Y., Frolking, S., Yao, R., Qiao, Z., & Sobrino, J. A. (2019). Satellite Remote Sensing of Surface Urban Heat Islands: Progress, Challenges, and Perspectives. *Remote Sensing*, *11*(1), 48. doi:10.3390/rs11010048

Zhu, W., Yu, Q., Tian, Y. Q., Becker, B. L., Zheng, T., & Carrick, H. J. (2014). An assessment of remote sensing algorithms for colored dissolved organic matter in complex freshwater environments. *Remote Sensing of Environment*, *140*, 766–778. doi:10.1016/j.rse.2013.10.015

Zoological Survey of India. (2005). *Fauna of Sambhar Lake*. Zoological Survey of India.

Zou, Z., Dong, J., Menarguez, M. A., Xiao, X., Qin, Y., Doughty, R. B., Hooker, K. V., & David Hambright, K. (2017). Continued decrease of open surface water body area in Oklahoma during 1984-2015. *The Science of the Total Environment*, *595*, 451–460. doi:10.1016/j.scitotenv.2017.03.259 PMID:28395260

About the Contributors

Athar Hussain is working as an Associate Professor in Department of Civil Engineering and Head, Civil and Environmental Engineering Department at Ch. B. P. Govt. College of Engineering, New Delhi. Dr. Hussain completed his Ph.D. in Environmental Engineering from IIT Roorkee in year 2007. Thereafter he joined as an Assistant Director with Environment Management Group at National Productivity Council, New Delhi till Oct, 2010. Thereafter he worked as an Assistant Professor, civil engineering department in School of Engineering at Gautam Buddha University till December 2015. His research area includes industrial wastewater treatment and solid waste management with special focus on anaerobic treatment of wastewater and solid waste. Dr. Hussain has more than 15 years of experience in the area of teaching, research and consultancy. He has also guided more than 60 M.Tech dissertations. He has published more than 125 papers in international referred journals and international conferences. He is also an Associate Member of The Institution of Engineers (IE) India. He has also handled various national and policy level projects with government and private organizations.

Chandra Tiwari is a professor in the Department of Civil Engineering, Delhi Technological University, Delhi, India. He is an ex-army officer and his research interests primarily include detection of small and buried objects using microwave and hyper-spectral remote sensing. His other areas of interest include detection in different bands of the electromagnetic spectrum, wave interaction with various media, and neural network applications in remote sensing.

* * *

Muhammed Sulfikkar Ahamed is an urban planner having received his master of planning from the National Institute of Technology, Calicut, India in the year 2020. He has worked in several organizations on various subjects such as disaster risk management, transportation, and other fields of city planning. Muhammed's current research interests include resilient cities, disaster risk reduction, and risk-informed planning. He is currently working for RIKA India, as a project officer.

Shyni Anilkumar is an assistant professor in the Department of Architecture and Planning, National Institute of Technology Calicut, Kerala, India. She has over a decade of experience in the field of urban planning, disaster risk management, Architecture, and so on, and has guided academic projects as well as field projects on disaster resilient planning and mitigation.

Dimple Behal is an Urban & Regional Planner assisting in the development of carefully designed communities. Dimple is a development enthusiast actively involved in Spatial Planning through strong community engagement and Inclusive Urban Development. Core strength in bringing a different perspective for Environmental Planning, Comprehensive Planning, Research, and Policy formulation. Dimple thrive as an Advocate where she Plans better with public participation and by enhancing institutional governance. Extensively worked in the area of Climate and Disaster Risk Resilience, City Sanitation Plans, Environmental Impact Assessment by incorporating Natural and built environment, and she helps in promoting cities that are Environmentally sound and socially inclusive.

Parul Bhyan earned her P.G. Degree in Master in construction and real estate management, DCRUST Murthal, Haryana. Professional member of Council of Architecture, Indian Institute of Architects, HR Chapter and GRIHA certified Professional. Presently, pursuing her PhD in the Department of Architecture and Planning, Malviya National Institute of Technology, Jaipur, India. She has been working on institute assistantship during her P.h.D handling the following subjects, such as Housing, Barrier free Architecture, Building Construction, Architectural Design, etc., at the Postgraduate and B. Arch level. Her specialisation serves Housing, Sustainability, MCDM Analysis. Her major publication serves as Sustainable Housing, sustainability assessment and Customer Satisfaction. Her Scopus Author ID: 57211507117.

Tarush Chandra is a Professor in the Department of Architecture and Planning at Malaviya National Institute of Technology (MNIT), Jaipur, India. He completed his Bachelor degree in Architecture from Government College of Architecture, Lucknow in 1993. After his post graduation in Urban and Regional Planning in 1996 he received his PhD also in 2016 form CEPT University, Ahmedabad. He has been in academic profession since 1996. In last 25 years he has worked on various research projects in collaboration with Indian and foreign Institutes/ Universities. He holds his research interest in rural urban transformation, integration of passive techniques/ practices for improving human habitats and resource optimization for improved urban environment, etc. He widely writes on affordable housing, sustainable built environment and management of solid waste in municipal towns.

Şeyda Emekci, MSc, PhD, received her bachelor's, master and doctoral degrees from Middle East Technical University (METU). Worked as an associate at the Development Agency and the General Director of Nature Conservation and National Parks, in Turkey from 2014 to 2017. Postdoctoral researcher at East London University, in the UK in 2019. Currently teaches in the Architecture Department at Ankara Yıldırım Beyazıt University (AYBU). Has been involved in various change projects at the different architectural levels in public and private sectors and with civil associations in Turkey since 2011. Specialization areas include sustainability, energy efficiency, housing economics.

Peeyush Gupta is a 1st degree connection Real-Time Information Specialist, National Mission For Clean Ganga, Ministry of Jal Shakti.

Kazi Saiful Islam is a Professor of Urban and Rural Planning Discipline of Khulna University, Bangladesh. He received both Masters of Engineering and Doctor of Engineering degrees from the Department of Urban Engineering of the University of Tokyo, Japan. His research interest includes the application of quantitative geospatial methods in urban complexities. Apart from being an academician, he is a protuberant professional urban planner and geospatial expert. He has contributed to many vital planning projects in Bangladesh (in different capacities).

Vishal Kamboj is a young researcher working in the field of environmental pollution, aquatic biodiversity, and environmental impact assessment. He received his Ph.D. in Environmental Science from Gurukula Kangri (Deemed to be University), Haridwar. He received his master degree in Environmental Science from Gurukula Kangri Vishwavidyalaya, Haridwar with gold medal. He has awarded INSPIRE Fellowship funded by the Department of Science and Technology (DST), New Delhi, India during their Ph.D. programme. He has published 14 research papers in national and international peer reviewed journals. He has 17 conferences abstracts, and 14 book chapters in his credit. Recently, he has obtained Young Researcher Award (2020) organized by Institute of Scholars (InSc), India.

Saloni Khandelwal is an Architect and Environmental Planner. Author has solid foundation in planning including governance, policy analysis, capacity building, climate resilience, water resource and natural resource management, sustainable development, smart cities, and land use.

Nand Kumar earned his Ph.D. and P.G. Degree in Urban and Rural Planning from the Indian Institute of Technology, Roorkee, and has 22 years of academic and professional experience. He is Professional member of Council of Architecture, Institute of Town Planners, India and Indian Institute of Architects, India Presently, Working as an HOD & Associate Professor (Planning) in the Department of Architecture and Planning, Malviya National Institute of Technology, Jaipur, India. He has been handling the following subjects, such as Housing, Urban Planning, Planning Theories and Techniques, Building Construction, Town Planning, Architectural Design, etc., at the Postgraduate and B. Arch level. His major publication serves as an analysis of energy consumption in residential buildings and transportation systems.

Asia Lachir is a professor at the National School of Architecture in Agadir, Morocco and an associate scientist at the Laboratory of Processes for Sustainable Energy & Environment at the Science Faculty of Cadi Ayyad University in Marrakech, Morocco. She graduated in civil engineering and hold a PhD on environment and climate change. Her research activities are related to sustainability in architecture and urban planning and focus on building energy efficiency, bioclimatic design, urban climate and particularly urban heat island.

Richa Madan has completed her bachelor's degree in Chemistry (Hons) from Zakir Hussain College, Delhi University in 2014. She did her masters from Kurukshetra University, Kurukshetra, Haryana in Environmental Science in 2016. She is currently pursuing PhD in Environmental Sciences from Department of Zoology and Environmental Science, Gurukula Kangri (Deemed to be University), Haridwar, Uttarakhand on the research topic "Textile-Dye Industry Effluent Treatment using Aerobic Moving Bed Biofilm Reactor and Low-Cost Biochar". She has research papers and chapters in journals and books indexed by Scopus. She also has research papers in conference proceedings. She has attended many workshops and seminars related to research methodology and academic research paper writing. She has teaching and training expertise in water and wastewater analysis.

Ankush Mishra is a graduate in Civil Engineering from SRM University and Post Graduate in Environmental Planning from School of Planning and Architecture, Bhopal. He has a passion for working in the development sector and solving problems from grassroot level. He has worked under Madhya Pradesh Government as a Young Professional - Consultant and currently employed as a Green Fellow, Associate for the Deputy Commissioner of Dumka district by Swaniti Initiative.

Satya Prakash has interest in GIS and Remote Sensing and has been working in the field for the last 30 years. He has published several research papers on varied topics such as disaster management, environmental planning, urban heat island modeling, carbon sequestration, displacement studies using SAR, besides many others. He is also on board of a few GOI projects which is using GIS for water resource planning, urban planning and school mapping.

Kiranmayi Raparthi is an Architect and Urban planner. She earned her doctorate degree in Urban Planning and Public Policy from The University of Texas at Arlington, U.S.A. Her doctoral research was aimed towards assessing the impact of urban planning policies on carbon-dioxide emissions through empirical evidences and examined the significance of urban planning policies as an effective tool in climate change mitigation. Her research interests deal with sustainable urban development, climate change studies and an interdisciplinary understanding of climate change resilient, urban development practices in developing nations. Kiranmayi has more than six years of research and teaching experience in various reputed universities in India and U.S.A. She has taught both undergraduate and postgraduate students of architecture and planning. Both in her teaching and research, she consistently strives for opportunities that allow her to explore sustainable and ecological approaches which offer new ways of enhancing Urban Resilience and Sustainable Communities.

Uma Shanker is a Trainee Engineer at Delhi Pollution Control Committee.

Bhavna Shrivastava earned her Ph.D. from the MANIT, Bhopal and has 15 years of academic and professional experience. She earned her P.G. Degree in Master of Planning with specialization in Urban Development and Planning (M.U.D.P), MANIT, Bhopal, 2010. Professional member of Council of Architecture and Indian Institute of Architects, MP Chapter. Presently, Working as an Assistant Professor in the Department of Architecture and Planning, Malviya National Institute of Technology, Jaipur, India. She has been handling the following subjects, such as Housing, land economics and housing, community planning, housing finance, building construction, town planning, architectural design, etc. at the Postgraduate and B. Arch level. Her specialisation serves Housing, Architecture, Design Pedagogy and Urban Planning. Her major publication serves as Housing Delivery Methods, Housing Quality levels and Customer Satisfaction.

Garima Toor is a Research Scholar in the Department of Architecture and Planning at Malaviya National Institute of Technology (MNIT), Jaipur, India. She received her Bachelor degree in Architecture from DCRUST, Murthal (a state university) in 2015 and Master degree with distinction in Environment Planning and Management from School of Planning and Architecture (SPA), Vijayawada, India in 2017. Prior to pursuing full time research, she worked for 2 years as contractual faculty in the same department. Her research interest lies in environmental planning, monitoring, natural resource management and spatial analysis. Her current research deals with rejuvenation of ecological sensitive areas by exploring impacts related to ecology, biodiversity and environmental degradation due to urbanization with focus on applications of geospatial tools and techniques.

Rubeena Vohra was born in Ludhiana, India, in 1981. She received the B.Tech. degree in Electronics and Communication Engineering (ECE) from Punjab Technical University, Jalandhar, India, in 2003, the M.Tech degree in ECE from Guru Gobind Singh Indraprastha University(GGSIPU), New Delhi, and Ph.D. degree in Civil Engineering from Delhi Technological University (DTU), New Delhi, India, in 2011 and 2020, respectively. In 2009, she joined the Department of Electronics and Communication Engineering in BharatiVidyapeeth's College of Engg. (BVCOE), as a Lecturer, and thereafter became Assistant Prof. Since August 2009, she has been with the Department of ECE, BVCOE, New Delhi. Her current research interests include digital image processing, remote sensing in satellite imageries and geo-informatics. She was awarded the topper certificate in B.Tech in ECE, LLRIET, PTU, in year 2003. She received Gold Medal in 2011 from GGSIPU, New Delhi for being the topper in M.Tech. She has published 07 research papers in referred international journals, 01 research papers in referred national journal, 04 research papers in international conference and 04 research papers in national conferences.

Index

H

I

K

L

M

N

O

P

Q

R

S

T

U

V

W

Y

Z

www.ingramcontent.com/pod-product-compliance
Lightning Source LLC
LaVergne TN
LVHW062139300926
849161LV00010B/2572